D1483749

Fire in the Hole

Fire in the Hole

Miners and Managers in the American Coal Industry

CURTIS SELTZER

THE UNIVERSITY PRESS OF KENTUCKY

The University Press of Kentucky
Scholarly publisher for the Commonwealth,
serving Bellarmine College, Berea College, Centre
College of Kentucky, Eastern Kentucky University,
The Filson Club, Georgetown College, Kentucky
Historical Society, Kentucky State University,
Morehead State University, Murray State University,
Northern Kentucky University, Transylvania University,
University of Kentucky, University of Louisville,
and Western Kentucky University.

Editorial and Sales Offices: Lexington, Kentucky 40506-0024

Library of Congress Cataloging in Publication Data

Seltzer, Curtis.
Fire in the hole.

Bibliography: p.
Includes index.
1. Coal-miners—United States—History. 2. Coal trade
—United States—History. 3. Trade-unions—Coal-miners—
United States—History. 4. Industrial relations—United
States—History. I. Title.
HD6976.M6152U67 1984 331.7'6223324'0973 84-11805
ISBN 0-8131-1526-4

For John Nicholas, who always liked books
but won't get a chance to read this one

Contents

Tables

Preface

It is the writer's need to have his say about his subject—to say something new, something useful—that prompts the reader to read defensively, to ask as he reads, What ax is the author grinding? What is left out? What reach is exceeding the author's grasp? I have never read a totally objective book—certainly not one on labor relations in the American coal industry. Authors are not totally objective; it is not in the nature of the beast. But it is reasonable to ask an author to be fair, to say what he has to say clearly, and to interpret events with care and precision.

In this book I try to explain and interpret the historical relationship between miners and managers in the American coal industry. That relationship has played a central role in the industry's history, and as long as coal is dug from the ground, it will shape coal's role in America's energy future.

Few readers could be expected to read all that I feel a need to say about coal, and space does not permit a description of every strike or an examination of every facet of coal's political economy. Many aspects of the industry are necessarily given brief treatment, and some—mine safety, air pollution, synthetic fuels, transportation, federal coal leasing, federal tax policy and subsidies, environmental impacts and public regulation, and new mining technologies among others—are not discussed at all.

Inevitably, a book of this sort ends up with a lot of sentences that say something like "Miners thought this" and "Coal operators thought that." Such sentences trouble me because I've come to appreciate the diversity of opinion in an industry that is all too often described simplistically and one-dimensionally. Miners and managers are individuals with all the complexities and variations that characterize any group of individuals. When they act as a group, I drifted into the convenient author's shorthand of assigning them a group mentality. I apologize to all of the miners and managers with whom I've spoken over the years for failing to convey their individuality in my effort to describe the broad shaping of events.

"Fire in the hole!" was a warning that miners shouted a few seconds before blasting the coal loose from the bed where it lay for millions of years. The phrase is an apt one for describing the relationship between labor and management over the years.

Dozens of people have had a hand in helping this book to take shape over the years. I've benefited from conversations with many persons mentioned in these pages. A number of years ago, the Center for Community Change helped in a time of need. The Energy Group at the U.S. Office of Technology Assessment gave me a deeper understanding of the complexities of coal development when I was one of several authors of *The Direct Use of Coal: Problems and Prospects of Production and Combustion*. The Institute for Policy Studies provided a congenial environment for several years during which I wrote the first drafts. Institute interns helped with the research, and my thanks go to Kelly Branham, Laura Cooley, Leah Goldman, Dana Handley, Ronald Hodges, John Houseman, Barbara Kritt, Julia Leighton, and Andrew Warner. Peter Lambropoulos was an able and resourceful research assistant for the better part of a year. Carole Collins, Maria Enrico, and Anne Wilcox-Staats scrupulously typed draft after draft. My wife, Melissa Ann Dowd, paid her dues during the last frantic stages of editing. The six years I spent in West Virginia were essential to my grasp of the subject. Finally, the following people have enlivened my investigations in one way or another: Willie Anderson, Thomas N. Bethell, Mike Burdiss, Mary Coleman, Alan Crane, John P. David, Nancy Debevoise, Fred Decker, Bill and Mil Duncan, Howard Fineman, Anise Floyd, Walter Burton Franklin, Si Galperin, William Greider, Robert Guerrant, Harold Hoffman, Pershing Henry, Gibbs Kinderman, Spencer Klaw, Kate Long, Linda Nelson Lucci, Davitt McAteer, Darrell McGraw, Warren McGraw, Arnold Miller, Joseph Mohbat, Dan Moldea, Robert Nelson, Robert Payne, Frank Powers, John D. Rockefeller IV, Warner Schilling, Carl Schoffler, Steve Shapiro, Barbara Ellen Smith, Cosby Ann Totten, David Underhill, Jim Weeks, and Helen Winternitz.

John Duray of the *UMWA Journal* generously provided photographs from the union's files. Each of the brilliant photographs of Jack Corn, Rick Diehl, Earl Dotter, Robert Gumpert, Lewis Hine, Russell Lee, Marat Moore, Cathy Stanley, and Ted Wathen tells more than a thousand words.

The people at the University Press of Kentucky have treated the author and his manuscript with kindness and competence, for which we both are thankful. Despite their patience and professionalism, errors of fact and judgment will inevitably slip through. These visitors are unwelcome but come anyway and stay forever. Such is life.

Introduction

Coal began to be made in North America 250 to 350 million years ago, long before people got a notion to mine it, to burn it, or to write about it. America's coal story has something for everyone. Some are most interested in its statistics—tons mined, cars loaded, parts per million, dollars made. Others concern themselves with the science and technology of mining and burning coal. Many focus on a particular current problem that coal poses. I've chosen to look at the people who have dug coal in America over the past two hundred years—the mine owners and those they hired as miners.

The relationship between these two groups has dominated the politics and economics of this industry as no other. For well over a century they have locked themselves in a constant struggle over how much each deserves for its part in wresting coal from the ground. The framework for this struggle has been the private enterprise system. To a great extent, then, this book looks at how that system functions and what its consequences have been for the people of coal. The constant mismatch of supply and demand together with the continuing turmoil of labor's contest with management has led almost every other coal-rich country in the world to replace private ownership with some form of public or mixed ownership. Faced with the same problems, the American political system never altered the basic private sector framework from which our coal industry sprang in the early nineteenth century. How well has private enterprise treated mine owners and their workers? How has market competition shaped the relationship between owners and miners? What pressures has the system of private enterprise brought to bear on each side, and how has each handled those pressures? Finally, how were miners and operators changed by their struggle?

Coal has aptly been called a casket for ancient vegetal material. Cellulose, the basic pulp in plants, is a carbohydrate. Sunlight synthesized carbohydrates from carbon dioxide, water, and other elements in primordial vegetation. The carbon, which gives coal its value, originated in the photosynthesis of billions of plants millions of years ago.

Coal formed only where ancient marshes and swamps lay next to salt water.[1] Over thousands of years, plants and other organisms lived, died, and partially decomposed there. Organic mass piled up in these watery areas and eventually became giant peat bogs. Low-lying to begin with, these bogs either sank or were inundated with salt water and were compressed as the water, sand, and minerals swept over them.[2] The earth heated the peat at low temperatures for centuries, gradually transforming it into better coals with higher carbon contents. The salt water retreated, leaving material that would eventually become the clay and rock that now cover coal seams. Without these two sequential processes—the buildup of organic material and the application of heat at 100 to 200 degrees C over a long period of time—coal did not form. There are as many coal seams in a particular area as the number of times this sequence of biochemical and geochemical processes occurred.

Neither process is neat and clean. The plants themselves, and the ocean, contained dozens of substances that worked their way into coal's basic hydrocarbon molecule. Sulfur, ash, nitrogen, radioactive material, and trace minerals—arsenic, beryllium, cadmium, fluorine, lead, mercury, and selenium, among others—contaminate U.S. coals in varying degrees. This geological legacy creates a substantial environmental threat where coal is burned in large quantities with less-than-adequate pollution controls. Another legacy of coal's origins is methane, a highly explosive gas, which is the product of vegetal decomposition. Geological forces trapped methane in coal seams under great pressure. Methane explosions have killed an estimated ten thousand American coal miners since 1900, about one of every eleven fatalities. Sulfur is the most persistently troublesome curse of coal's messy beginnings. That which was part of the original plant and animal matter has been passed down as organic sulfur, and no way has yet been invented to remove it before combustion. Inorganic sulfur, which is not part of the original molecular structure, can be stripped from coal before combustion. When sulfur-bearing coal is burned, sulfur dioxide (SO_2) and other oxides are produced and spread through the earth's atmosphere. These sulfur compounds, especially SO_2, and other combustion products threaten human health and the physical environment.

The coals that are mined today vary in quality according to the completeness of the biochemical and geochemical processes that have taken place over the past several hundred million years. Coals are classified into a hierarchy of ranks. Peat is the poorest, with a low heat value, much moisture, and many impurities. Brown coal, lignite, and subbituminous are low-quality coals that were formed during comparatively recent geological ages. Bituminous coals are the most common and useful. They vary in heat content from about 11,000 to 16,000 Btus (British thermal units) per pound. Anthracite, the highest-ranked coal, has a high carbon content and burns cleanly. Some bituminous and fewer anthracite coals melt and fuse when burned, forming coke, which is used in the manufacture of pig iron. These are commonly known as "metallurgical" coals. Most coals are burned as fuel to provide heat or make steam. Some are used as feedstock for products such as tars, drugs, and dyes.

Most of the goods and services that developed countries now take for granted were invented and made available to the mass population within the last hundred years. Industrialization based on hydrocarbons and mass production accelerated economic growth and made possible a consumer cornucopia. This model has counted cadence for every country we consider "developed," regardless of its ideological drumbeat. It is a model whose first energy source is coal. Coal for heat; coal for steam; coal for electricity; coal for steel.

As societies industrialize, they gradually climb an energy ladder. Five hundred years ago, every society depended on human beings and animals to do most of its work. Three hundred years ago, wind and water began to be harnessed. Wood was the overwhelming source of heat for homes, smelting, and other manufactures. As recently as two hundred years ago, wind, water, and wood were the principal sources of energy.

In the last half of the nineteenth century most of today's developed countries began to industrialize. The wood-fired steam engine was tinkered into existence in the early 1800s. Thomas Newcomen applied one of the first to pumping water from British coal mines. Not many years later, steam engines pulled passengers and freight over fixed metal rails. Coal soon proved better than wood for making steam. Pound for pound, it takes up less physical space and produces twice the amount of heat. Industrialization also required better metals in larger quantities than the iron and lead of the late 1700s. Metallurgical coal was burned in air-limiting furnaces to make coke, an essential ingredient in making steel.[3]

Coal was a familiar but lightly tapped global resource a century ago. Where crustial faults poked it through the surface, men and women scooped it out and burned it. The Chinese did so more than four thousand years ago, as did Bronze Age Europeans. Aristotle's student Theophrastus noted "earthy . . . coals; they kindle, and burn like wood coals [charcoal]."[4] Roy writes: "The primeval Britons, those savage and roving clans who inhabited the island at the time of its invasion by Julius Caesar, a people possessed of perceptive faculties of a high order, were doubtless acquainted with the existence and properties of coal."[5] A shipment from Tynemouth to London in 1269 establishes the beginning of the coal industry in Britain, Wieck tells us.[6] The growth of English cities around the time of Magna Carta began to sap the kingdom's principal energy supply, wood. An energy crisis befell the English in the mid-thirteenth century. Woodlots were stripped and inflation began. Commoners and royalty alike switched to coal to warm their stone rooms. London's fog was as much coal smoke as Thamesian mist, prompting Edward I and Parliament to ban most coal burning in the city in 1306 "to avoid the sulphurous smoke and savour of the firing." At least one violator was executed.[7] Demand for coal and the shortage of wood negated the act in due course.[8]

Protests against coal combustion and evidence of its ill effects appear in step with increased coal use. An early environmentalist, Sir Kenelm Digby, sent a memorial to the Crown in 1661 complaining about coal fires:

> Being thus incorporated with the very air which ministers to the necessary respiration of our lungs, the inhabitants of London . . . find it in all their expectorations: the spittle and other excrements which proceed from them, being for the most part of a blackish and fuliginous color; besides the acrimonious soot produces another sad effect by rendering the people obnoxious to inflammations, and comes in time to exculcerate the lungs, when a mischief is produced, so incurable that it carries away multitudes by languishing and deep consumptions, as the bills of mortality do quickly inform.[9]

Britain's swing to coal advanced her industry ahead of the continental countries, and her ascendancy in the nineteenth century was based in large part on mastery of coal—from "necessity rather than choice," as Roy puts it.[10]

The American experience with coal may have begun with the Arizona Hopis, who fired their pottery with lignite as early as A.D. 1000.[11] In 1760, a Jesuit missionary reported that he saw the Algonquins "making fire with coal from the earth."[12] Joliet and Marquette noted they had seen "Cole-Pits" in 1673, and Father Hennepin made a map of the Illinois River that pinpointed "mines of coal, slate, and iron" near modern Peoria.[13] Virginia's William Byrd wrote of "new settlements of ye ffrench Refugees [Huguenots]" about twenty miles above present-day Richmond who had discovered that "great Raines hath washed away the Banks [so] that the Coal lyes bare."[14] The "mines" referred to were protruding outcrops of the mineral. A mine in the sense of an excavation was called a "pit"; thus, early writers noted the presence of many mines and a few pits. The earliest actual use of coal in the Atlantic colonies was recorded in 1702, when one of the Huguenots petitioned the governor of Virginia for permission to burn coal in his blacksmith's forge.[15] The small coal basin near Richmond, Taylor wrote, contains "the oldest worked collieries in America."[16] According to one story, a woodsman called Necho Allen discovered Pennsylvania's anthracite when he spread his blankets one night near what is now Pottsville and was awakened several hours later after the embers of his campfire ignited the coal on which it rested.

By the time of the American Revolution, the political and economic elites in Pennsylvania, Maryland, and Virginia had commissioned mapping expeditions and knew of vast coal deposits in the hinterlands. Pittsburgh area deposits were the best known of any in the wilderness, owing in some measure to George Washington's explorations in 1770. Alexander Hamilton wrote of "fossil-coal," and Thomas Jefferson reported findings of "mineral coal" in 1781.[17] The Richmond field in Virginia was the main source of American coal at the time of the Revolution. Without it, revolutionary munitions makers would have been hard pressed. Pennsylvania's anthracite, which is hard to ignite and keep burning, was used at a few forges. It was not commonly used for space heating until after 1820. If it had been, Washington's army could have spent a very comfortable winter in 1777–78 at Valley Forge, just a few miles from the spot where Necho Allen is said to have burned his blankets. William Penn and his family began buying land in the anthracite as early as 1749. In 1768, the Penns bought most of Pennsylvania's bituminous coal west of the Susquehanna River from the resident Indians for ten thousand dollars as part of a treaty.[18] They bought the rest in 1784. Philadelphia investors locked up 484 square miles of anthracite coal in the Lackawanna and Wyoming valleys in northeast Pennsylvania.

Eighteenth-century American quarry pits were small; transportation limited the marketing possibilities; the mechanics of burning anthracite had not been mastered; wood was still plentiful and cheap. Most of America's rich bituminous coal lay west of the Appalachian mountains. The lack of cheap transportation for this low-cost commodity meant that coal had to be near markets. Early coal entrepreneurs puzzled over how to cut it from the earth. A pit could be dug in an outcrop or stream bank, but soon the hole would become too big to manage. Knowledge of underground mining was increasing rapidly in Britain—shaft sinking started in the middle of the fourteenth century, and horse-driven water pumps were used as early as 1486[19]—but this technology migrated slowly to America. The Crown claimed ownership of all minerals in the colonies, and royalties due the king discouraged local entrepreneurs. American mining was also slowed because of England's coal-export monopoly with the colonies. Later, protective tariffs limited British coal imports but never eliminated them until after the Civil War, when our rail network was able to haul coal cheaply from the inland mines to the Atlantic Coast. The clearing of Atlantic woodlands finally turned Americans toward coal as a ticket to energy independence.

American mining expanded steadily in the nineteenth century. By 1860, two distinct coalfields had emerged: the bituminous fields in eastern Virginia, western Pennsylvania, and western Maryland; and the anthracite field in northeastern Pennsylvania.[20] Americans were digging more than six million tons of bituminous and eight million tons of anthracite coal annually by the time the Civil War began.[21] In the early 1800s, Virginia supplied most of the bituminous coal, shipping it up and down the coast. Gradually, the bituminous fields around Pittsburgh became a major point of supply. Some of Pittsburgh's coal was shipped down the Ohio, but most was burned within a hundred-mile radius, in the metal mills and factories that were soon built nearby. The iron and steel needed for transportation—rails, locomotives, Conestoga wagon hardware—and for armaments made Pittsburgh a nineteenth-century boom town. By the

1860s, the Pittsburgh area's four hundred collieries were mining almost two million tons of coal annually, about one-fifth of the nation's total—and about half was burned or converted into coke locally.

Government promoted the fortunes of American coal entrepreneurs from the beginning. Government aid was deeply rooted in the general economic philosophy of the day. Americans saw businessmen as public servants in whose work the commonwealth participated as regulator and promoter.[22] Laissez-faire was not the rule through most of the nineteenth century; rather, "from Missouri to Maine . . . , governments were deeply involved in lending, borrowing, building, and regulating" business enterprise.[23]

Land that was taken from Indians or foreign powers was incorporated into the public domain, and much coal-rich land was distributed to war veterans.[24] Washington, for example, claimed more than thirty thousand acres under his soldier's bounty arising from the French and Indian War. He picked out tillable bottomland along the Kanawha and Ohio rivers that contained valuable coal deposits. After the Revolution, Washington persuaded the Virginia General Assembly to create joint-stock companies to make the Potomac and James rivers more navigable, in part, for coal traffic. The assembly gave Washington shares in each venture.[25] Land speculators engineered valuable war-service grants or purchase of thousands of acres. "In the early years of the Republic," Bogen wrote, "when such leading financiers as Robert Morris and Benjamin Rush were buying largely of Pennsylvania wild lands for speculative purposes, the state sold large tracts in this area [the anthracite fields in the Lackawanna and Wyoming valleys] for a small consideration."[26] Wealthy politicians—Albert Gallatin, De Witt Clinton, John Beckley, Wilson Cary Nicholas—took "the most extravagant parcels of western land during this period [turn of the nineteenth century], parcels that ranged in extent from 100,000 acres to 1,100,000."[27] Speculators knew that resources bought cheap would sell dear someday. The investments of politicians in Appalachian land was undoubtedly one factor in the federal policy of severing these mineral resources from the public domain.[28]

Government also promoted coal development—and commerce generally—through subsidizing transportation systems.[29] State governments built much of the inland canal system on which coal commerce depended in the first part of the nineteenth century. The states gave land to private railroad companies, lent them money, guaranteed their bonds, granted them immunity from taxation, and set up public lotteries for their benefit. Maryland chartered the Baltimore & Ohio, the first American line, in 1827, and bought $500,000 worth of stock. Pennsylvania built the "Main Line" of canals and railroads to link Philadelphia with Pittsburgh, and funded canals and railroads into the anthracite region. It lent $800,000—representing more than half of the private sector stock issue—to get the Delaware & Hudson's system of canals and railroads started in 1829.[30]

In 1831 New Jersey chartered the Central Railroad of New Jersey, which figured prominently in the anthracite traffic. The Central needed to pay no taxes until net earnings returned 7 percent of the property cost, whereupon the tax was 0.5 percent of that cost.[31] Pennsylvania even built and operated the Philadelphia & Columbia line in 1828. Direct state participation in rail operations petered out by the mid-nineteenth century as private investors, seeing profit opportunities in the now-tested rail technology, campaigned against public ownership.

The states did the most to underwrite early transportation systems, but the federal government was also involved.[32] By the 1830s, Washington had donated about four million acres of public land to promote canal construction in Ohio, Michigan, Indiana, and Wisconsin. It also bought more than $3 million in canal company stock. The Bank of the United States started the Philadelphia and Reading by subscribing to about 25 percent of its stock. During the Panic of 1837, the bank advanced money to this company on loans guaranteed by the line's own obligations![33]

With public land and money, a transportation infrastructure was reasonably well established by the time of the Civil War. By then, however, the general public had witnessed one scandal after another of politicians and developers cutting deals for their mutual enrichment. The public generally supported government aid for internal improvements in the first part of the nineteenth century, but sentiment shifted following disclosures that legislators had been given stock in transportation ventures to gain state charters. By 1860 eighteen states had prohibited themselves from aiding public works companies.

The American coal industry was well established by the end of the Civil War. Its markets were growing. A transportation infrastructure was being built. Investors were coming forward. Government had smoothed the path in the first part of the century so that private industry could stand alone for the second half. Later, government would again step in to aid coal suppliers through transportation subsidies, research, tax breaks, and labor legislation. At the same time, the first feeble efforts at public regulation of coal began in the area of mine safety. The character of growth in the postwar industry differed in scale from before the war. As hundreds of mines and thousands of miners were organized to meet the coal demands of the late nineteenth century, conflict between labor and management emerged as a fact of industrial life. Growth and conflict, intimately related, characterized the industry for sixty-five years after the Civil War.

1. Mines and Camps: The First Hundred Years

Before 1840, quarrying, or "trenching," was the basic mining technique, a primitive progenitor of open-pit surface mining. When a quarry pit was exhausted, it was abandoned. Digging was done by hand. No particular skill was needed to shovel dirt and coal out of an excavation. Hundreds shoveled—farmers, slaves, westward wanderers seeking a grubstake; hundreds more carried the coal from the holes in sacks, on mules, or in small wagons. Inasmuch as coal outcropped in dozens of places, geology gave rise to many self-employed quarriers. Coal gathering was a cottage industry, much like the yeoman farming of the day. But growing demand soon required more efficient mining systems. As the quarries were exhausted, mine owners began tunneling underground, importing the know-how from Britain.[1] The new technique transformed the organization of work and the structure of the industry.

Underground mines began to replace quarries after the War of 1812, and within thirty years or so, produced most of America's coal. Drift mines were cut into hillsides where the horizontal coal seam emerged. By the 1820s shaft mines, which tapped coal seams as deep as 350 feet or more, were used in Virginia.[2] Coal was loaded onto boxed sleds and dragged by men or dogs to the shaft's hoist. Ponies and mules were first used in the late 1830s to haul small carts over wooden, later metal rails. One early cluster of shaft mines in Virginia, the Black Heath Pits, was boldly renamed the "Maidenhead Pits," perhaps to suggest its intactness following an explosion that killed forty-seven men.[3]

In 1837, Edmund Ruffin, editor of the *Farmers' Register*, visited a Virginia mine where two 220-foot vertical shafts had been sunk. At their bottom, horizontal tunnels bored into the coal in a grid that left blocks of the mineral, called "pillars," in place to support the roof. (Today this is known as the "room-and-pillar" method and continues to be our basic underground mining system.) A horse-driven winch hoisted the coal to the surface in a *corve*—a box or basket with iron runners that held between 85 and 100 pounds. Slaves pushed the corves along the tunnel to the takeout point. Ruffin wrote: "Each man has a chain fastened by straps around his breast, which he hooks to the corve, and thus harnessed, and in a stooping posture [the tunnel was less than four feet high] he drags his heavy load over the floor or rock."[4]

Another account, written in 1828 by a Virginia engineer, noted, "The mines are worked day and night except Sundays—when the water is drawn as often as necessary to keep the works below from being flooded. The negroes prefer this to labor in the field."[5]

The shift from open-pit quarrying to underground tunneling forced mine operators to employ radically different techniques. Deep shafts and long tunnels had to be excavated; thousands of tons of roof rock had to be propped, hoists constructed, water pumped from the depths, ventilation provided, and blasting done in an explosion-prone environment. Dozens of workers had to be coordinated so that the rate at which the coal was dug roughly matched the rate of hauling it to the surface, which in turn approximated the rate at which it was sold. Underground mining, moreover, required more capital and better management: capital to buy the resource and construct the mine; management to coordinate efficiently the flow of capital, equipment, and labor.

The underground mining model that Americans borrowed from Britain required increasing amounts of capital but was primarily dependent on human energy.[6] The tools were simple—picks, augers, shovels, and carts. Machinery was primitive: steam-driven hoists and water pumps. "Natural" ventilation was used in the early days. Sometimes, for ventilation, "furnace" fires were allowed to burn in worked-out sections of a mine or in a sealed section of the vertical entry shaft. The fire drew air through the mine, but rarely supplied sufficient quantities to its most remote corners where the miners tunneled. The fire hazard of this ventilation scheme is obvious.[7]

Britain was the labor pool for American coal entrepreneurs, supplying experienced miners, supervisors, and engineers. Many Welsh amassed in the Pennsylvania anthracite region, where today their descendants speak with more than a trace of their ancestral Cambrian tongue. Other British filtered into northern Appalachia, along the Ohio River Valley, and west into Iowa and Colorado. Most leaders of the earliest American miners' unions—the Bates Union in the anthracite in 1848, the Illinois-based American Miners' Association in the 1860s, as well as the United Mine Workers of America (UMWA)—were British immigrants or their descendants.[8]

A class structure quickly developed. There were those who provided the opportunity to dig coal through their investment and expertise, and there were those who dug. Underground miners were multiskilled entrepreneurs, often referred to as "contract miners." Supervisors assigned each a "place" to work. The miner was responsible for propping the roof in his place, laying track from the main haulageways to his work site, and keeping the tunnel in repair. This was "deadwork," for which he received no pay. Deadwork could frequently amount to as much as 25 percent of the miner's time.[9] He was paid according to the number of tons of rock-free coal he produced daily.

Digging the coal from the seam was a two-step process that only the most experienced workers handled. First, the miner lay on his side and picked out a four-foot-deep V-shaped incision at the bottom of the coal face across the width of the tunnel. He wedged down the overhanging coal with a metal bar and loaded the lumps, leaving the "slack" on the mine floor. By the 1830s, blasting replaced the laborious "barring down" of the coal. With the face undercut, the miner drilled holes into the hanging coal, loaded them with gunpowder or dynamite, tamped the charges with clay, backed away, shouted, "Fire in the hole!" and lit the fuse. When the holes were drilled properly, most of the coal would rive cleanly in lumps and blocks. If not, the coal fractured into smaller, less marketable fragments mixed with waste rock. The miner then shoveled the coal into a four-wheeled cart. He or his helper pushed it toward the main track, whence it was taken to the surface. Each miner-entrepreneur hired two or three "buddies," often including a son. They were responsible for getting materials, laying track, and pushing the carts. These apprentices learned the miner's craft while doing unskilled but necessary work.[10]

Miner-entrepreneurs reaped few benefits from the independence of their enterprise and suffered many penalties. Piecework allowed them to work at their own pace and use their own judgment about safety. But the pay-for-tonnage system forced them to choose constantly between time spent on getting coal and that devoted to unpaid safety and deadwork. The more a miner timbered, the more care he took, the fewer tons he could load. When the company cut the piece rate, he had to take chances just to stay even. When accidents happened—as they did with appalling regularity—the companies blamed careless workers. The contract miner was not allowed to sell his coal in a free market, like farmers and merchants. Rather, the mine owner bought all of it, at a price he set. The miner owned the means of production but ownership gave him control of the work process without control of the product. Each miner had to buy his tools, lights, fuses, blasting powder, and blacksmithing from his employer, who made a profit on each captive sale.[11] The piecework-entrepreneur system persisted until the 1940s, building exploitation, or the perception of it, into every phase of coal production. A mine operator who treated his employee equitably had trouble competing against those who didn't.

A miner's weekly paycheck depended on his expenses, the working conditions in the place he was assigned, his output, and the tonnage rate. Company decisions influenced each of these variables. Working conditions vary—some spots have a more stable roof; some are level, while others dip; some collect water; some faces are split by layers of clay rock, called "middleboys," that have to be discarded; some seams are simply thicker than others. The mine owner got about the same net daily production regardless of where he assigned individual miners. But miners would be richer or poorer according to the places they were assigned. A miner's output depended on empty mine cars being available to him. Managers could be slow giving him "empties" when coal demand weakened or to teach him a lesson. Mine operators could shave expenses by "shorting" the tonnage credited to each miner.

A ton was more a frame of reference than an immutable unit of weight. The miner's ton was usually the long ton—2,240 pounds—rather than the 2,000-pound short ton. Some companies paid long and sold short.

Payment was usually based on the number of one-ton mine cars brought to the surface each day. If a miner was paid a fixed rate per carload, a mine owner might enlarge the car to hold 2,200 or 2,400 pounds. Another approach was to require that each car be topped with extra coal instead of leveled to the sides, which added another couple of hundred pounds. Where miners were paid by the ton rather than by the carload, the company docked pounds if the weighing supervisor felt that a car contained too much incombustible rubble. Sometimes miners were paid only for the lumps they loaded. Their gross tonnage was passed over screens through which coal chips fell. The screen holes varied from less than an inch square to more than two. As much as half of a miner's gross output might fall through the larger screens.[12] Finally, each company fixed the tonnage rate in light of its profit margin and selling price. When slow demand trimmed prices, mine operators knocked down tonnage rates. Opportunities for cheating and misunderstanding occurred every day.

Over time, these practices marked several generations of miners as indelibly as the blue "tattoos" of coal slivers and specks embedded in the cheeks of many veteran miners. The hatred and suspicion that miners learned in this closed world of the operators' making have never dissipated. It is a factor in the industry's labor-management relations today. If nineteenth-century coal capitalists had hired Karl Marx to fashion a system that would inevitably create a class-conscious work force and the worst record of labor relations in American history, he could have done no better than they themselves.

Yet the entrepreneur system also lent itself to a kind of worker independence, "the miner's freedom."[13] The work demanded self-sufficiency. Turn-of-the-century mines often used a thousand or more men underground at the same time, but they were divided into dozens of separate places where the coal was cut. It was isolated and individualistic labor, totally unlike factory work, where the architecture of the assembly line defined the work. The early miner shaped his environment, controlled it, set the rate of its change. The mind-numbing mechanical movements that factory managers found so desirable were unsuited to mining, where individual judgment, above all, was needed to mine efficiently and safely.

This freedom led to what mine managers referred to around the turn of the century as "indiscipline." Producing coal could be—indeed, was—done with little supervision. The miner quit work when he determined he had done enough; he insisted on doing the work in his own idiosyncratic way.[14] Section bosses were unable to visit each place more than once a shift, and usually not that often. Coal companies employed relatively few bosses; one to every one hundred workers was a common ratio.[15] In the mine, one miner said, "They don't *bother* you none."[16] Supervisors acted principally as logistical coordinators rather than omnipresent overseers.[17]

As entrepreneurs, miners accumulated a set of work rights. Among the most important were the right to refuse unsafe work; the right to a "square turn," each miner getting the same number of empties; the principle of sharing equally whatever work was available; the right of seniority; the right of consultation over work methods; the right of miner-elected committees to have some say over working conditions; the right of the senior miner to choose his own "buddies"; the right of the miner to his old place; and the right of "irregularity"—the miner choosing when to quit work each day.[18] British miners transplanted customary work-place rights to America. The miners' claim to rights in the work place was boldly radical. In the late nineteenth century, management and government conceded to labor few if any rights. Coal miners insisted on—and made stick—rights that were inherent in the work itself. Coal miners' grievances today often involve rights established 130 years ago. These rights helped to create attitudes that led to unionization in the late 1800s. The UMWA eventually codified many customary rights in its contracts.[19]

What impresses the twentieth-century observer is how much coal early mining techniques could extract with so little machinery. The room-and-pillar method, used in Europe at least as far back as the sixteenth century, mainly required human energy and local materials. A big mine in the late 1800s might at most have two big pieces of equipment—a water pump and a hoist. Inasmuch as 70 percent to 80 percent of production costs went to labor, it was apparent that profits would increase if a way were devised either to accelerate the pace of mining or reduce the labor necessary to get the same amount of coal. Machines held the promise of doing both. Profit, more than necessity, mothered invention. Mechanization allowed the operator to extract more value—that is, more income from each worker—than before by making him more productive. A pick-and-shovel miner might produce three tons per day while a machine-assisted miner could cut and load half again as much at manageable extra cost.

The first procedure to be mechanized in the mining sequence was undercutting. Electric cutting machines, introduced in the 1890s, accounted for about 25 percent of total production by 1900.[20] These machines slashed by 75 percent the time that hand undercutting required, and the productivity of the average bituminous miner rose from 2.98 net tons per work day in 1900 to 4.52 tons in 1925. The cutting machine restructured the traditional work process. It vastly increased the need for handloaders, and, soon, for mechanical-loading machines.

Undercutting machines spread quickly, but mechanical loaders accounted for only 10 percent of underground tonnage by 1930.[21] Mine operators discovered that the saving gained through mechanical loading was somewhat offset by the machines' indiscriminateness. The handloader, paid by the *clean* ton, had to meet quality controls. Mechanical arms could not distinguish between coal and worthless slate; the machine loaded both. Consequently, some operators were forced to build expensive surface facilities to size and clean their coal. The use of dirty coal contributed to urban air pollution—the "noon-day midnights" of Pittsburgh—until the late 1940s.[22]

Labor and management soon learned that even the most primitive machines changed many relationships besides who did what and how. Machines slowly eroded both the "freedom" and the "indiscipline" of coal's early days.[23] The work became more standardized, less dependent on craft. As machines grew in sophistication, they were engineered for simplicity of operation. Workers who ran them were semiskilled, and those who handloaded—a task that mainly required strength and stamina—were even less skilled. Mechanization meant that the cheapest, most desperate workers could be hired with no loss in production.

Machines imposed a new job structure on mine labor. Each machine created a new job title. The old generalist saw machines break his job down into separate tasks, each with a different wage rate. Machines did not lend themselves to piece-rate measurement. Miners did not want to be penalized for machine breakdowns, and mine operators also pushed for hourly wages. The old uniformity of a mine-wide piece rate increasingly gave way to a ladder of hourly wages, and management controlled access to each rung. Cooperative workers could be rewarded with better-paying jobs, others penalized.

Hourly wages destroyed the custom of irregularity and often lengthened the time a miner had to work to meet his income needs. At first labor tried to adapt piece rates to machine mining, but abandoned the effort by the 1920s.[24] As hierarchy replaced democratic uniformity in job tasks and pay scales, labor demanded a rigid seniority system that gave the individual a measure of control over what he did and how much he received. Job access and "classification" are issues today, as they were ninety years ago, over who controls work.

Machine mining required more investment and greater work-place coordination per ton of output. Unlike men, company-owned machines were expensive to replace. To protect its capital and coordinate the work process, management expanded work-place supervision. Machines turned mining into something closer to factory work, where profit depended on efficient coordination of a sequence of machine-related activities. When one step went faster or slower than expected, workers sat idle. Management assumed that miners could not or should not coordinate work on their own, so supervisors were hired. The number of section bosses increased as the number of miners declined. Where there was one supervisor for every one hundred or so workers in the early 1900s, at least ten oversaw every 100 by the late 1960s. And the 1,000 workers who had mined a given amount of coal had dwindled to 100.[25]

Machines created a new hierarchy *within* the mine work force that reversed the proportion of skilled to unskilled labor. They brought a different kind of worker into American coal mines. No longer were skilled British miners recruited. Instead, the industry sought non-English-speaking immigrants from central and southern Europe and inexperienced, untrained labor, white and black, from the American South. These workers were thought to be more easily managed than skilled, homogeneous miners with well-established work rights. A West Virginia coal operator and mining engineer observed in 1894:

> Machine mining is coming into general use, not so much for its saving in direct cost as for the indirect economy in having to control a fewer number of men for the same output. . . . As the machine does the mining, the proportion of skilled labor is largely reduced, and the result is found in less belligerence and conflict; a sufficient inducement though the direct costs be the same.[26]

Black or foreign-born miners often got the worst jobs. Early cutting machines needed a worker who was able to stand the noise and vibration:

> The machine runners were generally great big negroes, who could stand the racket. When the pick of the machine would stick, the body of the machine would pound the runner. That was what the "coon" was there for, to get the licks the coal did not catch. Whenever a runner would "quit the biz" the other coons would explain it by saying "his guts wasn't nailed in."[27]

Foreign-born workers were thought to be impervious to unionization, sometimes even to ordinary feeling. George Baer, spokesman for the anthracite operators at the turn of the century, told a reporter: "They don't suffer. They can't even speak English." West Virginia's first constitution provided for an immigration officer to advertise in Europe and arrange transportation to the United States. These officials "often sold their services to the company which would pay the highest commission on imported laborers."[28] But pluralizing the racial and ethnic base failed to reduce worker militancy.

Working conditions continued to radicalize miners. Even the least class-conscious followed UMWA strike calls, which frequently lasted for months. The basic egalitarianism underground usually stayed intact above. Some European miners promoted their socialist ideas after the trans-Atlantic passage. Mainly, though, coal operators created their own labor troubles. Machines made mining a duller, faster, and more alienated form of labor. While mechanization ultimately allowed coal to compete with other fuels, it created conditions that stimulated unionization. Where unionization failed when mining was a craft, it succeeded roughly apace with machine-related industrialization of the work force. The UMWA established itself in the very decades when mechanization began draining the skill from the work and changing the social base of those who did it.

Coal operators hoped mechanization would weaken unionism among miners. Thomas A. Stroup, superintendent of the Utah Fuel Company, expressed a common industry sentiment in 1923:

> The first steps in checking the spread of unionism in the coal mines and in restoring the industry to a sound position, on labor at least, will be to abolish the contract system, to mechanicalize the mines thoroughly, to standardize every operation down to the minutest detail so that no responsibility of any kind will fall on the worker.[29]

Stroup assumed that machines would make miners brainless robots. Instead, industrialization further politicized miners, because it worsened the social and economic relationships of the work itself without altering many of the issues that had created conflict before, such as safety, decent working conditions, fair treatment, and job security.

Beginning in the 1880s, mechanization figures prominently in the coal industry's labor relations and its ability to compete in the market. Machines created new pressures in the work place, setting the stage for almost four decades (1890s to early 1930s) of bloody conflict. All of the changes that mechanization brought—taking the craft out of the work, recomposing the work force, reducing the price of the required labor, phasing in hourly wages, eventually creating a surplus labor pool—were factors in that struggle. In some instances, mechanization stepped up the intensity of conflict. The early machines accelerated mining without making it safer. They created new hazards—shocks, sparks that could ignite methane or dust, moving machinery, even electrocution—that for the first thirty years of this century contributed to an annual average of more than two thousand fatalities.[30] By raising productivity, machines created surplus labor that destabilized industrial relations for years.

Market forces demanded higher productivity and cost controls. The question is not whether mining had to be mechanized, but how. Did the market modernize the coal industry humanely? Did the market internalize the social costs of mechanization? Did the market restructure its labor force wisely as mining mechanized? Did the market approach to modernization produce the greatest good for the greatest number? The coal industry should have done better—much better—at modernizing its technology and adapting its managerial philosophy. The cost-cutting pressures of competition and the philosophies of managers prevented the industry from doing so.

A miner's life outside the work place was directly connected to his work. Mining created a near-total frame of reference for defining a worker's life. That framework transferred the class relations of the work place to the social relations of the community. Most coal mining took place in isolated—by both geography and design—communities where industrialization of the mining craft played out again in the industrialization of the mining camp. Mine management sought to exercise increasing control over both phases of their enterprise. Techniques of social and economic exploitation were common in the camps, techniques that trapped families in debt, stripped them of cash and dignity, and left them at the mercy of management even in their most private affairs. Conditions of life and work bred unionism in the coalfields. In the latter part of the nineteenth century and first decades of the twentieth, most mining was synonymous with a way of life that at best was only a step or two away from indentured servitude.

Three features of private sector development of America's coal reserves shaped the industry's relationship with its workers. First, entrepreneurs deliberately developed the coalfields as a *one-crop economy.* Mine operators discouraged other economic activity that might provide alternative employment and possibly a more benign business model. Where coal mines did coexist with other business—as in Iowa, the Ohio River Basin, and near Pittsburgh—coal companies tended to be more flexible than the norm. Coal unionism first sank roots in this northern tier because in part of the pluralism of its economy.

Second, much of America's privately held bituminous reserve came to be controlled by several dozen corporations with multistate holdings. Local entrepreneurs lacked the capital to assemble blocks as large as those of the out-of-state corporations with which they competed. Corporations headquartered in New York, Philadelphia, Pittsburgh, Cleveland, and Chicago bought the mineral rights to much of coal-rich Appalachia, the Ohio River Basin, and the Rocky Mountain states. Coal land was cheap at the time—especially in Appalachia, where hill farmers were chronically short of cash.[31] By the 1920s, absentee corporations had locked up most of America's coal save for that still owned or managed by the federal government.[32] The implications of absentee ownership were far-reaching. The political and economic fate of coal communities lay in the hands of businessmen and bankers who were insulated from the human consequences of their decisions.[33] These corporations developed America's coal like the crop of one massive tenant farm on which both capital and labor were considered the landlord's property.

Third, pioneer coal companies had to organize not only a mining operation but an entire community. Mining villages were occupational communities. They existed only to meet the needs of the town's one employer, the mine. Many "camps" were totalitarian societies molded by the mine operator to his goal of making a profit from his investment.

The basic model of the American coal camp came from Britain. There, before their gradual economic and political liberation in the nineteenth century, miners had lived as serfs or indentured "freemen" for centuries. The British coal camp had a rigidly structured class system that denied miners even the most minimal rights. The essence of the camp was monopoly—the company's deliberate suppression of all private sector activity other than its own. The company owned every building in the camp. Representative of a small operation is this list that Maryland's Cumberland Coal and Iron Company reported to its stockholders in 1853:

> There have been erected on the land of the Company 152 dwelling-houses for miners; two churches, one of stone; one large brick schoolhouse; a handsome stone dwelling and outhouses, for the residence of the engineer; 6 large stables, 3 granaries, 1 large machine shop, 5 blacksmith shops, 6 offices, 1 large stone house, 1 large lumber house, 2 engine houses, 5 weigh houses.[34]

The coal-camp system that began in the mid-nineteenth century remained virtually unchanged until the 1930s. Reporter Winthrop Lane analyzed its operation in the early 1920s during a West Virginia strike:

> The operators are in the position of power. Their power comes chiefly from their ownership of property. . . .
>
> The operators are not only the miner's employer; they are his landlord, his merchant, the provider of his amusements, the sanitary officer of the town, sometimes the source of his police protection and the patron of his physician, his minister, and his school teacher.
>
> It is paternalistic, in some ways a feudal civilization.[35]

Monopoly ownership distorted property and political rights. Ownership gave the companies plenary political authority; by virtue of being propertyless, miners were disfranchised. In many cases, miners were required to remain tenants as a condition of employment. Most operators were reluctant to allow miners to own private property in their camps because of the independence that might ensue: "In the Cumberland region the miners own their own houses and hence they have held out [on strike] as long as they have."[36] If a strike occurred, coal companies could evict tenant strikers.

The psychological, economic, and political leverage that camp ownership gave management is obvious. Miners were either discouraged or prohibited from living outside the camp if rental space was available. They lacked the most elementary tenant rights. Frequently, they were prohibited from having "outside" guests in their homes. Fourth Amendment protection against search and seizure did not apply. Consolidation Coal Company set forth the logic of the camp which, Harvey says, was "probably typical for Maryland coal operations in the 1870s":[37]

> In order to secure and retain good miners (men of family) it will be necessary to erect fifteen or twenty small tenements at one of our principal mines. . . . As these houses will yield a fair return in rent, the company will be fully reimbursed in interest on the outlay. By furnishing such tenements, skilled miners are secured, who become permanent residents of the region, while their sons as they grow up, become the faithful employees of the company which has given steady work and comfortable homes to their parents.[38]

This is the paternalistic side of the landlord-tenant relationship. More often, a harder side revealed itself. In 1864, three coal companies in the Blossburg District of Pennsylvania proposed that house leases be amended to allow the companies to evict on short notice for violations of housing rules or for disputes arising in the work place. (The operators' proposal was clearly in violation of existing state law, which allowed tenants to stay in their houses for up to a year as long as they paid their rent.) According to Wieck, the operators used the lease modification to precipitate a strike and break a fledgling union, the American Miners' Association, in their mines. The Fall Brook Coal Company wrote to its miners:

> The houses have been built for the accommodation of those employed and willing to do their duty; not for idlers or disturbers of the company's business. Hereafter no one can occupy a house except if he executes a contract defining his rights and duties. . . . There shall be no relaxation on this point. The company will maintain the right to control their [*sic*] property. Self-respect and justice require this. If the company had at any time denied you full, generous compensation for your services you would have had some reason to form combinations. As it has been, and is, your action is uncalled for, unreasonable and disorderly, as well as disrespectful to your employers and best friends. The continuance of such unjustifiable conduct cannot and will not be tolerated.[39]

Company housing was the rule in the Appalachian coal fields, where most mines were located. A 1924 survey found about 75 percent of the miners in West Virginia living in company housing, compared with 64 percent in Kentucky and 9 percent in Indiana and Illinois.[40] Some companies sold their rental housing in the 1930s. When a company laid off miners, the lack of rental income provided incentive to sell the property. By the 1960s, most company housing had been sold. The sellers, however, retained ownership of the minerals beneath these houses and towns; this allowed them to force the sale and demolition of the dwellings if they needed the land for mining or mine buildings. Coal companies generally refused to sell undeveloped land suitable for housing to miners, forcing them to buy the old company houses.

The company town was an all-inclusive system intended to control behavior by making residents totally subject to the authority of the coal operator. The company provided all town services (such as they were) in most camps. Some services were financed through wage deductions. One of the worst features of the camps was the mine-guard system, under which company-hired gunmen patrolled the camp as civil police.[41] Sometimes operators hired detective agencies for police work; in other cases, sheriffs were elected but were paid by the company. The privatization of the police power transformed union struggles into battles against public authority. The distrust miners often express today for "the government" is rooted in their families' experience with company police during the mine wars and strikes between 1900 and 1940.

Education, religion, health care, and cultural activities were company functions and reflected its thinking. The official God of the coal camp was the company's God, and the prescripts of His church dovetailed with those who risked their capital. The commingling of God and company was established first in the British camps. When miners in Durham and Northumberland struck in 1844, for example, they were evicted from their housing and forced to camp on the moor. Lord Londonderry, a leading coal operator, had a clergyman counsel the strikers in this fashion: "You are resisting not the oppression of your employers, but the Will of your Maker—the ordinances of that God who has said that in the sweat of his face shall man eat bread, and who has attached his penalty to the refusal to labour, namely, that if a man do not work neither shall he eat."[42]

Medical care was no less impartial. Mine operators deducted a fixed amount from each miner's paycheck to provide medical care and insurance.[43] Even though miners financed these services, they had no voice in how they were provided, or in the selection of the camp doctor. Medical records were not confidential. Mine operators provided health care at a small profit, or, at least, at no extra cost. The doctor was often given the use of a clinic. Some doctors built for-profit hospitals financed by a regular wage deduction, called a check-off. Such hospitals had little incentive to prolong care, as miners usually had no way of paying for lengthy treatment. The thrust of the "company doctor" system was to keep the labor force just healthy enough to continue work.

The company store, where miners traded as a condition of employment, was another legacy of British mining—the "truck" system.[44] In both countries, the store defined what a miner did with his pay as if the company never lost its grip on those wages. For all the industry's talk about the virtues of market competition, miners saw how free enterprise in coal went hand in hand with the denial of free enterprise in the camps. A miner caught buying goods outside the camp might forfeit his job. Not without reason did miners dub company stores "pluck-me" stores—they frequently charged higher prices than those where competition existed. Often, miners were paid not in dollars, but scrip. Each company had its own, which was worth something only in the camp. Miners paid as much as two dollars in scrip for every dollar in U.S. currency on the black market. The stores frequently encouraged credit buying, often with interest. Once a miner was looped into the circle of credit, he had to continue working for the company to which he was indebted. Only those who were free of debt could quit and move away. After the company deducted for rent, house coal, schooling, medical care, utilities, and other compulsory services, a miner's due was matched against his store account. The company subtracted his debt from his earnings, and if he had anything coming to him he was paid in scrip. Many went from paycheck to paycheck without ever pocketing a nickel in cash or scrip.[45]

The camps deliberately structured a class-divided society. Management lived apart, often on a hill overlooking the town. Everyone else lived in identical houses in the camp.[46] Where the two classes might meet—say, in the local church or school—self-imposed segregation was the norm. Companies frowned on miners establishing their own institutions, be they lodges, churches, or unions. They preferred to envelop miners in a total environment, the better to control their behavior. Mining became an occupational prison from which there was no escape, upward or outward. Most camps were run along the same principles, so it did little good to hopscotch from one to another. Miners who crossed their employers found themselves blacklisted and unable to work.

"Them and us" was the principle the coal camp laid down, and the principle miners upheld. Conflicts that arose in the work place engulfed the entire community. The hot-house environment of the camps intensified the perception of exploitation miners carried with them from the pits. Everyone, from women offended by bad drinking water to children self-conscious of their ragged clothes, felt aggrieved. The camps allowed no room to cool off work-place disputes. When a dispute triggered a strike, the entire community split cleanly into two sides. The decision to strike meant eviction, the end of medical care, uncertain food supply, and, of course, no income. Yet it was the very rigidity and oppressiveness of the camps that forced miners and their families to risk their lives to gain even the most marginal improvements. Dissatisfaction gnawed continuously at the residents, who blamed the companies. Had miners blamed themselves, or their stars, they might have become resigned and docile. Instead, the structure of their lives and work created anger.[47]

The coal camp was the basic community unit in America's coalfields for most of the industry's history. It was a "total institution" that defined the existence of its residents.[48] One central authority planned, scheduled, and controlled all camp activities. In most camps, no "public sector" existed in the way we use that concept today. Local government taxed at rates that mine operators said they would pay. The company, rather than government, decided how much attention would be paid to environmental protection, public health, and community services. This system permitted no competition for or division in political power. No independent middle class surfaced. In this way, mine operators created conditions that led miners to unionize. Operators correctly saw unionization as a challenge in the work place as well as a threat to an entire way of life. Consequently, from the 1880s through the mid-1930s, the relations between labor and management often turned into no-holds-barred class wars. Coal camps persist, if not in substance, certainly in form and sometimes psychology. Their legacy continues to shape collective bargaining.

The industry boomed from the 1880s to the early 1920s, but there was little praise worthy in its treatment of workers and their communities. Yet the operators, too, were caught in a system only partly of their own devising. It was not their fault that the nature of their business required them to build towns and provide services. True, mining was dangerous. But some of the hazard was in the nature of the enterprise and its primitive technology. Miners knew the risks. Operators could argue that had they not risked their capital, miners would have had no jobs at all—immigrants would still be impoverished peasants; blacks ensnared by sharecropping in the rural South; native whites chained to their meager mountain farms. Wages had to be cut and unions kept out, they said, because the brutal realities of the marketplace required it. Class war on the supply side was caused by war over markets on the demand side.

Both miners and operators were to an extent captives of forces they did not control. Neither could do much about the peculiar rhythms of coal demand. Neither could alter the inherent danger of boring tunnels unders millions of tons of rock. Neither could readily affect the simple technology that was available. But within this set of givens, coal operators concocted their own mean twists.

Why so? First, they didn't know any better. Their British colleagues had set up mining villages as prisoners-of-job camps, and Americans followed suit. Second, laissez-faire ideology in the first half of the nineteenth century contained no generosity for labor. Capital had rights; workers had none. Social Darwinism and the contempt that Yankee entrepreneurs held for workers of different color and ethnicity promoted at best paternalism, at worst repression. Finally, the pressure to make a profit increased the level of exploitation. Competition promoted cutting costs and corners alike. Mine operators used their company towns to recapture investment costs and increase profits. The company store system was so lucrative that mine operators often paid their workers to mine coal even when it sold for no gain. Repeated hundreds of times across the coal fields, this synergism contributed to the industry's chronic problems of oversupply and instability.

Journalists and public investigators publicized conditions in the coal camps with metronomic consistency after 1900—when striking miners and their families wintered in tents and caves upon eviction, when mines exploded, killing hundreds, or when some were killed in organizing battles. And nearly everyone lamented the squalor of coal-camp life—the open privies, the lack of potable water, the ramshackle dwellings, the scant concern for medical care and education. Yet coal operators survived these periodic exposés, because capital was mainly—perhaps only—responsible to itself. Legislation eventually prohibited child labor, scrip, company-paid civil police, and other onerous features of the system. But it took decades of suffering and turmoil to win each small change.

The American coal industry's application of the British coal model raises the question of whether alternative approaches might have been less destructive and more productive. As the U.S. industry matured, variations on the theme appeared. But, management and labor alike generated few other models of development. Little evidence appears of miners running mines cooperatively; Wieck notes only "isolated instances" in the 1860s of "small groups of miners leading and operating mines on a cooperative basis."[49] There is mention of a miner-run operation that failed because the local railroad refused to supply it with empty cars.

Resolutions calling for union-owned mines, stock ownership in coal companies, public management, and nationalization met with little success at most UMWA conventions. Labor—itself with British roots—accepted the basic model of private ownership and control despite its record of inequity. As the industry mechanized, there is little record of protest among miners of the change in their work-place status. The UMWA entertained few notions about worker self-management or joint management between workers and owners. Similarly, labor accepted for many years the basic theory of the coal camps, protesting only its abuses and seeking reform through legislative rather than contractural change. Those camps that were the exception to the norm—where housing was decent and might be purchased, where merchant competition existed, where politics was pluralistic—were the creations of company paternalism rather than worker pressure.[50] Union leadership never encouraged rank-and-file thinking about alternatives, and "practical" issues dominated worker concern. The private enterprise model defined the behavior of both victimizers and victimized.

The sixty-six years between 1860 and 1926 were coal's golden age. Total production grew from 20 million tons to 658 million. America's coal reserve was cheaply obtained and seemed inexhaustible. Mine operators had little incentive to mine carefully or efficiently. Expansion was the prevailing psychology, and enterprise was generally free from governmental regulation. No corporate income tax existed until 1910.[51] Labor unions had no legal right to exist. Coal operators could do whatever they pleased to the land, air, and water around their mines. Despite repeated congressional and presidential investigations into the industry's labor practices and economics, bituminous coal suppliers structured their mines and communities to suit themselves.[52] Only in the anthracite, where a handful of corporations controlled supply, did the federal government act—forty years after this "combination" began.

Coal's golden age was the product of two distinct forces: first, industrialization demanded steadily increasing coal production; and second, other fuels did not challenge coal's biggest markets. Just as coal supplanted wood, so too would other fuels supplant coal. All fossil fuels are hydrocarbons: molecules of hydrogen and carbon atoms locked in different proportions and to some degree interchangeable.

From 1870 to the 1920s, other hydrocarbons did not challenge coal because coal was cheap, accessible, and easy to use. It was cheap because it was plentiful and readily obtained. Coal's economic advantage lay in its labor intensiveness, as long as labor was cheap.

Market economics brought bituminous coal its golden age and kept it there *as long as demand increased steadily.* But expansion encouraged the opening of thousands of bituminous coal companies. Competition was intense. The diffuseness of America's coal reserve prevented the larger suppliers from concentrating production into an oligopoly. When coal demand stopped growing steadily in the 1920s, the industry found itself with too many suppliers and too many workers. Suppliers handled reduced demand by cutting labor costs, which allowed many of the least efficient companies to survive. Free-market competition drove the bituminous industry to its knees in the late 1920s and early 1930s. No mechanisms were available in the private sector to end the self-destructive circle of price cutting and labor exploitation.

Ever since the Civil War American coal suppliers have almost always been able to mine more coal than they could sell.[53] In the 1980s, this situation is called "demand constraint" and treated as if it were something new and temporary. Oversupply led to a particularly brutal form of supply competition where prices were cut—and cut again—to retain sales. Price cutting hit a miner squarely in his pocketbook—his piece rate or his wages were cut. Geology and free enterprise created conditions that gave birth to a miners' union, the one cloud over coal's golden age.

2. "Single Stones Will Form an Arch"

Relations between labor and capital have defined the public's image of coal. The contours of the industry's history show endless ridges of conflict separated by valleys of temporary truce. Disentangling historical fact from fiction is not easy. Labor writers have usually written of coal miners as either hopelessly exploited or mythically heroic. Miners, indeed, have been victimized, and many now-forgotten workers and their families have at times acted heroically. But as individuals they are as complex as anyone else. What defines their character as a group is their collective response to the circumstances in which they found themselves.

American miners' unions have British forebears. Until they were emancipated from serfdom in 1799, Scottish miners had been bought and sold as part of the mine property, the life's labor of their children "sold" at their baptisms.[1] A more subtle system of indenture replaced serfdom. It consisted of a contractual "bond," the truck system, and arbitrary fines for short tonnage.[2]

British miners organized local unions after 1825. The Miners' Association of Great Britain and Ireland was organized in 1841 as the first national union. It collapsed after several strikes, but miners had been able to modify the "bond." The association pressed Parliament to regulate coal operators. In 1842, boys under ten and women were prohibited from working underground. Until then, teenage girls, outfitted with girdle and chain, dragged 170-pound loads on their hands and knees and then up winding stairways for fourteen hours a day, naked from the waist up; and six-year-old Shropshire lads "hurried" coal for their miner fathers through three-foot-high tunnels. (American mines employed young boys until it was outlawed by child-labor legislation in the first part of this century, but there seems to be little evidence that girls and women worked in our mines.) The first mine safety legislation, the Mines Regulations Act of 1850, established a public inspectorate—of two! After five years the act was to expire, an example of what antiregulators now call "sunset legislation."[3] A second national miners' union, the Miners' National Association, formed in 1863, two years after Americans organized their first national union. The association urged Parliament to provide minimum ventilation standards, more inspectors, regulation of the truck stores, and public education for coalfield boys before they entered the mines. Mine safety legislation passed in response to headline-grabbing explosions, a gruesome politics of stimulus and response that America copied.[4]

Anthracite miners formed the first recorded American coal union in 1848, in Schuylkill County, Pennsylvania. Known as the Bates Union after its founder, Englishman John Bates, it won a three-week strike over scrip and other grievances, but fell apart a year later.[5] Bates vanished with what was left of the union's treasury. About twenty years later, John Siney, another anthracite man, helped to organize the Workingmen's Benevolent Association. He became president of a successor union—the Miners' National Association—in 1873. Siney opposed strike violence and demanded that union members give "strict adherence to social law and order." In his founding declaration of December 2, 1868, he stated that the object of the Workingmen's Association was to unite "in one band all who earn their bread by hard toil—more especially the miners and laborers in Pennsylvania. Without obligations, signs or password, we move along . . . determined to judge for ourselves in the future, having clearly learned that our labor is our own capital, that our company is a strong one, and that its stock is always worth its par value." The association organized 20,000 anthracite miners in an unsuccessful strike for the eight-hour day in 1870, and a second a few months later for reduction of stockpiles and adoption of a minimum wage. Siney urged conciliation and arbitration in place of strikes, yet he found himself embroiled in one after another over wage cuts in the 1870s.

The anthracite companies organized spies and vigilante groups to wipe out labor militants. Miners fought back with guerrilla warfare. The operators claimed that the Molly Maguires, a secret Irish brotherhood, masterminded the resistance behind the public front of the Ancient Order of Hibernians. Many Irish miners in the anthracite region were Hibernians, but the specter of a Molly Maguire conspiracy seems to have been mostly employer propaganda and media hysteria, although both sides engaged in terrorism and property destruction.[6] Nevertheless, Mollies were front-page news in the late 1870s, and the anthracite operators portrayed themselves as victims of working-class terror. A *New York Times* editorial of May 14, 1876 called for Molly blood:

The Molly Maguires have long been suspected of many violent crimes. It now seems that from burning "coal-breakers" and mobbing men who were willing to work for wages under the rate demanded by strikers, the rise to darker deeds was easy and natural. The trade unions of England have resorted to murder, assassination and arson, in order to strike terror into the hearts of those who oppose them. This root of evil has been planted on American soil. . . . The Pennsylvania authorities owe it to civilization to exterminate this noxious growth, now that its roots have been discovered.

Nineteen Irish miners were tried on conspiracy and murder charges in a series of trials after 1875, during a period of falling prices, wage cuts and sabotage. Franklin B. Gowen, head of the leading anthracite company, the Philadelphia & Reading Railroad, broke the 1875 strike and, as prosecuting attorney, won convictions against all nineteen. James McParlan, a Pinkerton detective, employed by Gowen's railroad, supplied incriminating testimony against the miners. He was the only witness who claimed the Molly Maguires existed. All nineteen were hanged.

Anthracite miners were the first to organize, but bituminous miners were the first to see the need for a union of all coal workers in a single organization. Bituminous miners were the victims of the cutthroat competition among employers, who tried to maintain sales by cutting wages. A national union held the promise of being able to equalize labor costs and end this practice. Mine-worker policy beginning in the 1880s demanded an elaborate system of pay differentials (which factored in mine location, seam thickness, and the prevalence of machines) designed to equalize production costs across the field. The UMWA had some success in getting employer agreement, but labor costs could not be equalized in practice and the UMWA was never able to organize all suppliers. Consequently, low-wage, nonunion companies continued to undersell unionized suppliers, forcing unionized operators to cut wages in response.

Immigrant British miners organized the American Miners' Association in Illinois in 1861.[7] The association organized industrially and excluded no ethnic or racial group. It built a democratic, representative structure, which the United Mine Workers of America inherited almost thirty years later. A convention of miners established policy. Local "lodges" had a measure of autonomy. The association's first constitution carried the motto, "Union is Strength—Knowledge is Power," and contained this poem:

> Step by step, the longest march
>   Can be won, can be won;
> Single stones will form an arch,
>   One by one, one by one.
> And, by union, what we will
> Can all be accomplished still.
> Drops of water turn a mill,
> Singly none, singly none.[8]

The association faded in the late 1860s in the face of demand cutbacks, internal dissension, and stiffening employer resistance.

Miners tried again a few years later, this time as the Miners' National Association with John Siney as president. Its founding convention in 1873 adopted moderate goals: to eliminate the causes of strikes and accept the principle of arbitration, obtain legislative improvements, shorten work hours, secure true weights, sue for compensation when employer negligence caused accidents, and establish a strike fund. The 1873 Panic prompted drastic wage cuts and precipitated uncoordinated local strikes. Despite much political repression, the association claimed more than thirty-five thousand members in 347 locals in twelve states and the Indian Territory (Oklahoma) in 1875.[9] Poor economic conditions and the lack of central discipline over strikes undid the association.

The 1880s saw other efforts to cobble together a national miners' union from the remnants of earlier organizations. The National Federation of Miners and Mine Laborers organized in 1885 in the northern coal tier from Pennsylvania to Iowa. Its first interstate joint conference negotiated a system of wages that ranged from $0.60 to $0.95 per ton in 1886. The conference provided the framework for multistate bargaining for the next four decades and established "minimum wage rates per ton as a means of setting some limit upon competition."[10] The Knights of Labor organized the National Trades Assembly No. 135 for all workers in and around the mines in 1886. The Knights advocated secret meetings and membership from all classes (including professionals), while the National Federation favored open meetings and organization according to craft. The two unions began to consolidate in the late 1880s. The first effort was the National Progressive Union of Miners and Mine Laborers (NPU), which embraced both anthracite and bituminous miners. NPU gave its national officers and executive board substantial power to regulate locals and districts. It advocated a minimum wage, but opposed government regulation of labor relations because "it would be unmanly for us as miners to ask either national or state legislation to exercise paternal surveillance over . . . difficulties in which we ourselves are supreme and control." [11] After losing a strike in the anthracite in 1888 and failing to work out a new contract in 1889, miners from the Knights and the NPU convened in December 1889 to make a new union.[12]

In 1890, miners—conservatives, liberals, and radicals—from several states reconstituted themselves as the United Mine Workers of America (UMWA), which wove together the organizational strands of its loosely knit predecessors. Coal production was by then 158 million tons, from mines in more than a dozen states. The industry was growing steadily and was intensely competitive. Management's largest cost—70 percent—was labor. Sales depended on cheap labor and steady production. Unionism was the inevitable result of that formula.

The UMWA's first presidents led miners through one tumultuous strike after another. Most were over wages and the UMWA's right to organize. Without bargaining rights, the union could not sustain itself. For strikes to be effective, UMWA presidents had to control their use. John Rae, first UMWA president, denied locals and districts the right to strike except when he and one member of the union's executive board consented. The UMWA stumbled when a market tailspin prompted operator wage cuts in the mid-1890s. Its treasury held only six hundred dollars in 1897 when UMWA president Michael Ratchford called a "national" shutdown to save the organization.[13] Miners responded, and they won their desperation strike. Wages increased four cents per ton, to $0.65.

Of the early UMWA presidents, John Mitchell, who took over in 1898, cut the deepest mark. A business unionist to his marrow, Mitchell combined philosophical conservatism with strong—although not always sound—leadership. His brand of unionism shaped UMWA policy but less so rank-and-file thinking. Mitchell believed: "Trade unionism stands for no impractical, speculative theories. It is a plain, common sense, business institution that insists upon labor receiving a fair share of the wealth which labor produces."[14] He described his philosophy in *Organized Labor:* "There is no necessary hostility between capital and labor. Neither can do without the other. . . . There is not even a necessary fundamental antagonism between the laborers and the capitalist . . . . Broadly considered, the interest of one is the interest of the other, and the prosperity of one is the prosperity of the other."[15] Mitchell believed in a confederative union structure. He favored local "autonomy" whereby each district elected its officers and handled its finances. He allowed each district to negotiate its own contract. But this system, the districts discovered, was self-defeating. Miners in a high-wage district found themselves out of work when their employers could not price their coal competitively. District bargaining also undercut the UMWA's effort to equalize wages. Socialists, Wobblies, Mother Jones, and UMWA radicals bitterly attacked Mitchell's decentralized approach.

Mitchell helped the UMWA grow from about 33,000 miners in 1898 to about 250,000 and a treasury of more than $1 million by 1903, but his incumbency from 1899 to 1908 established a pattern of financial manipulation that some later presidents followed in varying ways. With the advice and assistance of Frank S. Peabody, president of Peabody Coal Company, and Harry N. Taylor, president of the General Wilmington Coal Company, Mitchell invested in coal companies, coal lands, and other ventures.[16] He was the principal stockholder in the Egyptian Powder Company, which coal operators set up to sell blasting powder to miners. He seems to have been part owner of the Carbon Hill property, whose mine in Illinois employed his own members. He helped to organize a health-and-accident insurance plan for miners for which he received commissions, and tried to get anthracite miners to donate five thousand dollars to build him a new house. He also marketed a "John Mitchell" cigar, which he hoped would become a profitable sideline. For much of his UMWA term, he negotiated with and accepted payments from the National Civic Federation (NCF), a businessmen's group largely financed by Mark Hanna, Andrew Carnegie, and August Belmont. Mitchell left the UMWA to work for the NCF, but resigned when UMWA miners made his continued membership in the UMWA contingent on disassociation.

Mitchell mined his followers for cash; he manufactured conflicts of interest as if he were running an assembly line. At the same time, he was a shrewd strike leader and a reasonably competent union administrator. Mitchell is best remembered for "winning" the eight-hour day[17] and leading a strike in the anthracite in 1902. Radicals berated him for undercutting several other strikes and impeding united strikes of bituminous and anthracite miners in 1902 and 1906. Although Mitchell's UMWA represented miners from both fields, he argued that each should have a separate contract. (The UMWA continues this practice today.)

The anthracite strike of 1902 boiled into a national crisis and established Mitchell's reputation. The strike focused on union recognition, wages, and child labor. The deadlock lasted five months. President Theodore Roosevelt brought the two sides to the White House to seek a settlement. The leader of the anthracite operators, George F. Baer, refused to negotiate. Roosevelt then contacted J.P. Morgan, whose bank controlled the railroads that owned the anthracite companies. Morgan came to the White House and announced: "The operators will arbitrate."[18] Roosevelt convened the U.S. Anthracite Coal Strike Commission to investigate the issues and publish findings. This was the first time that the federal government had intervened in coal's labor-management relationship. The arbitration commission pitted Mitchell, Clarence Darrow, and Henry Demarest Lloyd against twenty-three corporation lawyers. Darrow asked to see the operators' books. He asked Baer about the 20,000 children who worked as trapper boys, mule drivers, and slate pickers. Baer said, "The unions . . . are corrupting the children of America by letting them join their illegal organizations."[19] Darrow replied that had Baer not employed children in the first place, they could not have joined the union. The UMWA emerged from White House arbitration with a 10 percent wage increase but without recognition. Radicals like Eugene V. Debs and Mother Jones denounced the decision.

Union radical Thomas L. Lewis succeeded Mitchell in 1908. He was the first UMWA president to advocate government control of coal: "He advocated a *Governmental Fuels Policy,* a long-run conservation scheme, wherein the government would regulate fuel prices to enable the mine owner to sell his coal to the consumer at a reasonable price."[20] Lewis coupled to his radicalism a belief in strong central leadership. He asked the 1908 UMWA convention to empower the president to suspend district officers for insubordination, place the union's executive board in charge of all strikes using union funds, locate responsibility for organizing drives in the national headquarters and require that all contracts be signed simultaneously. Lewis failed to persuade the rank and file to make all these changes. He did make organizing a headquarters function and its staff became the field workers in his political machine.[21]

When war erupted in Europe, America's was a coal-dependent economy.

Bituminous production would more than double between 1900 and 1920, from about 212 million tons to more than 568 million. Employment increased from 304,000 to 649,000, and the number of mines also more than doubled to 8,900. World War I boosted the price of a ton of bituminous from $1.04 to $3.75, which was to be its highest until 1945. More than 270,000 bituminous and 100,000 anthracite miners were UMWA members in 1920.

As demand grew, UMWA organizing succeeded despite legal impediments and industry resistance.[22] In the first quarter of this century, the coalfields witnessed epic strikes—"mine wars."[23] The sources of mine-worker militancy lie in the decisions made in America's boardrooms and the combat-like conditions of mine work itself. Like soldiers, miners dressed daily for battle with danger and hardship. Survival depended on using their wits, each pulling his own weight and not letting down his fellow miners. In the hostile environment of the coal mine, with its dust, darkness, crampedness, noise, and threat of explosion, a profound sense of peer-group loyalty blossomed. The operators became the enemy in this work-place battlefield. Strikes frequently involved gun battles between strikers and company guards.[24] State militia and federal troops were frequently ordered into the camps to maintain order. Civil liberties—freedom of speech, press, and assembly and the right to a fair trial—fell victim to the concern for labor peace. Mine buildings were dynamited and the tents of strikers burned. These early union battles became legends, and their hatreds, if not their details, were passed down to children and grandchildren on both sides. Those who today read the accounts of mining conditions and these strikes may be more amazed at the miners' patience than at their violence.

Coal's abysmal record of industrial relations in the first two decades of this century cannot be blamed solely on competition and cost cutting, which were major factors later. It is also explained by the industry's attitude toward labor and its steadfast unwillingness to treat workers with anything resembling fairness and decency. Thomas A. Stroup, a mine superintendent, expressed the common view:

> The old tradition of the American workingman, intelligent, efficient, alert, individualistic, inventive, capable of using good judgment and requiring a minimum of direction, dies hard: it, of course, was always largely a myth and today there remains no excuse for its persistence in the minds of reasonable men. The psychologists have shown beyond reasonable doubt that 50 per cent of our population is underdeveloped mentally, that it is incapable of self direction or of exercising independent judgment. Considering all the factors it is probable that from 80 to 90 per cent of the employees of the average factory or mine are . . . incapable of clear thinking or self direction.
>
> . . . This simple fact of the preponderance of the subnormal in our working population must be the point of departure for all successful schemes of employer-employee relationship.[25]

George Baer, the leader of the anthracite operators, is sometimes referred to as George "Divine Right" Baer for the letter he wrote to one of his stockholders in defense of his policies during the 1902 strike: "The rights and interests of the laboring man will be protected and cared for—not by the labor agitators, but by the Christian gentlemen to whom God has given control of the property rights of the country and upon the successful management of which so much depends."

Miners were less unified in their attitude toward management than might be expected. Radicalism has always been a factor in mine-worker politics, but the prevailing attitude among UMWA leaders—and back to John Siney as well—has been a business unionism that does not challenge the private-enterprise system. The business unionist tries to regulate management behavior through contracts and legislation, but upholds management's right to set policy and produce for profit. This philosophy allowed the UMWA to survive as an institution, whereas a more radical orientation might have led to the kind of political repression that wiped out other left-wing unions.

Rank-and-file socialism was an undercurrent in UMWA affairs since its founding. Miners ran as socialists for political office in the first two decades of this century. Laslett notes that UMWA socialists had "considerable influence in several of the district unions, most notably in Illinois" as late as the mid-1920s.[26]

The UMWA's 1904 convention rejected one resolution that would have had the union "buy or sink one coal mine in each state" and another that authorized the purchase of stock in coal companies.[27] The 1907 convention again rejected a proposal that the union buy mines of its own, but it endorsed a resolution that argued for the "necessity of the 'public ownership and operation' and the 'democratic management' of all those means of production and exchange that are collectively used."[28] The convention, however, rejected a resolution that called specifically for government ownership of the mines. A year later, the convention advocated that the American people should own and control the country's coal mines. Government ownership and democratic management of all public utilities including coal mines was endorsed by the 1914 convention. UMWA socialists succeeded in getting these philosophical statements passed, but the general thrust of UMWA political and legislative activity was centrist and pragmatic. Although the union had organized industrially, rather than by craft, from the start, it never supported the radical idea of a single union of all workers. Mine-workers conventions regularly dismissed resolutions endorsing a working-class political party until 1919, when delegates resolved that labor could only be represented in Congress and state legislatures "by the organization of a labor party that will be representative of and under the control of the workers of hand and brain of the United States."[29] John L. Lewis, acting president that year, ruled from the chair that this resolution was "simply a declaration of principles" and did not bind any mine worker. A more guarded resolution on the labor party issue slipped through the 1924 convention and then the debate ended. The 1919 and 1921 conventions called for the immediate nationalization of the coal industry, a clear shift in rhetoric from earlier statements. Yet even this, the high-water mark of radical thinking, was a working-class perspective couched in political moderation. The proposal envisaged full payment to current owners and public management rather than worker management of an expropriated industry. Several UMWA officials—including John Brophy, a democratic socialist, who would challenge John L. Lewis for the UMWA's presidency in 1926—were authorized to draft a detailed plan, which Lewis, who distrusted government and accepted the notion of public regulation reluctantly, subsequently discredited and mothballed.

Radicals failed to take over the UMWA for a number of reasons. At the time of their greatest strength just after World War I, the federal government cracked down on radical unionists, chilling the efforts of those who were spared trials and imprisonment. Second, an avowedly socialist union of mine workers could expect nothing but nonrecognition and hostility from both management and public authority. Mine workers saw how precarious survival was for even a middle-of-the-road union and doubted the viability of a radical alternative. Third, the radicals were never sufficiently united to overpower their political opponents within the union. Fourth, much of the rank and file never saw itself as socialist for whatever reason. Small gains in wages and working conditions persuaded some to steer away from radical politics. The risks of radicalism—unemployment, difficulties within the union, blacklisting, and so on—were high, and many considered them not worth taking. Finally, the institutional momentum of the UMWA flowed in a more centrist direction. Mine-worker organizations since the middle of the nineteenth century had been reformist rather than socialist. They had sought modest contract improvements and legislative reform, preferring arbitration to strikes. Radicals found little encouragement in their union's historical evolution and met hostility from most of the elected leadership.

Business unionism was less risky and more likely to achieve gains in the here and now. It prevailed within the UMWA over the years because it adapted to changing circumstances. While opposing strikes, business unionists like John Mitchell and John L. Lewis established reputations as strike leaders. While opposing nationalization, Lewis came to advocate strong federal regulation of the market to stabilize competition and protect labor. The very harshness of coal management allowed conservative unionists to maintain credibility among the rank and file.

These two currents confronted each other in the years following World War I. The federal government through the U.S. Fuel Administration had controlled wages and prices during the war. When the agency was abolished in January

1919, the UMWA demanded a 60 percent incease in wage rates, a six-hour day, and a five-day workweek. The Wilson administration and the coal companies argued that the UMWA was still bound by the war's wage controls because a peace treaty had not been signed. After a six-week strike in November, John L. Lewis ended the confrontation, accepting a small wage increase. A bituminous coal commission was established to investigate conditions in the industry. UMWA radicals berated Lewis for his unwillingness to confront the Wilson administration. Their anger was siphoned into a proposal to nationalize the industry, which Lewis correctly saw as a manageable risk and limited threat to his own political plans. The times turned against union radicals in the 1920s. Lewis became president in 1920 and spent most of the decade eliminating left-wing opposition within the union. At the same time, demand for coal nosedived and unionized companies broke their UMWA contracts. Socialism within the UMWA did not survive these forces.

3. Too Many Mines, Too Many Miners

The year 1920 marks a major change in the American coal industry. Coal demand, which had increased more or less steadily for a hundred years, generally stagnated from 1920 to 1970. America mined about 659 million tons in 1920, compared with only 613 million in 1970. The high production mark was 688 million tons, in 1947. Between 1920 and 1945, the industry mined more than the 1920 output only twice, in 1944 and 1947. During most of this half-century, annual coal production was 100 to 150 million tons less than in 1920. Anthracite production peaked in the decade before 1920 and fell to one-ninth its 1920 level by 1970. Not only did coal production plateau, but supply almost always exceeded demand. Overcapacity and oversupply crippled suppliers and poisoned the industry's labor relations. Supply-side competition impoverished both mining companies and their workers. Lack of demand framed the politics and economics of coal in this period. It implanted a depression psychology within management that narrowed corporate vision and created an opportunistic profit-taking "run-up-the-score" mentality during the years when production perked up. It also directly affected collective bargaining and the industry's practice of shifting production costs onto its employees and the communities where coal was mined. Van Kleeck summed up coal's predicament in 1934:

> The coal industry . . . presents the spectacle of an overdeveloped, undermanaged industry, in a continuous state of internal and external economic struggle. . . . Even if by the move of a magic wand the demand for coal should double . . . , then bankrupt mines will be reopened, and new mines will be opened and the competitive war will go merrily on. Then retrenchment, wage cuts, strikes, unemployment, cut-throat competition, bankruptcies, commence all over again. The old squeaky merry-go-round will commence again to cycle: too many mines, too many miners, too much equipment, too little management, no planning, no profits, no living wages.[1]

In the idiom of modern economists, coal was "demand limited." It floundered in the very decades when America's consumption of energy increased 242 percent.[2] Two kinds of factors explain this paradox: those related to other fuels and those related to the coal business itself.

Other fuels—oil, gas, and electricity—were easier to transport and handle.

Like coal, they were abundant and cheap, and often they could do work that coal couldn't. They burned more cleanly and presented fewer problems in maintenance and cleanup. And their production was less subject to strikes. While bituminous coal production rose only 1 percent (in Btus) between 1920 and 1970, crude petroleum output rose 662 percent and natural gas (wet) 2,589 percent.

As railroads switched to diesel fuel and residential consumers converted to oil and gas in the late 1940s, electric-generating utilities became coal's principal market. Demand for electricity required rapid growth in all the hydrocarbon fuels as well as nuclear energy. Oil, gas, and nuclear power were considerably less labor intensive than coal, and each had received major federal subsidies and tax breaks over the years. Coal's only hope for survival lay in cutting labor costs by mechanizing the mining process and limiting wages and benefits. Interfuel competition forced coal to mechanize, which resulted in the permanent loss of thousands of mine-worker jobs.

Factors peculiar to the coal business were a second reason for coal's protracted lack of growth. Coal's dispersed reserve promoted competition among thousands of small- to medium-sized producers. The biggest bituminous operators could not effect monopoly based on reserves or control of the railroads. In other industries, competition led to new production efficiencies. In labor-intensive coal, however, competition bred wage cuts, from which neither capital nor labor gained. Coal could not match the productivity gains of its competitors. Mining productivity lagged in the 1920s and 1930s for lack of capital. From 1929 through 1939, bituminous suppliers showed a *net* loss every year. (It was not until 1948 that the industry had more than $100 million in undistributed profits after taxes and dividends.)[3] Coal needed capital to meet its competition. Other energy industries evolved supplier oligopolies that regulated intraindustry competition. Coal never achieved the degree of concentration found in its competitors. Public regulation in the 1930s and 1940s stabilized competition among coal suppliers to a degree. In the 1950s, the major companies tried to limit competition through increased concentration and policies aimed at removing marginal suppliers. Lack of capital and the absence of stability among suppliers contributed to coal's inability to protect its markets after 1920.

Market competition was the main source of coal industry chaos after 1920. It impeded profitability and capital investment, destabilized healthy supply-and-demand patterns, created tumultuous labor relations, and produced chronic depression in mining communities. It is not surprising that coal's economic history after World War I was a search for ways to limit competition.

The coal industry was weakened by its inability to rid itself of its most inefficient suppliers. Between 1920 and 1970, the number of coal mines ranged from about 5,100 to 9,400 each year, the majority of which were small, very labor-intensive operations. In hard times, such mines might hang on by cutting wages. The more capital-intensive companies could not cut their production costs as much. Cut-rate coal from the small-mine sector undercut the productivity of the entire industry. The least efficient survived and the most efficient had difficulty growing. The overall effect was to hamstring coal's long-term interfuel competitiveness.

Conventional wisdom describes coal economics as the prisoner of a chronic "boom-and-bust" cycle. Within the 1920–1970 period, there were years when demand rose: in 1926, 1936, during the early war years, and in 1947. But these were transient, erratic surges created by special circumstances. Profits from the surges were distributed to dividend-starved stockholders rather than plowed back into the mining operations or research. When demand did perk up, it was the hundreds of marginal suppliers—the weakest and most unreliable sector of the industry—that boomed. When their production glutted the market, the surge flagged and coal prices plummeted. Rather than seeing coal as a boom-bust industry after 1920, it is more accurate to describe it as a chronically stagnant sector that was further destabilized by a couple of atypical production spikes. The psychology of stagnation resulted in management doing little long-term planning or investment. Had marginal suppliers not drained the steam from every demand surge, the more efficient companies might have had enough cash to deal more reasonably with labor and act more responsibly in their mining camps. The industry saw itself as relegated to a permanent third-class role in the American economy, a remnant sector, the material of liberal cheap shots and corporate head-shaking.

Although business was bad in the 1920–1970 period, it was measurably worse for coal miners. Coal employment fell 80 percent in a more or less straight line from the early 1920s to 1970. In 1920, 785,000 miners produced 659 million tons of bituminous and anthracite; in 1970, 146,000 mined almost as much: 613 million tons. The fifty-year decline in coal-mine employment was the direct result of mechanization. Productivity—the number of tons produced per worker per day—increased from 4 tons in 1920 to about 15.6 tons for underground mining and 35.7 tons for surface mining in 1969. Labor costs per ton fell from about 70–80 percent of the total in the early 1920s to 30–35 percent in the late 1960s.[4] Ironically, it was John L. Lewis and the UMWA's wage pressure that drove mine operators to invest in new equipment. Had Lewis not demanded higher wages, the industry would have modernized even more slowly. If the operators had invested in mechanization more quickly, they might have retained more of their market and ultimately caused fewer layoffs.

While mechanization was inevitable, the manner in which the industry carried it off maximized the social damage. The operators did nothing, for example, to help unemployed miners retrain or relocate. They never asked government to fill the gap. Neither did they make an effort to suppress the dust that new machinery produced. Machines can be blamed for five hundred thousand cases of black lung disease, but it was really the operators' disinclination to invest in dust control that is to blame.

As competition characterizes bituminous coal in the late nineteenth and early twentieth centuries, oligopoly—or "combination," as it was called—describes the anthracite industry. Geology was the difference. America's anthracite reserve is concentrated in one relatively small and isolated area. Those who controlled coal transport quickly came to control the mining companies there.

The social history of "the anthracite," as it is known in the coal industry, is in many respects a subplot to that unfolding in the bituminous fields. Three ties bound the two together. First, their markets overlapped. Anthracite began a long-term decline in the 1920s when it could not meet the competition from bituminous, oil, and gas. Second, the United Mine Workers of America succeeded in organizing both fields into a single labor union in the 1890s,[5] pulling all miners together and giving them a common institutional framework.

Third, a small group of Wall Street bankers financed almost all of the anthracite companies and railroads and many of the bigger bituminous companies by 1900. J. P. Morgan controlled the Reading Company, which owned about two-thirds of the anthracite and handled about one-third of the anthracite traffic. Morgan also influenced mining companies and railroads that hauled bituminous. The Rockefeller interests straddled both sectors. The Rockefeller family had by the 1920s invested in four anthracite companies—Glen Alden Coal; Delaware, Lackawanna & Western Coal; Lehigh Valley Coal; and Philadelphia & Reading Coal & Iron—while at the same time controlling Consolidation Coal and Colorado Fuel and Iron, two major bituminous producers.[6] The Rockefellers also bought into three major coal-owning steel companies—U.S. Steel, Bethlehem, and Jones & Laughlin—as well as more than a dozen coal-hauling railroads. The Rockefellers managed much of their investment through the City Bank of New York (predecessor to the National City Bank) beginning in the 1890s, and after 1930 through the Chase Manhattan Bank. Through control of investment capital and ownership of stock, the Rockefeller and Morgan interests shaped coal policy in both anthracite and bituminous. Bankers saw the coal industry in a different light than suppliers. Shifts in production from anthracite to bituminous were economic rationalizations that did not threaten their long-term interests. Bankers provided a kind of private sector planning committee for parts of the coal industry before the New Deal.

Five railroads controlled most of the anthracite's coal traffic by the Civil War. Beginning in the late 1870s, the five increased their ownership of coal properties, and by 1900 they owned most of the anthracite region.[7] Pennsylvania amended its constitution in 1874 to prohibit a common carrier from mining commodities for transport over its own line. The rail lines easily sidestepped the law by creating coal subsidiaries, all of whose stock they owned.[8]

Local capitalists began each of the railroads—Girard and others, Philadelphia & Reading; the Scrantons, Delaware, Lackawanna & Western; Edward R. Biddle and Asa Packer, Lehigh Valley Railroad; a group of investors from Elizabeth, New Jersey, Central Railroad Company of New Jersey; and William and Maurice Wurts of Philadelphia, Delaware & Hudson. In each case bankers came to control the roads—and the coal lands and mining companies they owned. The McCalmont brothers' London banking house backed the Philadelphia & Reading in its formative years. After William H. Vanderbilt of the New York Central bought a substantial share of the Reading, J. P. Morgan took it over in the mid-1880s. About the same time, Jay Gould and the National City Bank swept up the Scrantons' roads, which Rockefeller's Standard Oil Group soon came to control. The Vanderbilt-Morgan group and the Gould-Rockefeller group line battled each other for a while, but by the 1890s decided that greater profit lay in combination under Morgan's guidance than in competition.[9]

The "anthracite combination," as the consortium of five railroads was known, controlled the supply of about 80 percent of all anthracite production. The railroads needed steady traffic, so they mined at cost and made their profit on the haulage. The consortium set the baseline price for virtually the entire industry through its transportation monopoly. On April 1 of each year the coal subsidiary of the Reading line announced its price list, after which the subsidiaries of the other railroads fell into line.[10] By controlling production, pricing, and freight charges, the railroads were able to squeeze out most independent mine operators.[11] Each railroad was apportioned a share of the market that varied little over the years. New York bankers manipulated a cat's cradle of interlocking directors that kept the combination functioning to their benefit. The railroads fell out from time to time, but in one guise or another their monopoly hammerlocked the anthracite from the 1870s to the 1920s.[12]

The anthracite monopoly operated so brazenly that the 1922 U.S. Coal Commission, a presidential inquiry, minced no words:

> The fundamental evil in the anthracite industry is that of monopoly—the treatment of limited natural resources as if they were like other private property. Reliance on competition without supervision has resulted in . . . a permanent level of high prices above which extortionate increases were made whenever a suspension of mining or other disturbances gave rise to the phenomenon of premium coal. In the anthracite industry we have secured stability—which is desirable—but it has been at high cost to the consumer and has made anthracite a luxury fuel.[13]

After more than a decade of lawsuits, the five rail lines were told to "segregate" their coal interests in the early 1920s. They severed coal in the same way

American Tobacco and Standard Oil satisfied divestiture orders: the coal property was sold to a company having the same stockholders as the parent corporation.[14]

The U.S. Coal Commission thoroughly investigated the anthracite industry—its labor relations, working conditions, economics, and politics. The commission's work is remarkable for its breadth and its candor:

> Eight producing companies affiliated to some extent with the railroads, produce 74 percent of the total output and control 90 percent of the underground reserve. The remaining 26 percent of the output is contributed by so-called independent companies, but the largest of these companies retains a community of interest with one of the railroads, and nine others control 13% of the output.[15]

The commission found there was "no effective competition among 'railroad companies'" in the anthracite, and that "railroad ownership of coal-mining and coal-selling companies . . . was . . . the principal and underlying method of limiting competition." In its letter to the president, the commission stated:

> The fundamental fact in the anthracite coal problem is that heretofore these limited and exhaustible natural deposits have been in the absolute private possession of their legal owners, to be developed or withheld at will, to be leased for such royalties as could be exacted, to be transported and distributed at such rates and in such manner as a double-headed railroad and coal combination might find most advantageous from the point of view of private profit, to be sold at such prices as could be maintained by the restriction of output and the elimination of independent competitors, through such means as the maintenance of freight rates burdensome except to those who, owning both mines and railroads, could afford to be indifferent as to whether their revenue came from the one source or the other. [p. 43]

The commission came to view anthracite coal as a "public necessity," which, like the railroads, should be required to report to a public authority and no longer be guided by the principle of maximum profit to owners "but maximum service to the public." The commission said, "In the operation of coal mines, as in the operation of railroads, telephones, water companies or banks, the public interest must be respected and served, and . . . this requirement places limitation on the rights of owners of coal lands, operators, mine workers, carriers, and dealers."

The commission went so far as to suggest that "much might be said for the view that mineral deposits should have been held from the beginning as national rather than individual property." But it did not endorse nationalization of the mines. Instead, it favored a governmental authority with the power to require financial and operating reports, prescribe uniform methods of cost accounting, and determine conditions on which coal could be shipped in interstate commerce. The commission hoped its analytical indictment of the monopoly would force federal antitrust action and public regulation. The commission did not anticipate the decline of anthracite production in the ensuing decades, which removed the urgency of controlling corporate greed.

Since the late 1920s, anthracite production has fallen from the 80 million ton level to 6 million tons annually in 1980. Cheaper fuels stole its main market—home heating. No observer in the 1920s foresaw anthracite's decline.[16] Anthracite production eroded for several reasons. It cost more to mine than bituminous because its seams were more irregular and thinner. It was less suited to mechanization. It was also less convenient than alternative fuels. The oligopoly kept anthracite prices high on the assumption that consumers would not switch. When they did, the combination could not recapture them. If anthracite had been competitive, prices would have been lower and the long-run decline might have been less profound. But even a highly competitive supply structure would not have been able to overcome the geological irregularities that cursed the reserve. Nor would competition have improved the public welfare of a region.

Geology and oligopolistic capitalism created a depression that choked this region after the late 1920s. Chronic unemployment, capital depletion, and social anguish were its residues. Layoffs in the 1930s created such desperation that unemployed miners punched mines into seams they did not own and "bootlegged" coal to market. "Regular" coal companies were bankrupt, and local juries refused to convict bootleggers.

Compare how politicians responded to the needs of the anthracite at different times. Public agencies readily subsidized canals and railroads when private investors wanted to open the field. No state or federal agency effectively confronted the railroad-based monopoly for fifty years. The appalling conditions of life and work in this region were well documented, but politicians did little to force the private sector to do business under minimum standards of decency. Government policymakers boosted anthracite when demand was healthy, and abandoned it when it was not. In contrast, when coal markets nosedived in Britain and Western Europe after World War II, their governments initiated offsetting or compensatory measures. They believed that to victimize the victims further served no public purpose. The federal government could have stabilized anthracite output and planned a phased reduction. Public agencies might have *diversified* the anthracite economy through subsidizing the relocation of other industries or funding local initiatives. Instead, the decision of the market stood and the region fell.

Both the anthracite and the bituminous fields suffered from demand weakness after 1920. Both were victimized by competition from other fuels. Bituminous had a more diverse market—steam locomotives, cokemaking, industrial feedstocks, electricity generation and home heating—than anthracite, which was used almost exclusively for heating. That diversity cushioned the loss of bituminous sales to railroads and residential consumers in the late 1940s. Anthracite could not overcome its comparative expensiveness, a legacy of monopoly and geology. Market forces shrank the anthracite and paralyzed the bituminous. Some of that was inevitable, productive, and sensible. The consequences in terms of coalfield life were immense, and not the least of them was the way John L. Lewis managed collective bargaining during the forty years he was UMWA president, beginning in 1920.

4. John L. Lewis: A Paradox

What kind of person was John L. Lewis? Big, self-contained, and brooding,[1] he carried his 240-pound bulk on a large 6-foot frame, which made him seem larger. As a young man, he was well muscled in his arms and shoulders. His hands and head were disproportionately large, and he maneuvered both with dramatic flair. He came by a potbelly in middle age, but his dark three-piece suits camouflaged the fat as heft. His mouth turned down at the corners, giving his infrequent public smiles a slightly cruel or mocking cast. Wavy hair, reddish in his youth, turned black, then white in his later years, but stayed thick. His dark, deep-set eyes stared directly at the object of his attention. They did not flit. His heavy eyebrows grew thicker and wilder with age and had the ability to distract opponents—a coal operator once boasted to Lewis that he was not afraid of them! A long, thick cigar was a frequent prop. Its gray haze contributed to the aura of mysterious wizardry that Lewis liked to create. Lewis was not a friendly, likable person, though he could be charming in private. He was excessively formal and stuffy, even prim. He preferred florid language and used his stilted pomposity to relegate foes and subordinates to positions of intellectual inferiority.

At root, Lewis wanted to control people and events. This moved him to eliminate political opponents, to initiate organizations that he could dominate, like the CIO and the UMWA's Welfare and Retirement Fund, and to establish a relation of greater to lesser with as many people as possible. He respected those who had greater knowledge or greater strength, but would defer only to the latter. His ego fed on veneration, and he would settle for sycophancy. He saw himself as a larger-than-life person, an actor in history. He allowed himself no close male friends, and derived whatever emotional support he needed from his wife and, later, Josephine Roche, a colleague and social confidante. He insisted on space, physical and psychic, between himself and those around him. Space established the monarchal terms he felt best defined his proper role in the world.

Had he been born to the purple rather than the denim, Lewis's natural intelligence and guile might have pushed him to the heights of Wall Street. But unlike some other gifted working-class children around the turn of the century, he never connected with a powerful businessman as a patron. It was Samuel Gompers, leader of the American Federation of Labor, who discovered him and channeled his talents. Lewis brought his irrepressible entrepreneurial drive to bear on the raw material available to him. Human labor became his capital, and he managed it with a combination of care, cunning, and boldness worthy of the most successful robber baron. (Indeed, Lewis once described himself on the back of a 1917 photograph as "Jesse James in disguise.") He never took himself less than totally seriously. His self-righteousness enabled him to bull his way through opposition. Lewis rarely admitted to having made a mistake—although he made many. Mistakes were parasites that fattened off of lesser figures. Because he felt himself always right, he had few qualms about ends not justifying means. If not ends, he might ask, what else?

Lewis directed truth or falsehood at his enemies according to which would wound more seriously. He was an institution builder, a man who could take the dreams of idealists and turn them into programs and organizations. Some of his architecture was timely and magnificent: he helped to rescue the UMWA from near oblivion in 1933; he sparked the organization of the CIO; he masterminded a cradle-to-grave medical system for UMWA miners. Other constructions were warped, either through their original purpose or by his misuse: he bought the National Bank of Washington in 1949 and misapplied its assets; he formed joint ventures with coal management that were at best suspect; he made certain collective-bargaining alliances that did not serve his members' interests. No public figure as powerful and as long-lived as John L. escapes making errors of judgment. These, the historian can understand and forgive. History has a harder time resolving the errors that flowed from the deepest pit of John L.'s personality. These may be easier to forgive than to understand.

Born in 1880, Lewis came from a family of Welsh miners. With his brothers, he moved to Illinois to begin a UMWA career. He and his family built a political machine in his local. He was appointed UMWA lobbyist in Springfield, and later Gompers made him an AFL organizer. After serving a short stint as UMWA headquarters' statistician, he was appointed its acting vice-president in 1917. He became acting president when Frank Hayes, a disintegrating alcoholic, resigned in January 1920. Lewis "schemed to grasp the union power steadily slipping through Hayes's palsied fingers."[2] Some believe that he encouraged Hayes's drinking. Lewis won election in his own right later that year against Robert Harlin, an advocate of nationalization. As Hayes's vice-president, Lewis built a political machine based on appointed organizers. During the campaign, Lewis denied Harlin's slate access to the *UMWA Journal* and milked his incumbency for every ounce of advantage. Many of the "dirty tricks" that W. A. (Tony) Boyle was to pull fifty years later in his campaign against Joseph (Jock) Yablonski came directly from John L.'s bag.

In Lewis's first thirty years as UMWA president, three threads tie his story together: his relationships with the UMWA, with the coal industry, and with the federal government.

First, and in his mind foremost, was the need to turn the UMWA into an expression of his own personality and politics. By 1933, Lewis had stripped the UMWA of dissidents. As he demonstrated skill and militance in the 1930s and 1940s, the union's rank and file accepted his autocracy. Every nickel Lewis wrung from "the companies" was a nickel that each miner felt he personally had retaken from those who exploited him. Every time Lewis stymied management executives with a Shakespearean reference, each miner felt as if he had bested his own boss. Their loyalty allowed Lewis to cut a wide swath through the coal executives and politicians who ran life in the coalfields. Lewis converted the union's rank and file from cynicism and hatred in the 1920s to a dog-like trust by the 1940s. He encouraged unquestioning dependency, rewarding those who kowtowed and purging those who didn't.

As Lewis was winning his power struggle within the UMWA in the 1920s, he was losing his wars with the operators. Lewis vowed: "No Backward Step!" from the standards of the 1924 Jacksonville contract. Rather than continue to pay the Jacksonville scale, unionized operators abrogated their UMWA contracts. Lewis and the UMWA weren't strong enough to maintain the scale and keep their contracts. He had little bargaining leverage when demand softness and excess capacity created a desperate army of unemployed surplus miners. The vicious strikes of the 1920s were replaced in the 1930s by a bristly collective-bargaining relationship in which most of the industry reluctantly negotiated with him. Their relationship is best remembered for Lewis's mastery of bargaining tactics and vituperative rhetoric. By the late 1930s, Lewis had become the landlord of coal politics rather than its tenant. Lewis defended the free market and unregulated competition until the late 1920s, when the anarchy of the marketplace decimated his union and forced him to advocate public planning. After a decade of defeat, he began advocating legislation that would give unions federal protection and impose federal management over the coal industry. His success as a negotiator came only in the 1930s, after the federal government began to regulate the coal industry. New Deal legislation turned coal politics into a tripartite relationship in which Lewis often struggled with presidential power as much as with management. The balance of power shifted against him in the late 1940s with the passage of Taft-Hartley, the antiunion undertones of the Cold War, and the testy hostility of the Truman administration. When political circumstances changed after World War II, Lewis abandoned the tripartite mode. The coal industry threw off federal market regulations. Coal labor and management would try again to settle their affairs privately. The last ten years of Lewis's presidency—the 1950s—saw neither the class struggle of the 1920s nor the collective-bargaining battles of the 1930s and 1940s. Labor-management collaboration took their place. Lewis, as much as any other single actor, defined events in this sector of the American economy for most of these forty years.

The UMWA entered the 1920s as the spokesman for more than 500,000 miners in an industry that had just fattened off the Great War. By 1930, the end of Lewis's first decade, only 20 percent of U.S. coal output was mined under UMWA contracts.[3] Perhaps 75,000 or so still paid UMWA dues. Lewis had lost the battle with management as he vanquished his political opponents. The two were not unrelated.

UMWA dissidents attacked Lewis from two directions. Some, like Illinois district president Frank Farrington and Kansas's Alexander Howat, a socialist, were ambitious politicians. Farrington left after Lewis showed reporters a three-year contract that Farrington had signed with Peabody Coal Company to be its labor representative for $25,000 a year.[4]

UMWA radicals like Howat opposed Lewis's heavy-handed leadership and his suppression of union democracy. Before Lewis became president, the UMWA was a reasonably democratic organization. Democracy came directly from the work place, where danger was shared equally and everyone did the same work. Piece rates were uniform throughout mine and craft.[5] The rank and file preferred "equalization" of work when weak demand forced production cutbacks.[6] The most direct expression of rank-and-file democracy was the UMWA's district and national conventions, where delegates elected from local unions established policy. One of Lewis's first political goals was to take conventions away from the delegates. Some miners supported Lewis's bid for central decision making after witnessing the disaster of district-authorized strikes. Most understood that local "autonomy" had to be juggled with the need for central authority. Lewis struck no balance between the two.

Radicals also split with Lewis over his prescription for coal's economic ills. Lewis hoped an industrywide UMWA wage scale would equalize labor costs and stabilize competition. Recalling their thirty years of experience with such efforts, radical miners felt that the UMWA wage could never impose order over thousands of suppliers. They proposed nationalization, to which they linked opposition to mechanization and support for a labor party. They felt public planning would phase in mechanization while distributing the available work equally and softening job losses. Democrats and Republicans opposed nationalization and supported market competition. Radicals therefore saw a labor party as the union's only vehicle for dealing with irregular work, unemployment, and depressed wages.[7] Lewis opposed each link in their chain of logic. His handling of the nationalization issue early in his presidency derailed what appears to have been a burgeoning rank-and-file sentiment.

The 1919 convention established the Nationalization Research Committee under John Brophy to prepare a plan for nationalizing the mines. Lewis accommodated the union's left wing and, it might be argued, gave Brophy rope for his own political hanging. The committee's 1922 report, "How to Run Coal," called for federal purchase of the coal mines for $4.5 billion. A cabinet post, secretary of the mines, would operate the industry through a federal commission of mines. The eleven-member commission would have five representatives selected by industrial, technical, and engineering organizations; the other six would be presidentially appointed. The UMWA would enjoy collective-bargaining rights with the commission. The mines would be operated by regional councils of miners, technicians, and consumers; these councils would appoint mine management.

Lewis denounced the Brophy plan through a statement by Ellis Searles, editor of the *UMWA Journal:*

> The plan . . . was prepared largely by a bunch of Greenwich Village reds who do not belong to the UMWA and who have not the slightest right or authority to represent or speak for the miners' union. This crowd has long made it a practice to butt in on the affairs of the miners' union. These Greenwich Village parlor coal diggers, who might better be designated as gold diggers, represent nothing.[8]

Red-baiting and Lewis's stonewall stopped the plan.

The radicals challenged Lewis for the presidency in 1926. Brophy, a Fabian Socialist who was District Two president, organized the "Save Our Union" (SOU) movement. He advocated labor education, nationalization of the mines, support for a labor party, organizing the unorganized, and internal democracy.[9] Brophy saw the Communist Party's interest in mine workers as an anti-democratic threat to the UMWA. But during the SOU campaign he accepted help from all anti-Lewis elements—Farrington, Howat, Frank Keeney of West Virginia; socialists like Powers Hapgood and Adolph Germer; and the Communist Party. When the UMWA announced that Brophy had lost, 170,000 to 60,000, Brophy declared, "Lewis would not have won without stealing votes."[10] A number of issues that Brophy raised in 1926—internal democracy, honest bookkeeping, fair play for districts, and open conventions—surfaced more than four decades later when first Jock Yablonski and then Arnold Miller challenged Lewis's heir, Tony Boyle, for union leadership. Brophy's challenge cost him his district post, and two years later Lewis expelled him from the UMWA. SOU became a "dual" movement in that year after the Communist Party took it over. Brophy opposed dualism "on the general principle of waste . . . [and because] there is no solid group of support for a new union in the mining industry at this time."[11] Brophy did not join the National Miners' Union when the Party set it up in September 1928.

Brophy's campaign affected Lewis more than either man probably realized. Although Brophy was out of his hair, Lewis soon conceded that market forces were ruining coal suppliers, who in turn were taking it out on his members. In 1928, he began supporting federal legislation to impose government controls over coal supply and demand. During the next four years, Lewis supported a public-regulation plan for coal that was short of nationalization but more interventionist than antitrust policy. At the same time, Brophy was planing down his sharp-edged militancy. He deemphasized parts of his political program in return for the chance to implement others. Shortly after the New Deal began, Lewis tapped Brophy to aid UMWA organizing. The two effected a rapprochement, and Lewis appointed Brophy director of the CIO.

Lewis's treatment of Brophy is significant for what it reveals about UMWA politics. Lewis peeled articulate opponents from the payroll. He demanded that miners place loyalty to him above loyalty to the union. He turned the UMWA into an organization of sycophants. Independent rank and filers could not move up in the hierarchy. It was a bureaucracy of anvils on which only Lewis's hammer beat. When Lewis retired in 1960, the UMWA had few capable miners-turned-administrators. John L.'s legacy was a union stripped of talent.

As the 1920s unfolded, the bituminous coal industry collapsed. World War I had left coal terribly overexpanded. Coal production, both anthracite and bituminous, fell 38 percent by the end of the decade, employment 18 percent. About 141,000 miners lost their jobs, and those who still worked were down to 187 days a year. The number of operating mines fell from 8,921 in 1920 to 5,891 in 1930. The price per ton was more than cut in half, from $3.75 to $1.70. Competition intensified for remaining markets. Operators cut wages first, then capital investment. Instead of tightening the structure of the industry, the free market allowed everyone to hang on—and hang together. The vastness of U.S. coal deposits and the elasticity of labor's selling price thwarted efforts to rationalize competition.

Brophy and later Lewis realized that market forces created a self-destructive cycle in which too many suppliers produced too much coal. This insight was widely shared among miners. As one local union secretary expressed it in 1919: "It is precisely unrestricted competition which keeps the coal industry in its present state of demoralization and trouble for the coal operators and of disaster for us every year."[12] Lewis once voiced a solution to this dilemma: "Shut down 4,000 coal mines, force 200,000 miners into other industries and the coal problem will settle itself."[13]

Nonunion operators in the southern fields who paid below UMWA scale ate into markets traditionally held by unionized northern companies. Surplus miners—what one might call excess labor capacity—forced union labor to cheapen itself. In 1924, Lewis negotiated the "Jacksonville agreement" with the northern tier of operators (from western Pennsylvania to Illinois) with a daily wage scale of $7.50. Nonunion operators in the southern Appalachians paid less, which prompted the unionized operators to subvert or break their UMWA contracts. About 60 percent of the mines were nonunion by the end of 1925, compared with 40 percent a year earlier. The Coolidge administration extended preferential freight rates to nonunion companies. But Lewis would take "No Backward Step" from the 1924 standard, even though standing pat meant losing ground. The longer Lewis stood firm against wage concessions in a depressed market, the more union labor surplus he created, making the UMWA's future even more precarious. The unionized companies, led by Consolidation Coal and Pittsburgh Coal in 1925, broke their UMWA contracts.[14] Lewis and UMWA miners learned through long, painful strikes they could not prevent them from leaving the fold. If union miners struck, their employers would hire cheaper, nonunion labor. Lewis saw the small, nonunion mine as the bane of his effort to stabilize competition and secure better conditions for his men.

Government reinforced the mine operators' efforts to destroy the UMWA in the late 1920s. All unions functioned without federal protections, and coal companies continued to use blacklists, yellow-dog contracts, private police forces, and company stores to keep their workers in line. In 1920–1921, for example, West Virginia requested federal troops four times to police Logan and Mingo counties, where the UMWA was striking for recognition. When union miners marched en masse to "liberate" Logan County from the reign of the local sheriff and coal companies, a gun battle was fought at Blair Mountain, where company loyalists threw a homemade bomb onto the battleground from a borrowed airplane. (A squadron of U.S. Army pilots tried to get into the action, but got lost in the flight from Washington and crashed six of their planes in the mountains.) After five days of fighting, federal troops separated the warring sides. Local grand juries indicted almost 600 miners, a number for treason and murder.[15] Coal operators and government officials lined up again during the terrible strike of 1927, when more than 450,000 bituminous and almost 160,000 anthracite miners stopped work over a 21 percent pay cut. Benson Hough, a federal judge, issued a broad antistrike injunction in August that stipulated "no interference with 'business as usual' at any coal mine, no threats to strike-breakers, no gathering near mines, no 'vile or opprobrious names or insulting words,' and no use of words like 'scab' or 'rat.'"[16] The injunctions, which Lewis obeyed, crippled the strike. Injunctions had handicapped UMWA organizing campaigns and strikes for decades, but Judge Hough added a new twist when he issued the following warning to UMWA miners: " 'Any man, . . . who cannot talk English or is not an American citizen who dares to take part in picketing, shall be immediately deported.' "[17]

Brophy, Lewis, and other UMWA observers understood that the operators' antiunionism was partly a reaction to the competitive tailspin in which they found themselves. Brophy proposed nationalization and labor protections to stop the downward spiral. Lewis took a different tack. He wanted to stabilize competition by having a single coal operator impose order on suppliers—men like Richard Grant in the anthracite and George Love in the bituminous. He preferred partnership between labor and capital to federal regulation. But partnership assumed a dominant operator who read coal's tea leaves the same way he did.

Lewis found a leader in the anthracite in Richard F. Grant. Anthracite production peaked during World War I, and by the mid-1920s significant competition appeared from cheaper fuels in the residential heating market where most anthracite was sold.[18] Demand fell, and the anthracite monopoly shut down mines and cut wages.[19] On September 1, 1925, the UMWA struck as Lewis demanded no change in the existing wage scale and adoption of a checkoff, whereby the company would deduct union dues from a miner's pay and send them to the union.[20] The strike lasted 170 days. Only after Grant, president of the Lehigh Coal Company, assumed leadership of management did the sides settle. Monroe writes:

> A forceful man who badly wanted a settlement, Grant was also tactful and clever. While disdaining publicity and loud rhetoric, Grant was able to gain Lewis's trust through a series of private meetings and between them, they quickly worked out a contract that would allow both sides to save face, and ended the strike that neither side could win. The significance of this episode is less the contract these two men hammered out than the precedent they established for future negotiations. Lewis and Grant temporarily reached a consensus . . . as to how labor/management issues should be handled in the anthracite fields. Meaningful negotiations were to be carried on in private and potentially explosive issues were to be avoided. Their objective was to get the industry producing, so that it might again provide the profits and jobs which were the heart of the region's economy.
>
> The difficulty with such a solution, however, was that for all of the good intentions which Lewis and Grant possessed, the system was now designed to virtually exclude active participation in important decisions by the bulk of the miners. They were isolated from the negotiations. . . . In the ensuing decade, this arrangement had a significant effect on labor relations in the anthracite fields.[21]

This arrangement in the anthracite foreshadowed by twenty-five years one Lewis concluded with George H. Love of Consolidation Coal in the bituminous.[22] The similarity is eerie—deals cut in backrooms; exclusion of the rank and file from collective bargaining; autocratic leadership and suppression of dissent; promotion of "labor stability" and a strike-free environment; and union acquiescence in layoffs.

Anthracite miners rebelled against the job losses, and against Lewis. They demanded equalization of work, emergency relief, and union democracy. After unsuccessful wildcat strikes, directed against Lewis as much as their employers, the insurgents formed the United Anthracite Miners of Pennsylvania (UAMP), with Thomas Maloney as president, in August 1933.[23] The UAMP struck UMWA-organized companies in an effort to protect jobs and preserve their union. The operators, Lewis, and the Roosevelt administration joined forces to end this rank-and-file uprising, which threatened the industrial stability they were trying to achieve. Ironically, the Lewis-Grant accommodation did not work. Despite their efforts, anthracite's markets slipped away. The monopoly that had ruled northeastern Pennsylvania coal since the mid-1870s disintegrated from weak demand.

National energy planning might have softened anthracite's decline by reducing demand in a controlled way and regulating supply accordingly. Anthracite employers, miners, and communities needed protection from the market, which was achieving efficiency at substantial human and economic cost. A federal agency with the power to shape multiple-fuel demand need not have tried to bottle up the long-term efficiencies that market forces bring. Rather, it might have cushioned the market's verdict and given the region the opportunity to find other sources of employment. Brophy's nationalization plan might have functioned in this manner, but something far short of nationalization would probably have worked. Neither Lewis nor Grant considered federal planning in their discussions, and without it, the anthracite faltered.

When Lewis turned to the bituminous in the late 1920s, he saw economic chaos. Monopoly could not work in the widely scattered bituminous fields. The industry was too divided and fragmented. The crucial division within the industry was between operators in the northern field, running in a broad swath along the Ohio River basin, and those in southern Appalachia. The northern mines tended to be larger, more mechanized, and more unionized. Wage scales were higher than in the southern mines, which were generally smaller, less efficient, and nonunion. Over the years, the operators, the union, and the federal government tried various schemes to adjust wage and transportation costs regionally to control price-cutting competition among the various coal-producing districts. These efforts would succeed for short periods of time and then fall victim to the cascading trend of individual operators underselling to maintain sales. The southern operators compensated for their remote locations and inefficient mining systems by paying labor far less than northern operators. The competition between regions, between union and nonunion operators, between mines close to markets and those at a distance, between mechanized and less-mechanized operations, between mines with advantageous freight rates and those without, kept bituminous coal cheap for consumers but in a chronic state of instability, subject to strikes, bankruptcies, undercapitalization, and inefficient use of resources. The industry was also divided between "commercial" and "captive" mines: the former sold in the open market; the latter only to their parent companies in steel, railroads, manufacturing, or electricity generation. The coal companies owned by famous entrepreneurs like the Mellons, Fricks, Rockefellers, and Hannas never managed to discipline the supply side through concentration.

Price cutting and overcapacity were the related evils that plagued coal suppliers in the 1920s and 1930s, but they were basically symptoms of a structural change in energy supply that weakened coal demand in these years. Petroleum and natural gas ate into coal's traditional markets in the 1920s, and coal demand fell further during the Depression. Coal's demand depression began in the early 1920s and prompted various stabilization proposals that sought to stop price-cutting competition by fixing minimum prices across the industry (adjusted for regional differences) and, in some cases, by limiting production itself. Although some of the ideas went as far as nationalization, all were primarily concerned with stabilizing supply rather than shaping the structure of energy demand. The crisis in coal brought some Democrats and a few Republicans to the point where they saw no alternative to government coordination of supply, but federal allocation of demand was apparently too radical for even the most enlightened operators. The experience in the 1930s suggests the difficulty of trying to change producer behavior to stabilize supply while market forces were driving down demand.

During World War I, the Wilson administration imposed controls on the industry through the Lever Act, which authorized the federal government to establish maximum coal prices in the war-heated economy and even to require producers to sell coal to the government. The Fuel Administration imposed order on coal suppliers during the war's demand surge, which left the industry terribly overexpanded and subject to competitive warfare. Federal coordination and controls ended when different economic circumstances presented an even greater need for them. Senator William S. Kenyon (R., Iowa) proposed an "industrial code" for coal in the early 1920s that would have declared coal a public utility and established basic wages and working conditions, collective-bargaining rights, and dispute-settlement procedures. Herbert Hoover, Harding's secretary of commerce, proposed a plan to have operators sell coal through district cooperatives and reduce production capacity through a system of penalties and incentives to less efficient operators. Lewis attempted to use the Jacksonville wage scale to impose uniform labor costs across the industry. Nonunion competition undercut his effort, and unionized operators broke their contracts, nearly destroying the UMWA. As the situation worsened, Lewis eventually came to embrace federal stabilization as the only way that coal capitalism might be saved from itself. Henry Warrum, UMWA chief counsel, and W. Jett Lauck, union economist, developed legislation in 1928 that drew on the Kenyon plan. Its theory was based on the idea that reasonable wages and price controls would stabilize labor relations, provide decent conditions to miners, and eliminate price-cutting competition among suppliers.[24] Lewis convinced Senator James E. Watson (R., Ind.) and Representative Henry Rathbone (R., Ill.) to introduce a bill to suspend the Sherman Act to give bituminous operators permission to combine in selling pools and fix prices through marketing agreements under federal oversight.[25] Coal operators would be required to bargain collectively, and miners would have the right to a check-weighman, free speech and assembly, and the abolition of scrip. The bill died in the Senate. Coal producers in southern Appalachia set up Appalachian Coals, Inc. in 1931 to market their coal and allocate production. Lewis went back to Congress in 1932 with a measure similar to the Watson bill. Introduced by

Senator James J. Davis and Representative Clyde Kelly, both Pennsylvanians, the Davis-Kelly bill would have regulated "interstate and foreign commerce in bituminous coal, require[d] the licensing of corporations producing and shipping coal in interstate commerce and create[d] a Bituminous Coal Commission."[26] The commission would have encouraged mine operators to form marketing sales agencies and set minimum prices. The operators supported price fixing and joint marketing, but refused to agree to the labor protections Lewis wanted. Their opposition was sufficient to kill the bill in committee.[27]

Though he voted for Hoover in 1932, Lewis recognized that Roosevelt's mandate created a new political context for labor and coal. Maryland Senator David J. Lewis and Arizona Representative Carl Hayden introduced a bill that would have fixed minimum coal prices, set mine-by-mine quotas, marketed coal through producer cooperatives, and set up a committee of operators to oversee the system. The Black-Connery Thirty-Hour bill would have spread employment and equalized the calamitous effects of the Depression by adopting the old mine-worker principle of sharing work during hard times. Roosevelt, however, favored the National Industrial Recovery Act (NIRA), which set up the National Recovery Administration (NRA) with the power to regulate business practices and guarantee labor the right to organize and bargain collectively.[28] The NIRA had its roots in the Kenyon proposal of 1922 and the UMWA-sponsored stabilization plans. The NRA persuaded industries to set up and administer codes to regulate competition, raise prices, and extend labor protections. The NIRA imposed the responsibility for stabilizing industry on the private sector through the NRA's codes of fair competition, which ruled out certain destructive business practices. The federal government would step in only where an industry could not agree to regulate itself. Northern and southern coal operators could not agree on a single code, the Southerners objecting in particular to the collective-bargaining provisions. Finally, Roosevelt gave the industry a twenty-four-hour ultimatum in September 1933, warning that theirs was a doomed resistance and radicalism would win if they persisted. Within a few days the Bituminous Coal Code emerged, and the operators signed the Appalachian Joint Wage Agreement with the UMWA. It established a minimum daily wage; an eight-hour day, forty-hour week; union recognition and collective-bargaining guarantees; a union-sponsored checkweighman; and abolition of the requirement that miners reside in company housing, accept scrip, and trade at company stores as a condition of employment. Coal producers could not sell below a "fair market price" set by the code's marketing agency. Anthracite and captive mines were exempted from the first code, but a Lewis-led strike in the summer of 1933 forced Roosevelt to broaden its scope.

The NRA's Bituminous Code helped coal. Production increased from 310 million tons in 1932 to 372 million in 1935; employment from 419,000 to

462,000; days worked from 146 to 179; and price from $1.31 per ton to $1.77. Days lost to strikes declined from 5.9 million to 3.0 million. Coal was not bursting with vitality in 1935, but the NRA had scraped it off the floor. The NRA allowed the UMWA to bring close to 90 percent of the industry under contract in the 1930s, which gave Lewis the opportunity to use collective bargaining to impose uniform labor costs on the industry as a way of controlling competition that resulted in price and wage cuts. The code's minimum price structure raised prices and helped to establish a floor of return to the operators capable of supporting better wages. To the extent that higher prices disadvantaged coal's position with other fuels, the code's decent-wage policy shaved demand and encouraged investment in labor-reducing machines. The coal code had some success in fixing prices and controlling unfair trade practices, but it could not overcome the divisions within the industry that set region against region. The coal code was the product of UMWA and federal pressure, but the degree of control it exercised did not go far enough to stabilize supply, nor did the code address the problem of demand. Price fixing and unfair competition were expressions of deep-seated problems in the industry that the industry itself was incapable of fixing through the tools of the code. The U.S. Supreme Court declared the NRA unconstitutional in 1937 (*Schechter Poultry Corp. v. United States*, 295 U.S. 495), which promised to leave the coal operators on the same barren island where they had foundered in the 1920s. As a partial replacement, Senator Robert Wagner of New York drafted the National Labor Relations Act, which strengthened federal labor protections and virtually prohibited the company union. This act, which FDR supported at the last minute, established the National Labor Relations Board to shield labor from unfair practices and determine union representation. Wagner folded in the NRA's labor protections Lewis had had written in 1928. The Supreme Court upheld the Wagner Act in 1937.

Lewis went back to Roosevelt and Congress to reestablish a NRA-type code for coal despite the *Schechter* ruling. The Guffey-Snyder Coal Conservation Act of 1935 was stronger than the NRA code it replaced. A national "impartial" board would establish its competition rules, set prices, regulate wages and hours, and parcel out production quotas among mining areas. The NRA had generally negotiated codes with industry groups, relying on rules by which they themselves would live. Guffey-Snyder, instead, established a 25 percent tax on all tonnage and promised to rebate 99 percent of the payment to those operators who agreed to play by the rules. A portion of the remaining tax would be used to retrain displaced miners, a truly useful idea. Guffey-Snyder, like the NIRA, did not address the problem of coal demand and intrafuel competition. The Supreme Court cut down Guffey-Snyder. It was a "sad commentary upon our form of government," Lewis said, "when every decision of the Supreme Court seems designed to fatten capital and starve and destroy labor."[29] Lewis secured a third coal-stabilization bill—the Guffey-Vinson Act of 1937—in alliance with Roosevelt (to whose reelection he had contributed half a million dollars from the UMWA) and the northern operators (who saw that the code rules benefited them in their struggle with the southern mines). Guffey-Vinson set up a national commission and twenty-three operator-controlled district boards to set prices, which resulted in a system coordinating several hundred thousand approved prices. The act slapped a 19.5-cent-per-ton tax on noncomplying operators and a one-cent-per-ton tax on all tonnage. The Supreme Court approved the act, but its own conceptual and administrative inadequacies and the competitive pressures within the industry reduced its effectiveness. Congress did not renew Guffey-Vinson in 1942 when wartime fuel needs erased the need for minimum prices. Each of the stabilization acts had the effect of muddling the suppliers through the Depression by carrying everyone—miners and managers—along on a demand-sharing basis that rested on rigging higher prices. The acts were a temporary fix, a patchwork rather than a structural approach that allowed the private sector to live with its basic flaws. The acts failed because they did not go far enough in extending some sort of central control over energy supply and demand, not because they went too far.

The significance of coal's economic adaptations in the 1930s becomes clear in retrospect. The no-holds-barred class struggle that was the industry's "labor relations" since the 1890s evolved into collective bargaining. Related to this new balance of power was the reconstruction of coal's economics. Through most of the 1930s and much of the 1940s, coal was a federally regulated industry on matters of supply and price even though private investors continued to own the companies. While industry grumbled over Lewis and the New Deal, they had pulled the chestnuts of private capital from the fire of the market.

When we recall the 1930s and 1940s in the coal industry, we remember John L. Lewis and his strikes. What irony! Here was a Republican conservative spearheading the nation's most militant labor union. Lewis, the superb strike tactician, led few truly successful strikes before the mid-1930s.[30] Lewis, the huckster of federal plannning, was a free-market capitalist as ideologically firm as any pinstripe on Wall Street.[31] Lewis, the organizer of the unorganized, had purged John Brophy, the UMWA's most articulate organizer, a decade earlier. Lewis, an inveterate red-baiter, was stuffing the fire-breathing CIO with every left-wing organizer he could afford. Lewis, the miners' cosmic hero, had done his utmost to deny them genuine participation in their union's affairs. Lewis, the free-enterprise advocate, bore the wrath of the very capitalists he was trying to rescue.

John L. Lewis saw special opportunities in the 1930s and took them. His bedrock, unalterable principle was the acquisition and use of power. His conservative political philosophy was sacrificed to principle. When human welfare and political vision coincided with principle, he changed history, as with the founding of the CIO. Lewis moved toward public regulation only after the market left him no choice. He would reverse himself in the late 1940s when, for a short time, the market looked promising. Lewis never abandoned his business unionism and his belief in capitalism, but they did not stand in his way where they conflicted with what he wanted to do. Irony was the by-product of opportunism.

The coal industry's performance in the 1930s and 1940s mirrored the economy generally. New Deal stabilization measures laid a safety net under the industry. Bituminous production and price improved between 1932 and 1950. World War II turned mining from a perennial Wall Street loser into a temporary profit maker. As the average price per ton rose from $1.31 in 1932 to $3.44 in 1946, bituminous operators turned a net loss of $51 million into net gain of $75 million.[32] But war pumped coal demand to artificial and unsustainable heights. Mine operators were surprised in the late 1940s by the collapse of two traditional markets. In 1945, railroad and retail (residential and commercial heating) sales took 22 and 21 percent respectively of the 560 million tons of bituminous coal consumed. Five years later, they had fallen to 13 and 19 percent respectively of 454 million tons.[33] Coal consumers were switching to heating oil, natural gas, and electricity. American railroads converted to diesel, phasing out both coal and fuel oil.[34]

Lewis found himself again in the tightening coils of a postwar market. The specter of the 1920s appeared. Thousands of combat veterans roamed the coalfields looking for work. Days worked fell from 261 in 1945 to 183 in 1950. In the late 1940s, Lewis tried to regulate supply by suspending work. As part of the 1947 contract, he had won a right-to-strike clause, which said the contract applied to miners "during such time as such persons are able and willing to work."[35] This language enabled miners to strike even while the contract was in force. Lewis called these strikes "a stabilizing period of inaction." But they failed to stabilize production; indeed, they could not, because unionized operators were only part of the problem. Nonunion operators had saturated the postwar market with cheap coal. UMWA strikes merely helped nonunion mines overcome their technological and geographic disadvantages. More important, coal's crisis was one of demand, not supply. If demand had been strong, the right to strike could have been effective as both a threat and weapon. But during its brief existence—1947–1950—Lewis was fighting the wrong enemy with the wrong weapon. The most he could do was to push a bad situation to the point where political creativity would be welcomed. Lewis was maneuvering without his old allies from the 1930s. His endorsement of Republican Wendell Willkie in 1940 alienated him from organized labor and the Democrats. His wartime bargaining tactics had forced Roosevelt to seize the mines twice in 1943. His opposition to the Taft-Hartley Act's anti-Communist pledges embarrassed mainstream AFL and CIO unions. By the late 1940s, Lewis's world was again confined to the hilly boundaries of the eastern coal industry. The father of the CIO had waived paternity in a political snit. The man who had once been mentioned as FDR's running mate was almost completely isolated from Democratic decision makers. Other labor leaders—be they AFL, CIO, or radical—wanted to have little to do with him.

But over the years he had learned to become a savvy helmsman in that river of events where coal and politics flowed together. His negotiating ability had boosted the miner's hourly wage from $0.75 in 1935 to $1.94 in 1950, exceeding those paid to manufacturing and steel workers.[36] But after the war, he recognized that an industry in long-term decline could not afford to carry so many workers. He willingly accepted the inevitable loss of employment and union membership that mechanization produced. But, he said, miners who continued to work must be rewarded. Not only were they to be paid well, but a new benefit plan should be established—a plan that was to shape coal politics for the next thirty years.

Until 1945, UMWA miners had no pensions and only sporadic medical coverage. Still in use in mining communities was the company doctor system, which provided substandard care.[37] Fifty thousand permanently disabled coal miners—victims of their work—lay bedridden or housebound in 1945. They had no specialized rehabilitation program.[38] UMWA officers had been proposing a welfare plan for injured and retired miners at their conventions since 1938, during years when Congress was defeating proposals for national health care.[39] Lewis proposed a health plan in 1945 after the National Labor Board ruled that a sickness-benefit program not exceeding 5 percent of payroll cost was acceptably noninflationary in the war-managed economy.[40] Companies "burdened" with excess profits could deduct the costs of health benefits as business expenses with "little actual expense, since they would have had in any case to have paid much of it out in taxes.[41]

The idea of a group health plan for miners, either miner controlled or jointly controlled, goes back to coal's earliest days. In the thirteenth century, German governments bestowed corporate rights on their miners' unions to provide funds for injured miners and benefits to widows. British unions set up insurance funds in the nineteenth century. Mine workers of Pennsylvania's Philadelphia & Reading Coal & Iron Company in the 1870s established a relief plan by which the company contributed the yield of one day's production and miners the pay of one day's labor. This fund paid $30 for funeral expenses, $3 per week to the widow, and $1 per week for each child under twelve, and "more liberal provision has been made for the females whose heads [men] have been crippled or killed in the mines."[42] UMWA conventions regularly considered various medical and pension plans, such as the one in 1906 that would have funded an accident compensation program with a 1 percent tax on each ton of coal, or the 1919 resolution calling for state-run compulsory health insurance. What was new in 1945 was that the UMWA had a chance to win a welfare plan of some sort. Lewis raised the demand during the only time from 1920 to 1970 when coal production was truly booming. If miners were to win health care, he had to establish the principle while production was up.

The struggle over what came to be called "the Fund" was fought in many rounds of collective bargaining after 1945.[43] Two central issues were the cost and quality of benefits miners would receive, and who would control the Fund's operations. Both surfaced as soon as Lewis proposed a union-run industrial health plan in the spring of 1945, to be financed by a nickel-per-ton royalty on all UMWA-mined coal. Lewis's original proposal did not include pensions.[44] The mine operators rejected all UMWA demands that would cost more money in March 1945. Labor Secretary Perkins sided with the operators against Lewis's health plan. Lewis settled without a plan. He raised the issue again a year later in a proposal that included hospital care, medical services, life and health insurance, rehabilitation for injured miners, economic help for hardship cases, and "cultural and educational" work for miners if any money was left.[45] Pensions were not part of the package as it was explained in the *UMWA Journal*. Lewis put the need for a medical plan in these blunt terms:

> We accuse, by the record, that the industry does not bury its dead or bind up the shattered bones and the mangled flesh of its victims in any adequate, humane, or modern sense. . . .
>
> We accuse, by the record, that the industry extorts annually from the pay envelope of the mine workers $60 million for pseudo-, hypothetical, and sub-standard medical service, hospitalization and insurance of an actual value of less than one-third of the aforesaid $60 million.
>
> We demand abatement of this slaughter. We demand cessation of the accompanying extortion.[46]

The operators were to finance the system through a 7 percent *payroll* tax, and Lewis proposed that the union manage all monies and programs. The choice between financing by tonnage or payroll tax was significant. A tonnage tax tied benefits to the level of unionized production. That link gave Lewis an incentive to promote mechanization, which promised to boost production. A payroll tax, on the other hand, tied benefits to the number of working miners and their wage scale. Depending on how it was calculated, a payroll tax could act as a drag on mechanization. In 1945, Lewis argued that the Fund should be financed by a tonnage royalty; in 1946, he switched to a payroll tax. He would reverse himself one final time.

The operators refused to negotiate in 1946, and 340,000 miners struck. Lewis put the miners' case to the industry negotiators in these words:

> When we sought surcease from blood-letting, you professed indifference. When we cried aloud for the safety of our members you answer [sic]—Be content—'twas always thus!
>
> When we urged that you abate a stench you averred that your nostrils were not offended.
>
> When we emphasized the importance of life you pleaded the priority of profits; when we spoke of little children in unkept surroundings you said—Look to the State!
>
> You aver that you own the mines; we suggest that, as yet, you do not own the people.
>
> You profess annoyance at our temerity; we condemn your imbecility.
>
> You are smug in your complacency; we are abashed by your shamelessness; you prate your respectability; we are shocked at your lack of public morality.
>
> . . . To cavil further is futile. We trust that time, as it shrinks your purse, may modify your niggardly and anti-social propensities.[47]

Under federal pressure, the industry's negotiator finally agreed "in principle" to a health and retirement plan. The operators insisted that the Fund be jointly administered, but Lewis argued that the companies would veto Fund programs. The industry then repudiated the agreement "in principle." They pointed to the payroll tax. Deadlock. Coal stockpiles fell, and steel and auto manufacturers cut back production. Antistrike legislation appeared in Congress, and two Virginia congressmen sought to outlaw employer-financed health, welfare, and retirement funds.

Truman ordered Secretary of the Interior Julius A. Krug to seize the mines on May 22, 1946. Seizure meant federal administration but not ownership in any sense. Profits were held in escrow until the mines were relinquished. Seizure simply meant that Lewis negotiated with federal bureaucrats. Less interested in preserving the privileges of the coal industry and more interested in getting coal mined for the postwar economy, especially in time for the off-year election, Krug quickly came to terms with Lewis. In addition to an 18.5-cents-per-hour wage increase, Krug agreed to separate welfare funds—one for pensions, the other for medical care—each to be financed by a nickel-per-ton royalty. The pension plan would be administered by a tripartite committee in which the union and the industry each appointed one trustee and the two together chose the third, "neutral" trustee. The medical plan was to be run by three union-appointed trustees. Krug also agreed to a federal survey of health conditions in the coal fields to reinforce the miner's case with the general public.[48] Admiral Joel T. Boone directed *A Medical Survey of the Bituminous Coal Industry*, a disturbingly forthright report published in 1947.[49]

The Krug-Lewis relationship soured when Lewis sought to reopen bargaining some months after his initial success.[50] Krug wanted to return the mines to the operators; Lewis wanted to continue bargaining with the government. A strike ensued, and the press scared the public with headlines like the one the *New York Times* ran in November: "25,000,000 May Be Idle If Coal Strike Is

Prolonged."[51] Krug obtained a temporary restraining order against Lewis; Lewis tried to fight it by hiring Henry W. "The Dutchman" Grunewald, a former FBI agent, to dig up dirt on the personal life of T. Alan Goldsborough, the federal judge who had issued the order.[52] Lewis asked Harry Moses of the Frick Coal Co., Cyrus Eaton, and Averell Harriman for counsel and assistance. To no avail. He and the UMWA were found guilty of civil contempt of court for continuing to strike in December 1946. The no-strike injunction was continued indefinitely; the UMWA was fined $3.5 million and Lewis personally $10,000. While the Norris-La Guardia Act protected private-sector strikes from injunctions, the Supreme Court ruled on March 6, 1947, that it did not apply to miners who had been made federal employees. Lewis called off the strike and retreated to Florida to lick his wounds. The court reduced the union's fine to $700,000 but affirmed his $10,000 fine, which he paid with UMWA money. The court upheld Truman's position that seizure allowed the government to order Lewis to negotiate with the operators (Lewis had argued that seizure meant negotiations only with the government). Lewis correctly interpreted Truman's actions as a signal that federal power could no longer be swung against coal management in collective bargaining.

The agreement Lewis signed with operators in July 1947 contained most of the original Lewis-Krug terms. The medical and pension plans were merged into a single Welfare and Retirement Fund, financed by a ten-cent tonnage tax and jointly administered by a three-member board.[53] But the Fund's income depended on output and productivity, and not necessarily on more workers.[54] In fact, the fewer the employees, the better care and bigger pensions they could receive. Financing medical care from production created a Fund interest in minimizing constraints on production and maximizing efficiency. Occupational safety and health came to be seen as constraints, and the UMWA and the Lewis-led Fund did little on either for most of the 1950s and 1960s. The Fund's production-royalty plan pushed Lewis into a partnership with the coal industry in the 1950s that ill served his members.

Between 1947 and 1950, Lewis fought to make the operators live up to their Fund obligations. They deadlocked over a proposed $1,200 annual pension. The operators' trustee sued four times to stop the Fund's operations. Strikes and injunctions jumped one another like checkers. The 1948 contract reaffirmed the Fund, but slumping sales pitched the industry into its familiar battle with Lewis over the royalty. The Fund ran out of cash in 1949 after the operators stopped payments. Tumult in coal's labor relations and falling demand finally brought the parties to the point where they were forced to do something different. The Spring of 1950 marks that point.

The Fund's shaky beginning foreshadowed the two failures of its later years. First, it promised benefits that it delivered and then canceled. This was not inevitable. Lewis could have negotiated higher royalties in the 1950s, but decided to spare the operators at his members' expense. The second was the dashed promise of the UMWA medical system. From its start, the plan incorporated progressive principles of medical service and coverage. Fund administrators sought to provide no-fee comprehensive services and total coverage.[55] The Fund paid for the complete rehabilitation—physical, psychological, and social—of thousands of disabled miners in its first few years. Policy making was centralized, but carried out through ten regional administrations. Health consumers—that is, miners and their families—were given significant influence over Fund programs at the local level.[56] To break the grip of the company doctors and their fee-for-service practices, the Fund built a network of group-practice clinics and hospitals. These facilities extended union-sponsored health care where it was needed most and removed the profit motive from medical treatment. The philosophical underpinning for these ideas came from a group of politically progressive doctors and administrators recruited from the U.S. Public Health Service and the Farm Security Agency in the late 1940s.[57] Although Dr. Warren F. Draper, the Fund's chief medical administrator, was a conservative with strong links to the American Medical Association, he allowed the visionaries to try out their ideas. He was generally willing to defend them against attacks from the medical establishment, which saw Fund cost-control measures and salaried medical staff as great leaps toward socialism. Lewis, no socialist himself, also generally supported the Fund's progressive medical ideas for the pragmatic reason that he saw no alternative.

Lewis served as the UMWA Fund trustee from the late 1940s until 1969, along with his close friend Josephine Roche—the "neutral" trustee—who was also director of the Welfare and Retirement Fund from 1947 to 1971. "Never in the entire time that Lewis lived did Miss Roche ever once vote contrary to his wishes," Finley says.[58] The tragedy was that although the Fund worked well when it was able to work as it was originally designed, it was able to do so for only a few years. Much of the Fund's history under Lewis and Roche is one of cutbacks in services and the denial of eligibility to thousands of rightful beneficiaries. The Fund was a soundly conceived medical system built on an imperfectly conceived financial foundation. Lewis had established a semi-socialized medical system for his miners by upholding capitalism within the coal industry—one of the truly rearguard defenders of the faith. Lewis and the Fund's planners had created a welfare institution to which miners were passionately committed. Not surprisingly, the first sparks of rebellion against the UMWA leadership were struck over denial of Fund benefits. Miners may not have been socialists, but they liked semisocialized medicine and were willing to fight for it.

5. The Lewis-Love Axis and the Great Shakeout

Clouds hung over the coal business in 1949 and the early months of 1950. Bituminous production had fallen 30 percent in two years, from 630 million tons in 1947 to 437 million in 1949. Workdays fell from 234 in 1947 to 157 in 1949, the fewest since 1932. The market signs were ominous. The days of the coal-burning steam locomotive were fading fast into Norman Rockwell's evocative pictures. Coal smoke was becoming less of an urban blight as business and individuals converted furnaces to other fuels. Lewis did what he could to minimize overproduction. In March 1949 he called a two-week strike to reduce stockpiles. They fell, but a sixty-day reserve remained.[1] UMWA miners stopped work again in June. Lewis ordered a three-day workweek.[2] In the fall and winter of 1949, Lewis had his 370,000 coal miners working intermittently. Union mines were shut as the new decade began.

Lewis and the industry feared government more than they feared each other. Lewis was affronted by the fines, injunctions, and Taft-Hartley strictures he had faced. The operators nervously observed our North Atlantic allies nationalizing coal and other industries. Public opinion seemed to entertain public management if coal couldn't set its house in order. Truman had the authority to seize the mines. Virginia had gone a step further; it "bought, mined, distributed and sold coal through local fuel commissions" totaling $1,162,472 over seven months in the unpredictable 1949–50 period.[3] A *Nation* editorial said flatly that "the . . . coal industry . . . is rotten-ripe for nationalization."[4] Other journalists were writing of the advantages of such a change.[5]

But the Truman administration was probably less inclined to get stuck in coal than either Lewis or the coal operators realized. Truman was philosophically more reluctant than New Dealers to meddle in the market. Federal officials raised the issue in terms of antitrust and competition. Some clearly favored anitmonopolistic energy policies.[6] Interior Secretary Harold Ickes and some staff in that department, which did most federal energy work, thought differently. Evelyn Cooper, a member of Interior's secretariat, proposed within the department in November 1946 that a public coal policy have three parts: "(1) the government's withdrawal of certain marginal mines from production through exercises of the right of eminent domain, a license system, or some other device; (2) public assistance to productivity increases; and (3) rehabilitation of miners replaced by mechanization."[7] Cooper's prescription was in keeping with the oversight and management functions that had evolved between coal and the government during the Depression and War. N. H. Collisson, who managed the Coal Mines Administration when Truman seized coal in 1946, wrote that the coal "problem far exceeds the ability of the industry to effect a solution," and that the industry had narrowly escaped "committing suicide as a result of ruinous competition between the various producers."[8] There was even talk of adapting the model of the Atomic Energy Commission, in some fashion, to coal. But those who succeeded Ickes at Interior, J. A. Krug and Oscar L. Chapman, did not share his philosophy of federal responsibility. Although sentiment persisted in the Interior Department and elsewhere for public planning and oversight, it declined in proportion to the increasing influence of the major coal operators within the department through its National Bituminous Coal Advisory Council. The council worked hard to persuade its advocates in the administration to take a hands-off policy and let the private sector work things out. The industry felt that its economic and political clock was ticking, that a crisis might force Truman into nationalization. Lewis and the unionized operators foreclosed federal management by fashioning a partnership to control competition.

Aside from demand, the major enemy coal suppliers faced was themselves. They were too numerous, too divided, and too anarchistic to discipline their affairs. Lewis waited for the bituminous industry to come up with a leader, someone like Richard Grant of the anthracite combination. Lewis had pleaded for such a person for years:

> When asked . . . what the major difficulty was in collective bargaining, Lewis replied: "In one sentence . . . the lack of leadership among those with whom we deal. . . . If you were to ask me today who could speak for the industry on any suggested idea that would be constructive in itself for the industry, I would say very frankly that there are no such men."[9]

The biggest companies in the industry had come to the same conclusion during the chaos of 1949. One industry executive, Harry M. Moses of U.S. Steel, described in 1955 how a "small group of operators" finally got together to lay a new foundation for industry economics and collective bargaining:

> All this [dog-eat-dog competition] led to a small group of operators getting together in 1949 "to do some long-range thinking" with these several conclusions: (1) "kleig-light" bargaining with the UMWA had failed; (2) government intervention led to settlement of economic problems on a political bases; (3) both the industry and the UMWA had leadership with "sufficient wisdom to solve their own mutual problems without making a public spectacle of their dealings or calling on politically minded government for help."[10]

Moses did not name the "small group of operators." Subsequent events suggest that Consolidation and U.S. Steel were the key companies, representing respectively the leadership of the commercial and captive sides. Their goals were stability among suppliers without federal intervention; their methods, secrecy and "summit diplomacy" with Lewis. The "small group of operators" also approved the emergence of one of their own, George Love of Pittsburgh-Consolidation Coal, to restructure relations with John L.

George H. Love became to capital what Lewis was to labor. He began his career in 1926 in his family's coal business—Union Collieries—following Princeton, Harvard Business School, and two years as a stockbroker. In 1929, Union Collieries merged with the M. A. Hanna Company, whose president was George M. Humphrey, the Cleveland industrialist. Humphrey gave Hanna's mine to the Loves in return for an interest and a directorship in Union. Love and Humphrey worked well together. Hanna bought a large block of stock in Consolidation Coal from National City Bank (the predecessor of Citibank) in 1942. Hanna's dominant interest in Consolidation put Humphrey and Love in key positions from which they effected a merger with the Mellon-owned Pittsburgh Coal Company. In 1945, Love became president of Pitt-Consol,[11] the most important coal company in America.

From this position, Love began to discipline coal's buccaneers. Coal suppliers divided into three main groups: the northern commercial operators, the southern commercial operators, and the captive mines, which were found throughout Appalachia. Love represented the dominant northern operators and forged an alliance with the steel-owned captives. A new organization, the Bituminous Coal Operators' Association (BCOA), was formed in 1950; its members accounted for about half of all coal mined in the United States. The biggest companies were coordinated by George Love, who operated "behind the scenes in the development of the BCOA . . . just as George Humphrey had . . . [done] during George Love's campaign to make Consol the biggest company in the industry." BCOA policy was determined by votes weighted according to the tonnage each company produced.

> Consol accounted for 15.5 million tons, but Love also served as representative of other companies with 37.5 million tons. The total tonnage of the BCOA's members was 110.5 million; at each BCOA meeting, therefore, Love controlled 52 out of 110. If by some exceedingly remote chance that had not been enough to give him control of the organization, he had only to join forces with his friend Harry Moses, who represented the 19.2 million tons produced by U.S. Steel and therefore had 19 votes. By no possible combination could the other members of the BCOA defeat Love's aggregate 71 votes with their 39; voting was by a simple majority, not by two-thirds.[12]

Some months after the small group of operators had gathered secretly in 1949, their bargaining goals emerged. BCOA would try to dictate labor costs across the entire industry. Love said his first goal was a "new stable and binding contract . . . that neither party have the right to cancel or suspend . . . during the intervening two and one half year period."[13] He wanted, second, the "elimination of the 'able and willing' [right-to-strike] clause and a provision binding on the union not to limit production nor the use of facilities." His third goal was "to continue payment of 20 cents per ton [to the welfare fund] . . . but . . . all payments for pension or welfare [should] be limited to the employees of companies which contribute to the fund." If Lewis was willing to guarantee labor peace and contain labor costs, Love was willing to give him control of the welfare fund.

To get Lewis to yield his power to strike at will would not be easy. The strike had allowed Lewis to be the rogue elephant in this particular jungle for twenty years. But federal courts created an opening for Love. John L. had organized a walkout on February 11, 1950, after federal officials obtained a ten-day temporary court injunction barring "further stoppages under the provisions of the Taft-Hartley Act."[14] On February 20, the UMWA was hit with a civil and criminal contempt citation for continuing the stoppage. The government petitioned Judge R. B. Kleech for a permanent eight-day Taft-Hartley injunction. Negotiations with the operators resumed under that threat as well as an earlier NLRB ruling on an unfair labor practices petition. Judge Kleech

> enjoined the UMWA from insisting on certain non-wage demands which he ruled as illegal under the Taft-Hartley Act. His order restrained the union from demanding a closed or union shop, limitation of the benefits of the welfare fund to union members, the "able and willing" clause, or the right to stop work during "memorial periods."[15]

Kleech's ruling on the nonwage issues apparently forced Lewis to concede the "able-and-willing" clause in the 1950 contract. But UMWA lawyers never challenged Kleech's decision. Lewis may have figured that strikes could serve no useful purpose in a prolonged demand slump. In any case, no UMWA contract after 1950 contained a right to strike.[16]

Lewis made the best of a bad situation. He announced on February 17 a thirteen-point list of contract demands, which was more smoke than fire, including:

> a guaranteed annual wage based on 200 working days; 7½ hr. work day, portal-to-portal; an increase in the basic wage from $14.05 to $15.50 a day . . . ; an increase in the welfare-fund payment from 20¢ to 35¢ a ton; a new board of trustees for the fund, with the guarantee that the operators would cooperate with the union and its welfare program; abandonment by the operators of all lawsuits against the union, now totaling millions of dollars.[17]

But he made it clear to the BCOA that all demands were negotiable. As the *UMWA Journal* pointed out *after the contract was signed,* "the UMWA had proposed to negotiate a contract without any previous commitments in any way, shape or form on either the economic or controversial legal issues, such as the union shop."[18]

Although Kleech had painted Lewis into a corner, Love still needed John L.'s *active* help to solve the long-term problems they saw in the coal industry. Breaking the UMWA, as the industry had done in the late 1920s, would only cause more instability. Local strikes would abound. Price competition would feed again on cutting wage costs and, in the end, profits. No, Love reasoned, the better approach was to ally with Lewis against common problems. In a speech to the Edison Electric Institute on June 7, 1950, Love outlined the problems the coal industry had to solve: "(1) labor relations; (2) competition within the industry; (3) competition from other fuels; and (4) maintaining earning ability to permit continued progress."[19] He and Lewis could stabilize labor relations. Intraindustry competition was also subject to joint "problem-solving." If Lewis and the BCOA could impose uniform labor costs across the industry, they could squeeze out the small operators and finance the new machines that would raise productivity. Competition against other fuels drew Lewis and coal management together in the late 1950s in a joint coal lobby—the National Coal Policy Conference—that opposed nuclear power and imported oil. Finally, Love saw Lewis as central to guaranteeing coal's earning ability. Big coal could get long-term supply contracts from big consumers only if supply was predictable and coal cheap. Lewis was the key to both requirements.

Central to the economics of the 1950 partnership was mechanization. In the late 1940s, continuous-mining machines began to replace the drill-and-blast methods that required several separate machines and much labor. By cutting the coal directly from the face, continuous miners shortened the time needed to obtain a ton and reduced labor requirements. A continuous miner with 10 men could produce three times as much coal as 86 men in a handloading section and half again as much coal as 30 men on a machine-loading section. The labor cost per ton was $3.28 for handloading, $1.64 for machine loading, and $1.01 for continuous mining. The total cost per ton was $5.28 for handloading, $3.79 for machine loading, and $3.16 for continuous mining.[20] *Fortune* linked the 1950 Lewis-Love contract to new underground mining technology: "In the coal industry 1950 may be remembered as the year of the great shakeout—and also as Year One of its technological revolution."[21] Continuous miners were only one element in the mechanization that occurred in the 1950s and 1960s. To keep pace with the faster mining machines, mechanical loaders were improved and in-mine haulage systems upgraded. Together, they eliminated the need for most handloaders. As underground mining made itself more productive, so was surface mining. Excavating equipment increased in size and speed along with haulage techniques. As an ever greater share of U.S. coal was mined on the surface, ever fewer UMWA miners were needed underground.

Although the 1950 contract did not mention mechanization specifically, Lewis's support for "labor-saving" machines was part of his understanding with Love, who had no worries on that score: "Lewis' union . . . has one great advantage over many others. It has never fought mechanization."[22] Lewis had made plain his philosophy on mechanization as early as 1925:

> The policy of the United Mine Workers of America will inevitably bring about the utmost employment of machinery of which coal mining is physically capable.
>
> Fair wages and American standards of living are inextricably bound up with the progressive substitutions of mechanical for human power.
>
> All will agree that fair wages are only possible because of the increased productivity per worker, secured largely by advanced methods and means of production. It is equally true that such wage rates are the principal incentive for the invention and installation of new devices.
>
> . . . The development of low cost operations [mechanized] will automatically eliminate the uneconomic mine.[23]

In 1952, Lewis claimed full credit for mechanization:

> The American coal operators never could have mechanized their mines and increased per-man-day productivity unless they were compelled to do so by the pressure of the organization of the mineworkers. . . . We want participation. We ask for it. We ask for it . . . to compel him to modernize. Otherwise, he wouldn't move.[24]

Even at the UMWA's 1952 convention, Lewis made no bones about his policy: "We stand for the usage of all modern equipment and scientific devices and forms that will take the raw materials which God gave our people and make them of increasing value in the marts of trade. We only ask and we will always ask our rightful participation in that new productive efficiency."[25] The UMWA's rank and file seems generally to have accepted mechanization. They did not sabotage the continuous miners or insist on featherbed contracts. They did, however, press Lewis for a strong seniority clause to provide some protection against the layoffs that over twenty years would eliminate three of every four of their jobs. Delegates at UMWA conventions did debate the idea of unemployment compensation, shorter workdays (to spread the available work evenly), better pensions, and the health and safety hazards of the new machines. Nothing came of their concerns.[26]

Although Love and Lewis did not publicize the competitive implications of mechanization, they were apparent to anyone who cared to look, as did *Fortune:* "Mechanization is now a means of squeezing out marginal mines and thus increasing the industry's profitability."[27] Lewis had always linked the ability to mechanize to changes in the structure of the industry itself. Leadership and capital, he thought, were preconditions to mechanization.

> Before the technique of the coal industry can be brought to a state commensurate with what the nation has a right to expect, there must be a change in the financial structure of the coal business.
>
> . . . It is obvious that only solvent corporations, sufficiently financed, can undertake the improvements that the times demand.[28]

"Solvent corporations, sufficiently financed" meant the same thing to Lewis and Love: bigger companies with stable markets that were able to finance modernization. It meant more concentration within the industry. It meant the imposition of a semioligopolistic structure over supply through which a handful of major producers established the terms of business for most of the industry.

The logic that wedded Lewis and Love was economic. To survive the competition of other fuels, coal had to remain cheap and free from the threat of strikes. Mechanization would raise productivity and keep coal competitive with oil and gas. It also promised to bring about changes *within* the coal industry. Those who could mechanize fastest would be able to get the best contracts and make the most money. The Southern Coal Producers' Association (SCPA), the organization of smaller companies with UMWA contracts, could not afford both to invest in mechanization and meet the new BCOA wage-and-benefit standards. Lewis and Love hoped that mechanization would lower the production costs of union-mined coal below what the small, nonunion suppliers could achieve by cutting wages. Machines, they hoped, would eventually suppress competition.

Joseph E. Moody, SCPA president, clearly saw how the 1950 BCOA contract put his members in a bind.

> The union carried out a very forceful campaign to establish a national contract, aided at times by some management groups of the industry. The Southern group has opposed this throughout the years in every way available to us. We believe that it is impossible to establish terms on a national basis that are fair and equitable to the various groups in the industry. . . . A rigid level of labor cost is ruinous in normal competition. . . .
>
> That is the story of industry-wide bargaining. The price may be high to avoid it but there is no price too high to pay. Industry-wide bargaining will sooner or later lead us down the road to socialism and nationalization.[29]

It was not socialism or nationalization that Lewis and Love had in mind; it was free enterprise for the BCOA oligopolists and bankruptcy for the rest. The intent of the Lewis-Love partnership was summed up clearly in Harry Moses' conversation with John L., as he related it to *Business Week:*

> Mr. Moses: Well, you know how Lewis is. He sometimes makes a point by parable.
> *BW:* What was his parable on this point?

Mr. Moses: He said that if a fellow wants to stand out in the middle of the street with the traffic going both ways, it wouldn't be precisely preposterous if that fellow got run over.[30]

By 1952, Bethell writes, Lewis would sound like Love's "alter ego."

"The smaller coal operators," he said, "are just a drag on the industry. The constant tendency in this country is going to be for the concentration of production into fewer and fewer units. . . . More of the obsolete units will fall by the board and go out of production."[31]

Lewis was a proud, resourceful man who Love was well aware was pivotal in maintaining labor peace at BCOA mines. Lewis had to keep his troops under control in the face of layoffs. Lewis wanted wages and control of the Welfare and Retirement Fund in return for labor stability. The BCOA gave control of the Fund to Lewis, possibly in exchange for deletion of the right-to-strike clause.[32] *Coal Age,* the industry's trade journal, buried the importance of the new arrangement in one sentence: "By inference the union gained control of the welfare fund and therefore complete responsibility for its operations."[33] Legally, the Fund was not supposed to operate under a UMWA thumb, but federal prosecutors never made it an issue. While Lewis gained control, he never had enough money to make the Fund's promise a reality. The history of the Fund has been the reach of imagination exceeding the grasp of its finances. Control, however, gave Lewis access to many millions of dollars in pension capital and royalties.

The BCOA-UMWA contract of 1950 financed the Fund with a 30-cent-per-ton royalty, which was raised to 40 cents in 1952. The contract spelled out what the Fund might provide:

(1) benefits to employees . . . , their families and dependents for medical or hospital care, pensions on retirement or death of employees, compensation for injuries or illness resulting from occupational activity or insurance to provide any of the foregoing, or life insurance, disability and sickness insurance or accident insurance; (2) benefits with respect to wage loss not otherwise compensated for at all or adequately by tax supported agencies created by federal or state law; (3) benefits on account of sickness, temporary disability, permanent disability, death or retirement; (4) benefits for any and all other purposes which may be specified, provided for or permitted in . . . the "Labor-Management Relations Act, 1947" . . . ; and (5) benefits for all other related welfare purposes as may be determined by the Trustees within the scope of the Act.[34]

Lewis envisioned a cradle-to-grave welfare system that might provide supplementary unemployment and disability benefits, as well as almost any other social service the trustees chose. It might for example have provided benefits for "illness resulting from occupational activity." It might have been used to reduce the work-place dust that produced thousands of cases of lung disease and used up millions of dollars in medical expenses.[35] It might have paid severance benefits to miners as employment fell from 415,582 in 1950 to 124,532 in 1969—it might have seeded retraining and reinvestment projects in depressed coalfield communities, or helped miners relocate to more prosperous areas. All of these services would have been possible if Lewis and later UMWA presidents had negotiated adequate royalty payments. But for twenty years, UMWA negotiators never asked BCOA for a royalty high enough to have allowed the Fund to fulfill its original scope.

Why did Lewis and Boyle fail to negotiate higher royalties for so long? The answer, I think, has two parts. First, BCOA would have fought any increase tooth and nail. To survive the sales stagnation of the 1950s and 1960s, BCOA operators had to keep tight control over their costs to maintain competitiveness with other fuels. With coal selling at between $4.50 and $5.00 per ton for most of that period, even a five-cent increase every two years might mean the difference between sale and no sale. On the other hand, the industry as a whole generated increasing profits in both decades, so it is reasonable to suppose the operators could have afforded some increase without catastrophic consequences.[36] The second reason may be more telling. UMWA presidents accepted the logic of industrial partnership in which they played an essential but basically subservient role. The industry's survival—as well as their own—depended on keeping labor costs down and ending strikes.

Lewis and Boyle might have been more assertive had the rank and file been able to generate pressure for higher royalties. That pressure didn't build until the late 1960s, because those who had lost their Fund benefits had also often lost their jobs and their standing as union members. And those who were lucky enough to have jobs and Fund benefits didn't want to rock their own boat. Fund benefits always bobbed at the end of a seemingly arbitrary line of decisions about who was eligible and for what. The Fund inevitably transmitted to its beneficiaries the consequences of BCOA-UMWA politics. As Lewis and Boyle drove a wedge between themselves and UMWA miners, so too did the Fund move away from the people it was supposed to serve. As the bond with the mine operators tightened, the gap widened between the rank and file and leaders of the UMWA and Fund. When rank-and-filers questioned policies, they were sealed off like bacteria in an airtight jar. No wonder, then, that grievances against the Fund built up and exploded into strikes and lawsuits.

When Lewis and Love unveiled their contract in March 1950, no one foresaw the radical changes its twenty-year framework would bring. Love did not quite come out and say that the new BCOA had switched its Lewis strategy from fight-him to use-him, but he was blunt enough about the labor peace he had achieved:

This 2½ year contract gives the industry its first real opportunity for stability in the last decade. . . . The operators definitely established the right to control their own production and their mining facilities. The union asked for a cooperative administration of the Welfare Fund and we are giving it to them. . . . The responsibility [for the Fund] is squarely on the shoulders of the union and if it fails, the public and ourselves will look directly at the union.

This coal business is not a sick industry as it has been said recently. This country is one of the very few where coal mining is still in private hands operating under a free enterprise system. I hope that from this contract will come such mutual understanding that we will do away with coal strikes in the future.[37]

Lewis, who had from time to time called mine operators "human leeches,"[38] responded with a statement that might have been drafted by Love's public relations department.

It is [a] reasonable assumption that for a substantial period of time the industry can abate its labor warfare and apply itself, both management and labor, to the constructive problems of producing coal in quantity for the benefit of the American economy at the lowest possible cost permitted by modern technique. The Mine Workers stand for the investors of the industry to have a return on their capital; they stand for the public to have coal at the lowest possible price consistent with the right of mine workers to live a free life on the same standards and with the same opportunities and under the same laws as other citizens under our flag.[39]

The 1950 contract was the single most important change in the coal industry since the New Deal. It established new terms of labor-management relations that held together until they could no longer contain the social forces they created. Shifts in demand and the turmoil of collective bargaining had forced Lewis and Love to stabilize the industry. Stability meant an end to strikes, reduction of the work force, supplier-managed price control, higher productivity, mechanization, and the elimination of federal intervention in coal affairs. There was little, if any, significant legal conflict between the UMWA and BCOA companies for twenty years.

Two groups did, however, bring the UMWA and the BCOA operators into court: small coal companies and the UMWA rank and file. Both were targets of the 1950 contract. The UMWA suppressed its own members in order to subsidize BCOA employers. Miners lost their jobs, their health, and, for a time, their union. Some of what Lewis and Love did in 1950 was inevitable. Mechanization had to come. Inefficient suppliers and "surplus" miners would have been trimmed under any economic system. The relevant point is not that cuts were made, but how they were carried out. Market economics and Lewis's policies brought about these changes with little regard for the massive human suffering they caused.

6. Industry Peace and Labor Repression

The partnership that Lewis and Love began in 1950 darkened the UMWA for three decades. The chilly logic of cooperation evolved into a cold-blooded conspiracy against both the union's rank and file and small coal producers. What began as Lewis's support for industrial concentration translated into UMWA "organizing drives" to bankrupt BCOA competitors. The conspiracy was the product of the dark genius of two dyed-in-the-wool capitalists who saw nothing wrong in restructuring an industry according to their own wishes. Federal courts finally declared that the UMWA and Consolidation Coal had conspired to restrain competition. Love and the BCOA cannot be blamed for pursuing their interests; business is business. Miners, on the other hand, will not forgive Lewis and Boyle for using the UMWA to victimize them. The lesson for labor in the UMWA story of these years is how easily circumstances can transform union leaders into double agents.

Collaboration was new ground to Lewis and the BCOA operators in 1950. Yet old habits of old men changed with surprising ease. New institutional relationships hatched quickly. Contracts were negotiated peacefully, secretly, and infrequently. Each reaffirmed and amplified the terms of the 1950 understanding.

The 1952 negotiations blended a wisp of the old with the new style and content. It was the last time the federal government was to play an important role in coal negotiations until 1978. The Korean War had resurrected federal boards to regulate the economy and approve wage or price boosts. Following about twenty secret meetings with Harry Moses of the BCOA, Lewis signed a new contract on September 20.[1] He won a wage hike of $1.90 per day (bringing the basic industry rate to $18.25 per day) and a 10-cents-per-ton increase in the Fund royalty (to 40 cents per ton).[2] With a BCOA contract in hand, Lewis pivoted toward the unionized southern operators. The Southerners capitulated to the UMWA-BCOA terms, although Joseph Moody, their leader, had said earlier that the BCOA contract would be "economic hari-kari" for them.[3] Both BCOA and the southern operators filed approval petitions with the Wage Stabilization Board (WSB), chaired by Archibald Cox. The board gave contingent approval to the increased Fund royalty but denied the $1.90 increase in wages in favor of

$1.50 per day. Coal miners struck in protest on October 20 and stayed out for nearly a week. Moses and Lewis jointly requested that Roger Putnam, WSB economic stabilizer, overrule the WSB decision. After a conference with Lewis and Moses at the White House, Putnam received a letter from Truman directing the WSB to approve the wage agreement "as negotiated." Putnam complied, and Cox resigned in protest. Industry members of the WSB then boycotted meetings, and eventually resigned en masse to protest Truman's directive. A reconstituted WSB authorized the $1.90 wage increase on December 8, and the Office of Price Stability approved price increases on December 11.

Lewis consistently focused his pressure on Moses and the BCOA rather than on Truman. He was not asking additional concessions from management but political support in what had become a UMWA battle with the Truman administration. He laid his case before Moses following the WSB's decision to disallow the $1.90 increase: "We have a contract. It is with your association. It is complete. It speaks for itself. You signed it. It was negotiated in the American way—through collective bargaining. It is as pure as a sheep's heart."[4]

The southern operators doubted the purity of the sheep's heart. They realized the implications of the BCOA-UMWA axis. Less than a year later, *Coal Age* reported:

> [Moody] warned that . . . the southern operators face "economic ruin." . . . He charged that coal's affairs have been dictated by a group of northern operators, including the captive mines. This brought a rejoinder from Harry M. Moses, president, Bituminous Coal Operators' Association, that such charges are "loose and unfounded."
>
> . . . Mr. Moody [cited] the closing of 103 mines in his association's area in the 17 mos ending May 31, with 14,000 men jobless and an annual production loss of 15,000,000 tons. "It is about time," he asserted, "to decide whether we are going to starve to death quietly . . . . or come to grips with our situation."
>
> Sole hope for the southern producers is lower cost. "We cannot hope to reduce costs appreciably through the purchase of more machines," said Mr. Moody. Therefore a demand wage boost this fall would leave many companies only the choice of closing down or trying to operate non-union.[5]

The southern operators authorized Moody to terminate their UMWA contract if warranted.[6] Lewis growled to Moody: "There will be no negotiated wage decrease in the mining industry. . . . On that rock I place my ebenezer."[7] Moody and his group reluctantly stayed in the UMWA fold.

Lewis and the BCOA turned their ratchet wheel on the southern operators another notch in the 1955 contract. BCOA conceded a two-dollar wage increase but no change in the Fund's royalty. It was the first time since 1946 that Lewis had failed to *ask* for a royalty increase. Lewis described the contract in his inimitable way: "[The agreement] will not oppress the coal consumers, nor yet expose the brittle bones of the operators to the icy blasts of the coming winter. It is devoid of Marxian babble, and contains no wind or water. . . . The agreement is a constructive instrument with edible virtues."[8]

The UMWA allowed certain BCOA companies to pay less than full Fund royalties through sweetheart contracts to help them preserve markets and finance mechanization. By holding the southern companies to the full wage and benefit package, the UMWA made them less competitive. As long as the BCOA could offset increased labor costs through mechanization and sweetheart Fund payments, they could afford the contract while the Southerners could not.

The UMWA's wage pressure on the unionized southern operators forced some out of business and some out of the UMWA. Least preferable from the UMWA-BCOA viewpoint was continued operation as a nonunion mine. By 1958, *Coal Age* estimated that nonunion mines had grown to about 20 percent of total U.S. tonnage.[9] Nonunion operators found a niche in spot-market sales to electric utilities. As long as these mines could substitute very cheap labor for very expensive machinery, they could survive. With several hundred thousand miners out of work, labor was incredibly cheap. Many loyal UMWA miners were forced to work below scale in nonunion mines to get by. Lewis succeeded in ridding the UMWA-BCOA of *its* labor "surplus," which meant cutting the UMWA's own ranks by more than two-thirds.

The Tennessee Valley Authority's policy of purchasing the lowest-priced coal for its steam boilers gave nonunion Appalachian mines a ready buyer. TVA became the single largest coal consumer in America in the mid-1950s as demand for electricity exceeded its hydroelectric generating capacity. Nonunion operators began strip-mining the steep slopes of Kentucky and Tennessee, a comparatively cheap and productive method in the absence of any laws to conserve the environment. Thousands of acres of mountain property were destroyed as each operator sought to lower his price by stripping coal with no thought of reclamation. In this fashion, southern Appalachia subsidized cheap electric rates to TVA's customers. BCOA and the UMWA tried to stem non-union sales to TVA by insisting suppliers comply with the union-wage standards of the Walsh-Healey Act.[10] By the time they made their case, nonunion coal had established itself in this market.

While the BCOA-UMWA axis battled the smaller operators, major BCOA "commerical" companies took to consolidating themselves. Several dozen undertook a merger movement in the mid-1950s that appears to have been a carefully planned effort to strengthen themselves for a prolonged market contraction. In the spring of 1954, "top-drawer executives of 35 to 40 major producing companies, representing some 150,000,000 tons of commercial output," met informally in Chicago " 'to take the industry's pulse' and give 'a second hard look to its economic plight.' "[11] The names of the executives attending the meetings were not made public. All discussions were off the record. The group had no permanent chairman, no secretary, and no treasury. The press sniffed the smoke, but failed to detect the fire that George Love was feeding to forge a new coal industry. The *New York Times* reported on May 22, 1954, that key executives from the top forty coal companies had gathered in Chicago and set up a committee, in *Business Week*'s words, to "spell out consolidation methods that could be used by any group of coal producers."[12] George Love said Consolidation Coal was willing to make a loan to help establish a potential competitor.[13] The group met three times: in New York City on April 23, in Chicago on May 21–22, and in Cincinnati on July 8–9. Committees were established to study "freight rates, federal taxes, imported residual fuel oil, inferior uses for natural gas, coal merchandising and fringe benefits in the labor contract."[14] Although no merger committee was established, *Coal Age* reported that the subject was discussed:

> Some 40 to 50 major bituminous producers met July 8 and 9 to continue their informal review of the industry's situation and to hear reports from committees named at earlier meetings in Chicago and New York. The producers met in closed session and issued no statement at the close of the meeting. *Speculation persisted, however, that at least one result of the meetings probably would be a series of mergers to be announced within the next year or two.*[15]

Shortly after the last meeting of what *Coal Age* called the "coal executive group,"[16] a series of mergers was announced. On April 22, 1955, stockholders of Peabody Coal, the third biggest producer, voted at a special meeting to recapitalize "to permit Peabody to issue additional common stock shares for sale or for 'acquisition or development of desirable operating properties.'"[17] Less than two months later, Peabody offered to buy the Sinclair Coal Company's group of eight companies. On June 7—a day after the Peabody announcement—Island Creek, the fourth largest producer, said that a merger vote with the Pond Creek Pocahontas Company would occur on August 17.[18] Both mergers were completed through an exchange of stock. A few months later, the West Kentucky Coal Company bought all the stock of the Nashville Coal Company, making it the third largest after Consolidation and Peabody-Sinclair. Cyrus Eaton, chairman of West Kentucky and its major stockholder, denied at the time that John L. Lewis or the UMWA had anything to do with the acquisition. Nashville Coal immediately signed a UMWA contract. *Coal Age* commented in this remarkable fashion:

> Generally, the merger was greeted in the industry with satisfaction. Pittsburgh-Consolidation congratulated Mr. Eaton praising him for doing the industry a good turn. Mr. Eaton's own remarks about mergers were that coal needs more of them. He said that

"there is an urgent need for a number of larger, well-financed units that can spur coal's potential as the fundamental and best fuel."[19]

The next year, Consolidation merged with Pocahontas Fuel Company, the eighth largest producer. When the dust settled, the top handful of commercial producers—Consol, Peabody, Island Creek, and West Kentucky (later to be taken over by Island Creek)—had each absorbed a major competitor.

George Love's mark can be seen in the increasing concentration of the coal industry from 1950 to 1970. The four largest bituminous suppliers produced 13.4 percent of output in 1950, 23.9 percent in 1960, and 35.9 percent in 1970. The eight largest accounted for 19.2 percent in 1950, 32.5 percent in 1960, and 46.7 percent in 1970. Suppliers of fewer than 100,000 million tons per year declined from 20.7 percent in 1950 to 15.6 percent in 1960 and 7.3 percent in 1970.[20]

George Love felt that the federal government would have reorganized coal had he not done it first. He told the Edison Electric Institute in June 1950:

[The] . . . achievements [of the bituminous coal industry] are symbolic of what private ownership can do under adverse circumstances. I know that regulation and price-fixing in coal can only lead to government ownership, and if such becomes the pattern for coal, it may inevitably become the pattern for your industry and many others.

And ten years after masterminding the mergers, he confided to an audience at the Wharton School:

I honestly believe mergers and consolidations saved this basic coal industry of ours and permitted it to go on under private ownership as opposed to government control or nationalization. [Consolidation] . . . has been able to take the leadership in working with the union . . . and it has set the standard for other mergers and consolidations in coal.[21]

Love was probably right. Had he and Lewis not restricted competition through mergers of questionable legality and labor practices of unquestionable illegality, the federal government would have imposed a public solution to coal's perennial crisis.

These mergers created a framework for cooperation in matters of prices and markets. Each acquiring company tended to absorb those that strengthened its position in regional markets. In effect, big BCOA companies had come to a gentleman's agreement about who was to sell coal where. Mergers among the big companies made life harder for the smaller companies and helped coal compete with oil and gas. The Eisenhower administration allowed coal's merger activity to stand.[22] The BCOA's self-managed mergers enabled the big companies to increase profits despite market softness. These companies had transformed themselves into proven money-makers by the mid-1960s. Their work was duly appreciated by the oil companies, who began taking them over in 1963.

By 1955, the business press saw that the style of coal bargaining had changed radically. Although *Fortune* sensed the new ground rules, its writers hadn't quite figured out what the new game was. *Fortune* described the 1955 negotiations in this manner:

> Collective bargaining in the soft-coal industry, once a colorful pageant in which political fireworks, editorial dragon slaying, and judicial jousting were variously mingled, has been altered beyond recognition. . . . During the past few years, Lewis and Harry Moses, president of the Bituminous Coal Operators' Association (the northern group) have developed a quasi-social relationship; they meet periodically for lunch at the Carleton, in Washington, and frequently drop in at one another's offices, only a few blocks apart. "From time to time," as Moses explains this relationship, "Mr. Lewis raises the question of expanded benefits (I never raise it) in a purely personal, friendly way. We argue about it back and forth."
>
> . . . On the morning of August 11 . . . Lewis appeared in Moses' office with a typewritten statement. He laid it on the desk with a solemn avowal that it represented the bare bones. Moses glanced at the paper and commented that the skeleton had a lot of fat on it. He also told Lewis that he would not recommend the proposal to his executive committee. The discussion, however, was amiable. "We never have rancor," says Moses. . . . "I think we conduct ourselves as two honorable businessmen trying to make a contract for services."
>
> The week that followed was notable for the absence of any public breastbeating or gauntlet-flinging. It took, in sum, just three more meetings to reduce Lewis' proposed figure to the $2 finally agreed upon.[23]

Fortune's focus on style obscured how radically the substance of the collective bargaining had changed. The significance of Lewis's approach was not what he settled for, but what he never asked for.

Coal in 1956 had a modest year, but its best since 1951. Production rose about 35 million tons. Operators sold about $320 million more than they had in 1955, and price per ton was $0.32 higher. Lewis sought a modest amendment to the 1955 agreement, and was granted a "two-step, $2 pay raise for 200,000 bituminous miners . . . during secret negotiations."[24] The contract, he said, "encompasses all that the industry is able to pay at this time."[25] He told his membership that "if you look at it long enough, you will find it is good."[26] Some miners, of course, didn't see it that way.

Although Lewis had not called a strike since 1950, miners were striking over the working conditions the Lewis contracts created. As layoffs increased, miners struck over seniority, job bidding, and other job-security issues. Lewis came down hard on wildcat strikers. He told the 1956 UMWA convention:

The UMWA cannot undertake to negotiate wage contracts . . . . unless we can offer reasonable evidence that we intend to observe that contract. Those of our members who have with impunity been violating this contract in increasing numbers will have to find that out the hard way if they will not abide by the constitution and by the agreement. I cannot speak too strongly of the seriousness of this problem. Between Jan. 1, 1956 and May 5, 1956 . . . 170 local strikes of major magnitude had taken place in the various districts.[27]

Lewis had consistently opposed wildcat strikes since the 1920s. He had long since eradicated his political opposition, but was unable to suppress spontaneous militancy that kept alive the faint embers of union democracy.[28]

In late 1958, Lewis approached BCOA with the idea of a "protection" clause that "the industry vow not to handle coal mined by workers not in the United Mine Workers Union."[29] The operators balked, fearing federal prosecution under antitrust law. BCOA signed a new contract in December 1958, giving another two dollar per day, two-step pay raise, a twenty dollar increase in vacation pay, and the "protective-wage clause." This said in part:

During the period of this contract, the United Mine Workers of America will not enter into . . . any agreement . . . covering any . . . employees covered by this contract on any basis other than those specified in this contract. . . .

The operators agree that all bituminous coal mined . . . or prepared by them . . . or . . . acquired by them . . . under a subcontract arrangement, shall be mined . . . under terms . . . which are as favorable to the employees as those provided for in this contract.

The obligation assumed hereunder shall not affect any agreement in effect as of the date of execution of this contract: Provided, however, that any operator signatory hereto shall not maintain such inconsistent agreement in effect beyond the first date at which such agreement may be terminated by him in accordance with its terms.[30]

This language appears to protect UMWA jobs by forbidding BCOA companies to do business with coal companies that did not meet the terms of the UMWA-BCOA contract. BCOA promised to boycott hundreds of small Appalachian mines who had been leasing and mining its coal, or having BCOA clean, load, or market their tonnage. Certainly the clause saved some UMWA jobs. But its other purpose was to erect barriers between small nonunion producers and their markets. For his part, Lewis promised to refrain from negotiating any new sweetheart relationships. Those that were already in place with favored BCOA companies would stay.

BCOA had doubts about Lewis's scheme. Some signatories would lose money because of it. The major BCOA companies hoped the clause would set up a logistical blockade that non-BCOA coal could not surmount. Large coal consumers would rather buy from a few reliable suppliers than negotiate with dozens of small companies whose ability to clean and arrange transport for their coal was problematical. Lewis had devised yet another way for the partnership to regulate supplier competition. This one, however, backfired. In 1963, the National Labor Relations Board (NLRB) ruled the protective-wage clause illegal and restrained the operators and the union from enforcing it.[31]

> The board concluded that "the purpose of permitting Signatory Operators to subcontract work to producers observing contract standards, but not to others, was to create pressures conducive to the extension of the union's contract to unorganized producers, rather than to preserve work for the employees under the BCWA [Bituminous Coal Wage Agreement]."[32]

The NLRB's finding was the first in a series of legal challenges to the UMWA-BCOA partnership. Several antitrust suits brought subsequently by small coal companies exposed how the partnership affected nonunion suppliers. The Norris-La Guardia and Clayton acts exempted unions from antitrust prosecution, except in certain circumstances. The U.S. Supreme Court said unions were exempt except where one combined with a company to suppress competition.[33] Throughout the 1960s, plaintiffs tried to fit the UMWA and the BCOA into this framework. In *Pennington v. United Mine Workers,* the Phillips Brothers Coal Company was awarded a $90,000 judgment (automatically tripled to $270,000 under antitrust law) against the union, the Fund trustees, and major coal companies for conspiring to monopolize the bituminous industry and bankrupt small companies in east Tennessee that were seeking TVA contracts.[34] The U.S. Sixth Circuit Court of Appeals in Cincinnati upheld this judgment, but the U.S. Supreme Court remanded the decision. The district court assessed the applicability of antitrust laws in these words:

> We think a union forfeits its exemption from the anti-trust laws when it is clearly shown that it has agreed with one set of employers to impose a certain wage scale on other bargaining units. One group of employers may not conspire to eliminate competitors from the industry and the union is liable with the employers if it becomes a party to the conspiracy.[35]

The second time around, the district court again held against the UMWA; the appeals court reversed, and the Supreme Court in 1965 declined to review the case.

After a second unsuccessful challenge, the Supreme Court allowed conspiracy verdicts to stand against the UMWA and Consol in March 1971. In *Tennessee Consolidated Coal Company,* the Court affirmed a verdict against the UMWA for $1.5 million triple damages. The circuit court found that the "union and large coal operators, through their National Wage Agreement and its Protective Wage Clause, conspired in violation of Sherman Antitrust Act to drive small operators out of business."[36] The court agreed with Tennessee

Coal's allegation that the UMWA and the large coal companies in the BCOA had

engaged in an unlawful conspiracy to eliminate and suppress competition and production in the coal industry and to control the Southern Appalachian and Southeastern Tennessee coal fields. The plan of the conspiracy was that only the coal operators who signed and complied with the National Bituminous Coal Wage Agreement of 1950, as amended, including the 1958 Supplement, the Protective Wage Clause (PWC), would be permitted to carry on operations in that territory. It was alleged that the small companies could not operate profitably under the provisions of the National Agreement; that those who did not sign were picketed by UMW with force and violence, were not permitted to operate, and were forced out of business; and that plaintiffs were damaged as a result of the conspiracy.[37]

Two months later, the Supreme Court refused to review a Sixth Circuit finding that the UMWA and Consol had conspired to restrain trade in violation of the Sherman Antitrust Act. In *South-East Coal Company v. Consolidation Coal Company,* the Court agreed that the union and Consol had engaged in a "conspiracy . . . designed to force South-East and other small coal producers in Eastern Kentucky out of the bituminous coal business."[38] The South-East plaintiffs charged that the "1950 Agreement between the union and the Bituminous Coal Operators' Association *was formed specifically* to eliminate smaller operators."[39] The damage award: $7.2 million.

In the *South-East* case, the company had made arrangements with George Love of Consol to market South-East's coal. When UMWA vice-president Thomas Kennedy refused to relieve South-East from full payment of the Fund royalty, Harry LaViers, its president, broke his UMWA contract. Consol then stopped marketing South-East coal under the protective wage clause.

In 1960 Consol sold 270,000 tons of South-East coal; the following year Consol's sales on behalf of South-East were 133,000 tons. In the first part of 1962, when Consol knew South-East was preparing to go non-union, sales fell to 443 tons. After March 1, 1962, while South-East coped with a strike at its mines and struggled to set up its own sales force, Consol sold not a gram of LaViers' coal.[40]

As the BCOA-UMWA partnership slid into conspiracy, Lewis used UMWA organizing drives and financial resources to help BCOA eliminate competition. UMWA "organizing" in the 1950s resembled guerrilla war rather than unionization campaigns. Most targets were located in central or southern Appalachia, and the campaigns leaned on intimidation and destruction of property. Their purpose was less to sign up new members and more to bankrupt their employers, who were invariably small companies competing with BCOA suppliers. Non-union suppliers were given a choice: sign a UMWA contract and face bankruptcy, or refuse and face an "organizing campaign." The major BCOA companies supported Lewis's "organizing." Harry Moses, in fact, publicly denounced Lewis for *not* organizing nonunion mines.[41] Not once in the 1950s did BCOA criticize UMWA terror tactics.

The Lewis organizing campaign began hard on the heels of the 1950 contract. Lewis set the tone in 1950 when he told his Virginia organizers: "Organize these small mines and damn the law suits. We'll take care of them."[42] Tony Boyle, assistant to Lewis in the 1950s, directed many of these campaigns and developed friendships with officials in UMWA District 19 in east Kentucky and east Tennessee. (Albert E. Pass and William J. Prater, two District 19 leaders, were Boyle's co-conspirators in the 1969 Yablonski murders which landed all three in jail with life terms.) Tipples were burned and railroad spurs dynamited in Preston County, West Virginia. At least sixty UMWA miners were indicted on charges of violence against nonunion miners there in the early 1950s.[43] A fifteen-month organizing drive against the Elk River Coal and Lumber Company at Widen, West Virginia, began in September 1952. Two railroad bridges were dynamited and one nonstriker killed in May 1954. The drive ended in December 1954, when the president of UMWA District 17, William Blizzard, called the strike a "lost cause."[44] In southeastern Ohio and central Pennsylvania, tipples were blown, strip-mine machinery was dynamited, and workers were intimidated. A Clearfield County, Pennsylvania judge permanently enjoined the UMWA from interfering with the mining and transportation of coal from nine nonunion stripping operations in July 1953, following two years of violence.[45]

Clay and Leslie counties in eastern Kentucky bore the brunt of UMWA efforts in that state. The UMWA filed suit on September 11, 1951, for $2 million in damages against 612 defendants—including coal operators, judges, and sheriffs—for interfering with the union's right to organize and assemble peacefully.[46] Tom Rainey, the UMWA international board member who was leading the organizing drive there, had his car dynamited in January 1952. Rainey claimed it was the third car to be blown up and the seventeenth dynamiting or machine-gunning in the area since the organizing drive began.[47] Charles Vermillion, a UMWA organizer, was murdered in his parked car outside Hyden, Kentucky, in August 1953. He and three other UMWA members had been ambushed and wounded in January of that year.[48] On June 24, 1952, three more union organizers were ambushed, prompting Lewis to ask Kentucky governor Lawrence W. Wetherby to use his authority to stop the "reign of terror."[49] A Louisville & Nashville bridge was blown up in response, cutting the only railroad line into Leslie County, and another wave of dynamitings erupted.[50] The UMWA abruptly withdrew its $2 million "reign of terror" suit at a pretrial conference in Lexington, Kentucky, on April 1, 1953.[51] At the same time, a federal grand jury returned indictments against thirty-six UMWA members, charging conspiracy in their organizing drive.[52] The grand jury indicted the union members on May 1, 1953, for conspiring to "deny certain citizens the right to refrain from joining a union."[53]

In Virginia, Charles Minton, a former UMWA field worker, filed a $350,000 damage suit against the UMWA in February 1952, charging that the union had fired and blacklisted him because "he refused to . . . murder two owners of the Gladeville Coal Co."[54]

> Lewis sent an aide to District 28 headquarters in Norton, Virginia, with instructions that the operators be given "a good taste of rough stuff. . . ." After the aide's visit, Minton and another organizer were . . . ordered to blow up an electrical substation. . . . Minton and the other field worker obeyed. . . . For their work, the two saboteurs were paid "a handsome reward."
>
> Later, Minton was summoned to meet with Tony Boyle, "agent extraordinary of the United Mine Workers and personal agent of John L. Lewis," in Knoxville, Tennessee. Boyle told Minton he had been chosen to murder C. P. Fugate and Harry L.Turner, co-owners of the Gladeville Coal Company. . . . Minton refused. He was abruptly transferred to another job. Not long thereafter he was fired. In its reply to Minton's suit, the UMWA made a blanket denial of the charges. Nevertheless, Minton's action opened the way for nine other suits, totaling about $1 million, by coal operators whose property had been damaged but who previously had no evidence on which to sue. None of the cases, including Minton's, ever came to trial. They were settled out of court in a matter of months.[55]

The terms of the out-of-court settlement were never made public. Although reports circulated at the time that the companies received about $200,000, union officials denied them.[56] Minton settled for $45,000. The legal papers on the case had disappeared from the Wise County, Virginia, courthouse when reporters, investigating Boyle, looked for them in the early 1970s.[57]

In 1959 Lewis launched his last "organizing" drive, which centered on east Kentucky and east Tennessee. Companies there sold to TVA but had wriggled out of Walsh-Healey standards by setting up dummy corporations to mine, prepare, transport, and market the coal. Only one—the one with the fewest workers—paid union scale. The same investors owned all the companies. Again, tipples, railroad spurs, bridges, and trucks were blown up. Kentucky mobilized its National Guard. Finally, in April 1959, the NLRB enjoined all UMWA organizing activities in the area. Damage suits totaling more than $15 million were filed against the UMWA by the end of June.[58]

Lewis's war against nonunion companies failed. There were too many targets in too many places. Too many miners needed jobs—and jobs that paid anything. The Eisenhower administration was neither helpful nor particularly hostile. Lawsuits stopped Lewis cold.

This is an ugly story, but the ugliness is not all on Lewis's side. The mining companies he ticketed for destruction were among the least responsible in an industry that had elevated irresponsibility to a business virtue. Their mines were consistently the most dangerous; mining practices among the most savagely destructive; labor relations among the most backward. Had Lewis genuinely wanted to organize these miners, labor advocates might interpret his terror tactics somewhat less harshly. Fighting fire with fire; breaking eggs to make an omelet; ends justifying means—those arguments would be made. But he really didn't want more miners in the UMWA fold. He said so openly: "From a policy standpoint, it is immaterial to us whether the Union has a million or a half million members." [59] The only justification for Lewis's policy is that, had he succeeded, UMWA miners might have had more job security. But to blame him alone is to miss the point that he shared the logic of terror with Love and the BCOA. Lewis became the BCOA's thug, doing its dirty work in the name of *his* members. That's ugliness of another hue.

Lewis could claim one "organizing" success among these many defeats. This he accomplished without violence; he simply bought the coal company. In the late 1940s, Lewis had the UMWA purchase about 75 percent of the stock in the National Bank of Washington, D.C. and appointed Barnum L. Colton president. Lewis began transferring all of the Fund's accounts to this bank in 1950. Soon, "there was more than $36 million in deposits in a [Fund] checking account in the National Bank, drawing no interest."[60] Lewis now had millions in liquid assets—the UMWA's treasury, the Fund's unexpended income, and the assets of the National Bank of Washington.

He arranged for Cyrus S. Eaton, chairman of the board of the Chesapeake & Ohio Railroad, to buy stock with UMWA money in West Kentucky Coal Company, a major nonunion producer for the TVA market.[61] Eaton, who began his career as a messenger boy for John D. Rockefeller, Sr., was by the 1950s well on his way to amassing a fortune in steel, rubber, railroads, coal, shipping, and banking, worth nearly $2 billion. The West Kentucky deal was small potatoes to the man who had put together Republic Steel and was a major force in Cleveland-Cliffs Iron Co., Chesapeake & Ohio Railroad, Detroit Steel, Cleveland Electric Illuminating, Kansas City Power and Light, Goodyear Tire, and Sherwin-Williams Co., among others.[62] Once Eaton had taken over, West Kentucky signed a UMWA contract in 1954. Then West Kentucky bought another nonunion operator, the Nashville Coal Company. Lewis directed the loan or investment of $25.2 million in these companies by 1958.[63] Financial mismanagement soon turned West Kentucky's profits into losses, and the UMWA sold out at an $8 million loss to Island Creek in the 1960s.[64]

Most writers frown on Lewis's venture into the coal business because it was done secretly and ineptly. An interesting question is whether Lewis should have done more of it, and done it openly. American unions have generally avoided becoming employers of their own members on the grounds that they would face irresolvable conflicts. Yet business, even within a capitalist framework, can be organized cooperatively or on a nonprofit basis. The UMWA might have set up union-owned yardstick mines to demonstrate better and safer ways to organize work and produce coal. But as Lewis was philosophically opposed to worker ownership, the thought probably never entered his mind.

Lewis's scheming with Eaton was one episode in his lifelong penchant for wheeling and dealing with union money. In the 1950s, Lewis used the National Bank of Washington as an easy-money reservoir to help favored companies finance new mining machinery. The bank made eleven major loans, totaling about $17 million, to coal operators between 1953 and 1963.[65] The UMWA secretly collateralized the loans. NBW president Colton swallowed any objections in consideration of his own $25 million interest-free loan from the UMWA treasury.[66] Qualms of UMWA lawyers were swallowed in the same way: Associate counsel Harrison Combs borrowed $1,006,875, according to *The New York Times*.[67] Lewis dipped into the UMWA treasury, Bank, and Fund to buy stock in coal companies, coal-carrying railroads, and coal-burning utilities; to make loans to individuals and preferred businesses; and to buy coal properties (West Kentucky Coal, Coaldale Mining Company, and, possibly, mines in Kansas under UMWA contract).[68] A number of these investments were bad, illegal, or both. Yet even after Landrum-Griffin requirements made Lewis disclose UMWA investments, no federal prosecutor during the Eisenhower, Kennedy, and Johnson years ever brought him or Boyle into court for misusing money.

Lewis retired as UMWA president in 1960 to BCOA's praises.[69] For good reason. He and George Love, who headed Consolidation Coal until the late 1960s, made it possible for BCOA companies to mechanize, consolidate, and prosper through a demand-limited decade. In Western Europe, postwar recession resulted in nationalized coal industries. Miners' unions there fought mechanization and governments eased the transition through subsidies. The Lewis-Love partnership was America's private sector alternative to nationalization. It succeeded, but at great cost to several hundred thousand miners and their communities and hundreds of small capitalists. Their handiwork showed in the condition of the coal industry in 1960. Even though production had dropped to pre-World War II levels, BCOA and the UMWA had eliminated much of the excess capacity that had sapped profits in the past. The number of bituminous mines had fallen from 9,429 in 1950 to 7,865 in 1960. Anthracite and bituminous employment was slashed by 61 percent, from 488,000 to 188,000. Coal was more concentrated: the eight largest bituminous operators produced 19 percent of total output in 1950 and 33 percent in 1960.[70] Productivity—that important indicator of industrial efficiency—had increased from 6.77 tons to 12.83 tons per worker per shift. Coal had lost markets, but it had linked itself to the soaring demand for electricity which made up the losses and then some. By

1960, electric utilities consumed 42 percent of U.S. coal, compared with 17 percent in 1950. Big BCOA companies negotiated long-term supply contracts with utilities. These contracts provided predictable sales and prices and protected output from competition for the life of the contract, as much as twenty-five years.[71] The most startling change in coal was the absence of strikes. Lewis did not call a contract strike after 1952.[72] He had once kept miners on the front page, but reporters forgot about them during the era of "labor peace." They were buried in the classifieds: Wanted—a job; Wanted—protection in the work place; Wanted—a democratic union.

7. Tony Boyle and the Black-Lung Rebellion

If you were a coal miner in the 1960s, you found yourself caught in a three-way vise. Your employer was bringing in machines and cutting jobs. More than likely you were laid off, with no prospects and few marketable skills. If you were lucky enough to be working, you quickly learned the new rules of the game. You kept union talk down. You didn't cause trouble. You didn't fuss over safety. You hunched your shoulders, did your job, and hoped the axe wouldn't fall on your neck.

Your union—the UMWA, which had helped to free you and your kind from mine guards and starvation wages—turned the vise from another angle. The contracts Lewis and his successors negotiated didn't cushion you against layoffs, or even include a solid seniority clause. The companies scrapped those they didn't like—the militants, the blacks, the old. Now, the UMWA in Washington, D.C., seemed to be helping the companies to eliminate your job. Somebody heard the union was giving money to the companies to buy continuous miners. You knew for a fact the mine over the mountain wasn't paying its royalties to the Fund. And that was another worry. The Fund was cutting folks off—disabled miners, widows, those who were out of work. It didn't do any good to complain to the union. Your district president and executive board member were appointees who had a constituency of one old man with bushy eyebrows.

Finally, the government squeezed you from a third side. When the Bureau of Mines pulled an inspection, someone phoned your superintendent in advance. The company generally ignored whatever citations it received. Save for some surplus-food commodities, it was hard to see how the politicians were doing much about the mine layoffs that were wrecking your community. You were on your own.

Appalachia's fate was that of its coal industry. Some counties there lost half their populations between 1950 and 1960.[1] The unemployed who remained were made into dependents. Mine layoffs ground over the coalfields like a glacier, scooping out pools of despair and self-doubt. They left each miner with a profound sense of individual powerlessness, of being unable to imagine, let alone achieve, some control over the forces that shaped his life. Who was to blame? Not John L. The companies? They couldn't help it if people weren't buying coal. The politicians? Maybe, but they weren't any better or worse than they had always been. So miners blamed themselves. Forced dependency created self-contempt; only later was anger directed outward. While many migrated and most turned inward, a few began to complain. The 1960s is the story of miners beginning to resist the vise.

By the early 1960s, the coalfields had hit bottom. Anthracite and bituminous production fell to 420 million tons in 1961, the lowest since 1938. Employment had fallen from 488,000 in 1950 to 166,000 in 1961. John F. Kennedy saw the coalfields as an industrial disaster area in 1960. He funneled economic development and social service money into the Appalachian states, but never confronted the industry and market forces that created depression there. As bleak as the coalfields were, the biggest BCOA companies were showing respectable returns. Consolidation Coal, for example, increased net profits from $12 million in 1955 to $20 million in 1960.[2] The UMWA as an institution was doing fine too. As Kennedy shook his head over the misery he saw in West Virginia, the union reported that it was the richest labor organization in America, with assets of more than $110 million.

John L. Lewis retired in 1960, at the age of eighty. Consolidation Coal's *Annual Report* in 1959 praised him as a "great statesman" . . . [who] long ago established as a firm policy of the United Mine workers the principle that productive efficiency was to be encouraged even at the expense of the loss of jobs by the introduction of labor saving machinery, and without this the coal industry could not have survived." His successor, Thomas Kennedy, was ill and left in 1963. Lewis, in retirement, transferred the presidency to W.A. (Tony) Boyle.

Boyle, born in 1901 in Bald Butte, Montana, was a quick-tempered and insecure man with a ninth-grade education who had spent much of his adult life on the UMWA's payroll as an organizer, district official, and Lewis's assistant. As president of the union's Montana district, he had impressed Lewis with his political and organizational abilities. Lewis brought Boyle to Washington in 1948 and counseled him: "Be anonymous." Boyle served the aging Lewis as an energetic assistant in the 1950s, emulating the old man's opportunism and disdain. He handled tricky jobs for Lewis and a father-son bond developed. But Boyle could never emerge in his own right, owing to Lewis's own skills and legendary presence. Boyle's ideas about union leadership formed as he watched and served John L. in the 1950s. He was loyal to Lewis and to the partnership Lewis worked out with Love. (Boyle's brother owned a coal mine in Montana.) Loyalty to the institution meant loyalty to the president, both men reasoned. Boyle learned from Lewis how to ridicule critics, impugn their motives, and eliminate them if they proved to be calculable threats. During his nine-year tenure, Boyle extended the worst features of his mentor's character. That Lewis lacked the good sense to anoint a better successor suggests that either the old man was guarding his place in history or age had dulled his judgment. Boyle lacked the great man's mastery of politics and the intuitive creativity to adapt. Like a child stubbornly insisting on jamming a square peg into a round hole, Boyle persisted in applying the model of industrial partnership and union autocracy beyond its time. Had he understood Lewis's brilliance, he would have taken steps in the late 1960s to defuse rank-and-file critics. But Boyle tightened the turnbuckles with the BCOA in proportion to the growing discontent within his own ranks. He was neither inclined nor able to break with the BCOA, whose passive support he increasingly needed. He had never been a rank-and-file leader and couldn't conceive of democratizing the UMWA from the top down, even though it might have saved his incumbency. What had he to fear from noisy miners? Successful rank-and-file insurgencies did not happen in organized labor.

Consolidation Coal and BCOA found Boyle receptive. Boyle opened negotiations in late 1963 only after rank-and-file dissidents in northern Appalachia agitated for replacing the 1958 contract. Most of the UMWA's twenty-three districts had lost their autonomy—the right to elect district officers—years before, when Lewis appointed "provisional" administrators who proved to be permanent. District 5 in western Pennsylvania was one of five that still elected some officers, and there the grumbling surfaced over seniority. Older miners wanted layoffs based on length of service. The 1958 contract's seniority clause was subject to different interpretation. Miners said it meant length of service at the mine; BCOA insisted it meant length of service in a particular job classification.[3] Seniority based on job classification was phony protection because an operator could easily arrange to group unwanted workers in his least productive section and then lay them all off at once. When District 5 miners complained, Joseph A. (Jock) Yablonski, the elected president of that district, told them that Boyle would give "serious consideration to your suggestions."[4]

The miners didn't strike, and Boyle reopened the contract in December 1963. Signed in March 1964, the new contract provided for a two-dollar-per-day wage boost, a twenty-five dollar increase in vacation pay, and the establishment of mine-wide seniority based on "length of service and qualification to perform the work." Both layoffs and callbacks were based on the employer's judgment about an individual's "qualification" to do a job. You could sail the *Queen Mary* through that loophole and have room on each side. But Boyle was simply extending Lewis's policy of allowing Big Coal to do what it wanted with UMWA workers. Boyle continued Lewis's 1952 $0.40 royalty to the Welfare and Retirement Fund even though inadequate royalty income had forced the sale or closure of its miners' hospitals and a reduction of services.[5] Boyle's showing in his first negotiations touched off wildcat strikes—the first since the late 1920s to protest a contract. Eighteen thousand miners in six states stopped work over "Two-dollar Tony's" contract. This was the first sign that industrial partnership would drive them to rebellion.

Some months later, a quixotic maverick, Steve "Cadillac" Kochis, ran against Boyle for president. He won 19,894 of the 115,978 votes in a race where the local-by-local breakdown was never revealed.[6] The Kochis campaign lacked money, coherence, a platform, and a skilled candidate. Its 15 percent share of the vote was another clue that miners were stirring. Wildcat strikes erupted a few months later, in September 1965, after Consolidation Coal fired several miners at its Ireland mine in northern West Virginia for illegal striking. Picket lines went up in the tri-state area and pulled fifteen-thousand miners for more than two weeks. The *Wall Street Journal* predicted that this "labor unrest" would bring "some sharp and fundamental changes in collective bargaining agreements and the structure of the United Mine Workers Union."[7] The *Journal* did not spell out the underlying causes of the unrest or suggest what changes needed to be made.

In 1966, Boyle negotiated his second contract. This one added one dollar a day to the miner's pay, continued the $0.40 royalty, and did not tighten the seniority language.[8] More than 50,000 miners struck for seventeen days in protest. In 1968, 130,000 miners struck as the contract expired. To head off a major intraunion conflagration, BCOA gave Boyle a three-dollar-a-day wage boost (with two-dollar add-ons in 1969 and 1970), which amounted to roughly a 25 percent increase in the standard daily wage. A $140 Christmas bonus was begun, along with a graduated vacation plan. But the Fund's $0.40 royalty did not change. Boyle signed in October 1968. Five weeks later, seventy-eight coal miners died in a Consolidation Coal mine in West Virginia. Three months after that, West Virginia miners stopped work for three weeks over coal dust and the lung diseases it caused. And a few months after that, Jock Yablonski announced he was challenging Tony Boyle for the union's presidency. Dominoes were falling.

A rebellion had started, using the wildcat strike.[9] From 1935 to 1952, Lewis led eleven industry-wide strikes that ranged from seven to fifty-nine days in length,[10] but UMWA leaders authorized no BCOA strike between 1952 and 1971. In the face of layoffs and weak contracts, some miners began to strike on their own initiative. Lewis had the UMWA's executive board ban unauthorized work stoppages in 1951, fine local unions, and suspend activists. Layoffs and UMWA policy suppressed most rank-and-file strike activity in the 1950s.

The willingness of miners to strike against Boyle's partnership contracts takes on added significance given the new legal penalties they faced. Coal companies could not get injunctions against wildcatting locals in the 1950s, although they could sue for damages resulting from such strikes. As late as 1962, the Supreme Court said the anti-injunction prohibitions of Norris-LaGuardia were still in effect.[11] Soon, however, the Court began to reason differently. In

Boys Market v. Retail Clerks International (1970), the Court said Norris-LaGuardia did not protect against federal injunctions granted during a wildcat strike where the contract included a mandatory grievance mechanism. "A no-strike obligation, express or implied," the Court said, "is the *quid pro quo* for an undertaking by the employer to submit grievance disputes to the process of arbitration."[12] *Boys Market* made individual strikers subject to fines and jail, and repositioned the federal courts behind management in strikes over working conditions and other matters. Miners saw neutrality stripped from the law. No matter how legitimate their grievance, no matter how blocked their grievance channels, the federal courts automatically issued injunctions against them. Injunctions transformed conflict within the private sector into a contest between labor and the federal judiciary. The Supreme Court had turned America's labor-relations clock back to the 1920s by making strikebreakers out of federal judges.

Despite penalties, miners increasingly turned to strikes in the 1970s to effect political and economic change. In some cases strikes worked; in others, they didn't. Although their union constitution, their contract, and their courts denied them the right to strike, miners believed withholding labor to be an inalienable right. While wildcat strikes are usually described as a breakdown of labor-management relations, it is clear from the circumstances of the 1960s and early 1970s that many were acts of constructive resistance. They reflected less on the character of the strikers and more on the policies of those being struck. What is most significant is that wildcats were often aimed at the UMWA, the Fund, politicians, or the federal courts. Intuitively, miners understood that striking the operators was their only way of influencing the forces that determined their fate.

Industrial partnership was big coal's way of limiting competition and suppressing labor's rank and file. Those whom the partnership targeted eventually formed bases of resistance against the circumstances in which they found themselves. One such base was the small operators who sued BCOA and the UMWA for conspiracy. The first miners to fight back were the disabled workers and widows who had been severed from the UMWA's Welfare and Retirement Fund. A second group was working miners who wanted fair play and democracy in the running of UMWA affairs. They believed that Boyle could not have sold them out had the UMWA been run democratically with rank-and-file voting on proposed contracts. A third group was miners who mobilized during the late 1960s over the issue of black lung disease, seeking financial compensation for their distress.

Finally, grass-roots groups in the 1960s and 1970s organized around the impact of coal mining on their lives and communities. Their issues were diverse—strip-mining, taxation, land ownership, air pollution, road damage from coal trucks, and dam safety. These issues were what economists call "cost externalities," the side effects of corporate decisions. Many felt that private sector coal treated Appalachia like a colony. Each issue was an outcropping of an embedded relationship between absentee corporations and the coalfields. These community-based, single-issue groups never coalesced around a single, coalfield-wide political agenda.

The first to protest the consequences of industrial partnership were disabled miners and miners' widows, who fell through the crack between public relief and UMWA care.[13] The UMWA's Welfare and Retirement Fund tried to spread a blanket of financial and medical protection over most union members. If a widow or a disabled miner did not qualify for a UMWA pension of $100 per month, a special $30 stipend was given, plus $10 extra for each dependent. The Fund estimated in 1948 that about fifty-thousand miners were unable to work because of mine-caused disabilities. If they were too young for their pension or Social Security, the Fund's stipend and medical coverage were their financial lifeline. Any accountant saw them as a nonproductive burden who didn't contribute to the Fund and drained its scanty financial resources. Nor did they vote in UMWA elections: widows lacked the franchise and the disabled miner often lost his if he failed to continue paying his union dues. When tight finances made the Fund choose among its beneficiaries, it excluded the most marginal first. It revised its pension rules in 1953. A miner now had to prove that his twenty years of service occurred within thirty years of his application—the oldest pensioners were chopped off. A year later, the stipend was cut. These pensions and stipends together had helped an estimated 80,000 miners, plus 30,000 widows and their children. About 55,000 were more or less totally dependent on this support.[14] Disabled miners lost their medical coverage in 1960 after the Fund ran deficits of $12 million and $21 million in 1959 and 1960 respectively. Faced with a $17 million deficit in 1961, the Fund ended medical coverage for miners who had been unemployed for more than one year. The trustees also stipulated that a miner's last year of employment before retirement had to have been in a UMWA-organized mine, regardless of how many years of UMWA service he had logged previously. This eliminated thousands of older miners who had been laid off in the 1950s and took nonunion jobs to survive. Then, in 1962, the Fund cut medical benefits for miners who worked for companies that had not paid full royalties. Lewis had approved these sweetheart deals. Now, as Fund trustee, he penalized his own members for his double-dealing.

The neediest fought back first. Two weeks after the Fund withdrew medical protection, "300 disabled miners gathered in Glen Morgan, West Virginia, to protest the cutback . . . , the first formal organization since the 1930s that was separate from the UMWA yet designed to reform its policies."[15] Leaders of the meeting went to Washington to meet with Lewis and Josephine Roche, but failed to alter Lewis's decision.

In late 1966 or early 1967, a group known as the Association of Disabled Miners and Widows began to meet in Beckley, West Virginia. They wanted to change the Fund's eligibility rules and its high-handed way of treating those whom it was supposed to serve.[16] They drew support from working miners, but had no political standing in the union or with the Fund. Their demands included retention of medical coverage after retirement or disability, a graduated pension system based on years of service, and pensions for mine widows. In February 1968 the association brought a class-action suit, *Blankenship v. Boyle,* against the Fund, UMWA, BCOA, National Bank of Washington (the UMWA-owned bank), and a number of individuals. The suit charged the Fund's trustees with "willfully defrauding" miners of $75 million, and misusing the Fund's assets for their own gain. The case exposed incredible mishandling of Fund monies under Lewis and Boyle. Between $12 million and $72 million had been kept in non-interest-bearing checking accounts in the National Bank of Washington between 1962 and 1967.[17] Mine operators were not prosecuted when they didn't pay their royalties. Harry Huge, a lawyer with Arnold and Porter, brought *Blankenship* during a stint with the Washington Research Project.

In 1972, U.S. district judge Gerhard Gesell found a "conspiracy among certain trustees, the Union, and the Bank through its president [Barnum Colton], and that all parties are jointly liable for the Fund's loss of income from the failure to invest." Gesell said of Lewis: "There is no indication that Lewis personally benefited, but he allowed his dedication to the Union's future and penchant for financial manipulation to lead him and through him the Union into conduct that denied the beneficiaries the maximum benefits of the Fund."[18] Gesell calculated the bank and the UMWA were liable for $11.5 million in lost Fund income. Specific breaches of trust included withholding medical coverage from miners whose employers had not paid royalties; investments in utility stocks that were intended to force consumers to buy BCOA coal; failure to collect full royalties, particularly with respect to West Kentucky Coal (the company Cyrus Eaton had purchased with UMWA money); and a financially unsound pension increase in June 1969 that Boyle engineered to gain pensioner votes in his race against Yablonski. Gesell removed Boyle and Roche from their trusteeships and severed the tie between the Fund and the National Bank of Washington. The UMWA also lost control of the Fund. The Taft-Hartley Act required that plans of this sort be separate from the unions whose members they served. Lewis and Boyle had so abused the Fund that Gesell finally enforced the law. Neither Republican nor Democratic administrations had ever tried to apply Taft-Hartley as long as miners were quiet victims.

While *Blankenship* wound through the judiciary, a disabled black miner, Robert Payne, took Fund policies into court in a different way. Six months after the 1969 Yablonski murders, the Disabled Miners and Widows of Southern West Virginia (DMW) formed around Payne, a twenty-seven-year veteran miner and a supporter of Yablonski's candidacy. Payne carried scars on his back from the mines. His hands, like those of many miners, were no pianist's: several of his fingers had been left underground. Payne's group wanted the Fund to restore pensions and medical benefits. They believed that direct action against the coal companies was the lever for shifting union policy. Payne and a committee went to Beckley, West Virginia in May 1970 to ask UMWA district officials to set up a meeting with Boyle and other Fund trustees. The Boyle-appointed officials stonewalled, whereupon Payne threatened a strike in five days if his group was not given "some consideration."[19] When nothing happened, he sent a telegram to Boyle requesting a meeting and threatening a "nationwide strike" on June 20 if Boyle did not reply.[20] Picket lines went up at midnight on June 20 in southern West Virginia. Five thousand were out after one day; nineteen thousand after two. One fellow miner summed up the frustration of the disabled miners: "Boys, we've been forced to beg for the crumbs off the table we built."[21]

The strike lasted for two months. Strikers carried guns on the picket lines. Payne, it is said, would pull a mine out by commandeering its communication system to announce that cars in the parking lot would be "shotgunned" if his fellow miners didn't stop work. When picket lines went up again on July 12 after the vacation break, the UMWA sent payrollers to break them up. Working miners were appalled at the sight of UMWA officials crashing through picket lines of wheelchairbound miners and old women. They struck in support of the pickets and in defiance of Boyle. The companies fought back through federal restraining orders that barred picketing. By the end of July, 160 mines were down. Boyle, feeling the heat, asked Congress to provide medical care to disabled miners until Social Security covered them at sixty-five. Senator Harrison Williams, chairman of a Senate Labor subcommittee, held a public hearing in Charleston, West Virginia to hear testimony from the picketers and their supporters. Dr. Donald Rasmussen, a lung specialist, told Williams: "The men who instigated this action were not pension miners, they were nonpension miners. They were not asking for liberalization, they were asking for just a scrap, just part of what was rightfully theirs."[22] The strike gained support; an estimated 25,000 miners respected the protest. Payne and Rasmussen were threatened: a murder plot—possibly linked to the UMWA—was foiled.[23] But Boyle refused to meet with Payne, and the strike wound down at the end of August. The coal companies were granted injunctions against picketing on September 8. Payne and others were jailed for contempt a day later for disregarding earlier orders.

Boyle prevailed by refusing to negotiate with the protesters, and the operators' injunctions intimidated Payne's followers. Disabled Miners and Widows never developed alternative methods of attack after the jailings. Its virtues—direct action, spontaneity, militancy, and rank-and-file support—were in some ways its flaws as well. The group lacked structure or resources to defend itself against legal attacks, although Richard Bank, a young lawyer, managed to free the picketers after short prison stays. DMW was the most authentically rank-and-file organization of several such groups that surfaced in the late 1960s. It accumulated no mediating layer of lawyers, coordinators, and propagandists. Its members were at the bottom of a very deep barrel, and its spirit was that of the Wobblies. As Payne put it: "We deals from the shoulder. And a deal from the shoulder is this: a lot of times you got to stop production—stop the coal from going into the cars. Then you can get some action done. That's the way we believe in operating."[24] Disabled Miners and Widows of Southern West Virginia took risks. The other organization, the Association of Disabled Miners and Widows, channeled its energy (some would say it *was* channeled) into lawsuits. *Blankenship* defined the association as the strike defined Payne's Disabled Miners and Widows. When *Blankenship* succeeded in April 1972, Payne's Disabled Miners and Widows were no longer much of a political presence.

Of all the protest groups in these years, Disabled Miners and Widows was the poorest, the most desperate, and the most militant. Payne's strike against the UMWA and the Fund legitimized the idea that union members could challenge UMWA leaders without being disloyal to the union itself. His courage made it easier for black-lung victims to organize the next protest.

Miners have lived with threats to their health from the earliest days of mining. As early as 1708, *The Compleat Collier* noted: "The Labourer . . . may loose his life by Styth [carbon dioxide or oxygen-deficient air], which is a sort of bad foul Air . . . , and here an Ignorant Man is Cheated of his Life insencibly [*sic*]."[25] Most medical authorities, however, did not believe that coal-mine dust produced lung disease until the 1960s, despite considerable historical evidence. Friedrich Engels described in the 1890s "the peculiar disease" of coal miners: "black spittle, which arises from the saturation of the whole lung with coal particles, and manifests itself in general debility, headache, oppression of the chest, and thick, black mucous expectoration."[26] When anthracite mining began to expand after the Civil War, several doctors noted the health effects. Dr. John T. Carpenter told the Schuylkill County Medical Society as early as 1869 that there existed a "very large percentage of [respiratory] disease among miners . . . [including] bronchial irritations . . . , a peculiar asthmatic character of cough . . . , emphysema and nervous distress in breathing."[27]

One hundred years after Dr. Carpenter's lecture, 40,000 West Virginia coal miners stopped work for more than three weeks to force the state legislature to recognize black lung as a work-related disease for which compensation should be paid. Why did it take a full century before anyone—doctors, miners, coal operators, politicians—did anything about black lung? An occasional doctor had written about how miners suffered from respiratory problems, some of which might be related to their occupation. Yet the burden of medical wisdom was that a unique, dust-related disease almost certainly did not exist, and if it did, it did not do much damage. Denial persisted until coal miners forced the issue from the back pages of the medical journals onto the front pages of the nation's newspapers in 1969.

Prevailing U.S. medical opinion in the twentieth century was that coal dust did not cause lung disease. The U.S. Public Health Service (PHS) studied coal miners' health three times between 1924 and 1945. Its first survey found that more than 40 percent of the bituminous miners X-rayed showed excessive fibrous tissue in their lungs (pulmonary fibrosis), but the fibrosis was not thought to be caused by mine work.[28] PHS studies in the 1930s found that coal miners, particularly anthracite miners, had lung disease, but the cause was thought to be silica rather than coal dust: when inhaled, "even in large quantities, [pure coal dust] produces little or no fibrosis."[29] In contrast, Great Britain had officially recognized in 1943 that coal dust caused coal workers' pneumoconiosis (CWP), a fibrotic disabling lung disease. British miners were compensated. In the 1950s, a few UMWA Fund doctors wrote articles that challenged the private sector and the Public Health Service. Joseph Martin, Jr., a doctor at one of the Fund's clinics, wrote in the *American Journal of Public Health* that coal miners he observed and those studied in Britain were victims of a "disabling, progressive, killing disease which is related to exposure to coal dust."[30] X-rays, he discovered, showed CWP, but that evidence did not match the degrees of disability he observed in individuals. The Pennsylvania Department of Health X-rayed sixteen-thousand bituminous miners in 1959 and found CWP in 14 to 34 percent of the sample, varying with age and length of exposure.[31] A few years after the Pennsylvania study, the Public Health Service found X-ray evidence of pneumoconiosis in 9.8 percent of working miners and 18.2 percent of unemployed miners it examined.[32]

These studies appeared in a political and medical context that denied that coal mining damaged worker health. Doctors who regularly treated miners knew they had special respiratory symptoms, called "miners' asthma," which included coughing, black spit, and breathlessness. Miners' asthma, the doctors said, was "normal" and nondisabling. Earl H. Rebhorn, a Pennsylvanian, wrote in 1935: " 'Miners' asthma' is considered an ordinary condition that need cause no worry and therefore the profession has not troubled itself about its finer pathological and . . . clinical manifestations."[33] Some doctors blamed bad sanitation and poor living conditions for these symptoms. A few said coal dust might actually help miners because it was quite possible that the dust had "the property of hindering the development of tuberculosis, and of arresting its progress."[34] Others opined that miners who claimed to be disabled from miners' asthma were actually seeking to trick compensation officials, or were suffering from a psychological fear of mining.[35] As evidence piled up in the 1950s and 1960s that these opinions did not explain the facts, mainstream medical theory shifted slightly. A single disease, coal workers' pneumoconiosis, did seem to exist, but it was uncommon, and disabling only in its most severe form. Disability compensation might be justified for "third stage" CWP as determined by X-rays, but something known as "black lung"—which referred to a variety of disabling symptoms—was beyond medical comprehension.

The primary reason for this medical obdurateness was that most coalfield physicians were still extensions of the coal industry. They had no interest in discovering—and a powerful interest in *not* discovering—that business-as-usual methods were routinely damaging the health of thousands of workers. Moreover, occupational medicine tends to look for a single, specific cause for a single, specific disease. As long as a distinct disease was not acknowledged, causes were not sought for general lung damage. When medical opinion first recognized a specific mine-worker disease—anthro-silicosis—the disease-causing agent was said to be silica, not coal dust. Once coal workers' pneumoconiosis was acknowledged in the 1960s, it was linked to "respirable" coal dust, invisible specks five microns or less in diameter.[36] This search for a single disease and a single cause obscured the obvious: exposure to a variety of unhealthy conditions and agents in the mines produced both specific and nonspecific lung damage. Miners called it "black lung." Most doctors were not so clever; they either could not or would not grasp the idea that years of coal-mine work frequently produced general breathing disabilities. Black lung was caused by the then-normal conditions of mine work and the decisions the industry made prior to 1969 not to improve them.[37]

Furthermore, the UMWA did little to educate the scientific community, coal operators, or its own members about occupational health hazards before the late 1960s.[38] Safety—the prevention of explosions and accidents—was the UMWA's focus. Lewis never asked BCOA for dust controls. It was only after the Fund set up a health-care system in the coalfields that medical pressure began to build for understanding miners' respiratory problems. The Fund assigned Dr. Lorin Kerr in 1951 to become expert in CWP and financed a few conferences, but never commissioned an independent study. Josephine Roche, Smith writes, "prohibited her staff from involvement in any political effort to gain improved workers' compensation coverage or occupational disease prevention programs."[39] Although miners had debated occupational respiratory problems at their conventions since 1924, UMWA leaders never acted on these discussions. When continuous miners were introduced in the late 1940s, workers soon recognized that they produced more dust than the conventional methods of undercutting, drilling, and blasting. No limit was placed on mine dust levels before 1970 and no regular sampling occurred. Even Tony Boyle suggested in 1963, "We may be mechanizing the coal industry past the point of safety. This automation and mechanization has gone far beyond what Mr. Lewis was talking about back in the 1920s."[40] Yet it was another five years before he allowed Kerr to tell the union's rank and file about black lung.

For much of the 1960s, black lung was a closet disease. After operating continuous-mining machines for ten years or more, miners discovered they couldn't catch their breath. They suffered their collective illness *individually*. Yet evidence of rank-and-file health consciousness was slowly building—at union conventions, at work, at the Fund's clinics, and in occasional medical studies. At the UMWA's convention in September 1968, black lung came out of the closet. Local unions submitted nineteen resolutions about compensation and a UMWA-negotiated standard to limit mine dusts. Boyle authorized his recently appointed occupational health specialist, Lorin Kerr, to address the concern. Kerr, formerly with the Fund, had dueled with the American medical establishment over coal workers' pneumoconiosis for more than fifteen years, but this was his first chance to bring his message directly to the miners. He didn't muff it. CWP, he said, constituted "the greatest medical and social problem in all industry." And 125,000 miners had it, he estimated.

> At work you are covered with dust. . . . You suck so much of it into your lungs that until you die you never stop spitting up coal dust. Some of you cough so hard that you wonder if you have a lung left. Slowly you notice you are getting short of breath when you walk up a hill. On the job you stop more often to catch your breath. Finally, just walking across the room at home is an effort. . . . Call it miners' asthma, silicosis, coal workers' pneumoconiosis—they are all dust diseases with the same symptoms.[41]

Kerr's words rang true. The convention unanimously resolved that all union districts should seek compensation legislation at the state level. This was modest enough. Boyle did not allow the convention to recommend any action at the federal level.

Alabama, Pennsylvania, and Virginia had already recognized CWP as a compensable disease. But the political watershed came in West Virginia, where a unique set of personalities and events intersected. On November 20, 1968, seventy-eight miners were killed in an explosion at Consolidation Coal's No. 9 mine near Farmington. The preventability of that disaster, Boyle's apologetic defense of the company, and the fatalism of public officials—all primed miners to fight back. And the battleground became black lung rather than mine safety.

Three local doctors—Isadore E. Buff, Donald Rasmussen, and Hawey A. Wells, Jr.—began educating small gatherings of miners about black lung. A few VISTAs—Craig Robinson, Rick Bank, and Bruce McKee—boned up on the medical literature and brought miners together with the physicians and Ralph Nader. A maverick Democratic congressman, Ken Hechler from southern West Virginia, took an interest. Buff suggested that the miners hire Paul Kaufman, a liberal ex-state senator, to draft and lobby a compensation bill through the legislature, which was to convene in January 1969. Charles Brooks became the first president of the Black Lung Association (BLA), which a dozen miners formed on a street corner in Montgomery after one meeting. He mortgaged his house to help meet Kaufman's fee of $10,600. With a name, a lobbyist, several lawyers and medical experts, and a single issue, the BLA entered West Virginia politics.[42]

Bank, Robinson, and Kaufman drafted a bill that was introduced in the first week of February. Three of its features stand out. One clause said that a miner with *any* breathing impairment was presumed to have been disabled by work-related black lung. This eliminated the question of work causation by which companies had avoided paying claims in the past. Second, the bill defined disease broadly enough to permit disability to be established by evidence other than X-ray. Doctors in sympathy with the miners argued that X-rays did not reveal the degree of disability. Third, a disabled miner would be allowed to sue his employer for damages. This was a truly radical change in compensation law.

For all its strengths, the BLA's compensation bill dealt with effects, not causes. It was *not* coupled with an equally tough dust-control proposal. BLA activists were for the most part men in their late forties or fifties who had been robbed of their health, the laboring man's capital—and they wanted capital in return. The BLA never broadened its focus to include prevention. This was a crucial political error. By focusing exclusively on compensation, BLA confined its base to disabled and retired miners—a base that time and legislative victories would shrink.

In the winter of 1968–69, such questions went far beyond what the handful of BLA activists were asking. They were weighing the very practical consequences that might befall them for their boldness. Coal had dominated West Virginia politics since the 1880s. Many governors and key legislators were coal executives, their lawyers, and their bankers. Coal lawyers drafted the state's laws concerning coal. Coal, the biggest industry in the state, paid an inconsequential gross sales tax and almost nothing in property taxes.[43] Coal towns were free-fire zones where private sector priorities determined the level of public good or ill.

West Virginia was a fortress of old-boy politics, but a few new boys were crossing the drawbridge. The Democrats split during the 1968 primary into three factions, two of which were reform oriented. John D. Rockefeller IV, a then-liberal, was elected secretary of state, although Republican Arch Moore nosed out a Democratic reform candidate for governor. The War on Poverty had seeded the state with community organizers, a network of college-educated question-askers unbeholden to local power structures. Their thinking was fired in the kiln of student protests. Their model of political participation was the local meeting, the demonstration, the direct demand for immediate political gratification, helping the bottom confront the top. The mobilization of miners on black lung in 1968–69 involved a clash of political paradigms. Had miners stuck to the established lanes of political appeal, the old boys would have routed them into an inescapable maze. By applying a different model—one that blended their traditional use of strikes with the new skills of the community organizers—they forced the legislature to accommodate them.

Throughout January, black-lung rallies were organized in southern West Virginia and drew the attention of national media. Buff, a large shambling man in his sixties, wrapped his warnings about lung disease in a populist analysis of coal-company exploitation and Rockefeller financial manipulations. He and Wells would often bring actual lung sections from autopsies to the meetings and crumble them into microscopic dust as their audience sat transfixed. "You're crazy if you let them do this to you," Wells would shout. Rasmussen, the most scholarly and expert of the three, would describe his findings on the lung diseases of miners in clinical terms, which added scientific weight to the more theatrical presentations of Buff and Wells.

On January 26, more than twenty-five hundred miners and supporters rallied in Charleston, the state capital. Hechler, Warren McGraw—a young, combative Democratic legislator from a southern coal county—and the three doctors stirred the crowd. Hechler accused Boyle of neglecting his responsibility to miners. Boyle responded with the accusation that the BLA was a "dual" organization, membership in which might justify expulsion from the union. He called his members who were making public demands for compensation "black-tongue loudmouths."

The UMWA tried to maneuver miners into supporting a weaker bill than the BLA's. By mid-February, coal lobbyists had bottled up all bills in committee. Then frustration popped the cork. Workers at two mines skipped a shift to attend hearings. On Valentine's Day, fifty miners marched to the capitol and threatened a national strike. Four days later, something triggered a walkout at a mine in Raleigh County. It was in the wind, nobody had planned it. In the next few days coal production stopped dead in mine after mine. News stories and roving pickets spread the word. Within a week, one-third of the miners in the state were out.[44] Some began to spend their days at the legislature. Their angry presence spooked the politicians, and state policemen were assigned to the cloakrooms. On February 26, more than two thousand miners marched on the legislature. Governor Moore intercepted them on the capitol steps, promising to introduce a new bill during a special legislative session. The crowd pushed past him, shouting "No! No! No! Now! Now! Now!"[45]

But the miners didn't know how to work the legislature once they got in. UMWA presidents had isolated them from firsthand knowledge of the political process. "Of all the people who got involved," said Arnold Miller, a black-lung activist and future UMWA president, "none of us knew a damn thing about the Legislature."[46] The strikers got a cram course from the capitol's janitors, many of whom had quit the mines because of black lung. The janitors sifted through committee wastebaskets for clues about the miners' friends and enemies.

When the House Judiciary Committee reported out a weak version of a compensation bill every coal mine in the state shut down. Forty-two thousand miners were out. As hundreds of strikers watched from the gallery, the House caved in and passed a stronger bill. Quinn Morton, president of the West Virginia Coal Association, called the House bill "galloping socialism in one of its purest forms."[47] It was "founded on sensationalism and irresponsibility, nurtured on emotionalism and passed . . . with total disregard of proven medical facts." The Senate, too, passed a weak bill, but under pressure, the conference committee worked out a stronger compromise that sailed through in the closing minutes of the legislative session.

The bill that Moore signed was a far cry from the BLA original. Maximum weekly compensation benefits were raised from $47 to $55, a small addition by any standard. The legislature did nothing to control dust in the mines to prevent future illness. The issue of occupational causation was left far short of a simple eligibility rule that related lung problems with years worked. If a miner could demonstrate that he was totally disabled by a lung disease "consistent with a diagnosis of occupational pneumoconiosis" following ten years of coal-dust exposure within the preceding fifteen years, his disability was presumed to be caused by his work. The presumption was not to be considered conclusive, however. The law did *not* compensate a black-lung victim who was not working in the mines as of July 1, 1969. Although consistent with legal precedents on retroactive liability, this proviso excluded thousands of retired miners and many of the most severely disabled. The act weakened the original BLA proposal by defining compensable disease as "occupational pneumoconiosis" rather than the more generic "black lung." Finally, disabled miners had no right to sue for damages.

Miners failed to establish the principle that underground coal mining inevitably damaged their lungs. Such a principle is not, of course, universally true. Many miners are not disabled when they quit the mines. Others have cigarette smoking to blame for their disability, in whole or in part. Disability is tricky to prove or disprove in any event. X-rays, which show the extent of pneumoconiosis, can't measure disability. As long as the principal means of determining disability was the X-ray, many would be denied compensation even though disabled. Causation is even more difficult to prove. Automatic entitlement based exclusively on years of work—a kind of industrial combat pay—would err on the side of miners, inevitably benefiting some whose disability was caused by something else. Experience rather than science suggested the essential reasonableness of such entitlement, if not its validity. The West Virginia legislature said only that mine work *might* have caused disability consistent with "occupational pneumoconiosis" after ten years of work. The individual still had to prove it.

In that way, the old boys snatched victory from the miners. Automatic benefits to the lung-disabled based only on years of exposure would have eliminated the inequity and errors created by the case-by-case system, in operation on the federal level since 1970. This system is costly, cumbersome, and inefficient, and incapable of making completely accurate individual determinations.

Observers at the time felt that the miners had won. They had liberalized the state worker's compensation law; miners could now qualify for benefits. But they had failed to establish uniform and collective treatment. Collective neglect of the coal operators, public officials, and UMWA leaders caused black lung in the first place. They had not drawn distinctions when they created dusty mining conditions. Treating cases individually made evaluation adversarial and contestable. Neither the government nor industry wanted to pay benefits, so each, in its own way, challenged many claims, resulting in a substantial legal business. If BLA had won automatic entitlement in West Virginia, Congress might have embraced the principle at the federal level when it took up black lung later in 1969. For all of these reasons, one prominer senator who had supported BLA from the beginning said the law was "not worth the paper it's written on."[48] Nonetheless, the sensations of victory that miners felt when Moore signed the bill were greater than their objective knowledge of its flaws. To have kept the strike going on behalf of a stronger BLA bill would have required a degree of leadership, cohesiveness, and discipline that the BLA did not have. Momentum had ebbed with the passage of the compromise bill. BLA's leadership was too fragmented and inexperienced to have mapped a political strategy to win a better bill in a special session. Had militants continued the strike, the movement would have split, and the new model of rank-and-file political action would have been discredited.

The black-lung strike did much to change how miners saw themselves and their union in the political process. They had exposed the Boyle administration as deceptive and management-oriented. The strike was focused on the industry-dominated legislature, but, indirectly, it was against the UMWA's historic failure to insist on dust control. Boyle himself had helped to set the movement rolling when he allowed the convention resolution calling for compensation legislation at the state level. But he baited and threatened his own rank and file when they began to act. Outsiders and Communists, he said, were behind it; the BLA was a "dual" movement. Justin McCarthy, editor of the *UMWA Journal*, said the union had not tried to win compensation legislation before 1969 because "we didn't know the disease existed," a claim that must have puzzled his staff colleague, Kerr.[49] The strike demonstrated how far the Boyle leadership had drifted from the rank and file.

The drama of the Farmington explosion and the black-lung strike dumped mine health and safety into the congressional lap. Congress had never addressed mining's health hazards and had done little on occupational health generally.

The safety bill President Truman signed into law in 1952 required that mines employing fifteen or more workers comply with certain safety regulations and permitted inspectors from the Bureau of Mines to withdraw workers from conditions of imminent danger when they observed them on their infrequent inspections. Truman called the law a "sham"[50]: it exempted small mines (which were often the most dangerous), expected the states to take primary responsibility for safety, "grandfathered" existing machinery from meeting the new standards, and imposed such procedural complexity on the administration of the act that enforcement was a matter of faith rather than fact.

President Kennedy established a Task Force on Coal Mine Safety in April 1963 to review the 1952 law and suggest improvements. Several months later the task force recommended that the administration come to terms with the four deficiencies Truman had identified eleven years earlier. But neither Kennedy nor Johnson was willing to confront the UMWA-BCOA alliance, which was committed to improving productivity at the expense of worker safety and health. In 1966, the UMWA and BCOA persuaded Congress to change the 1952 law to end the small-mine exemption so as to impose additional costs on BCOA competitors. Congress also requested a study of the federal safety effort. That report came in September 1968, nine weeks before the Farmington diseaster—the worst since 1952. The resulting bill suggested for the first time in American history that the federal government develop health standards and enforcement procedures for coal mines.

Congressional debate over black lung in 1969 revolved around prevention and compensation. Both involved spending money. One senator asked a group of black-lung activists why he, a politician from a state without a single coal mine, should vote for a compensation bill. A miner replied: "Because it's the coal miners that keeps your ass warm, that's why." BCOA did not, of course, like costly federal compensation and prevention programs. Even less, however, did the industry want a state-by-state arrangement that would impose different production costs in each locale. The operators also feared that miners would be able to persuade politicians to make the industry, not the taxpayer, pay for black-lung compensation. Therefore, the major BCOA companies, particularly Consolidation Coal, accepted a national dust-control program that imposed uniform costs across the industry and a compensation program funded almost entirely with public money. To prevent health damage to miners, the 1969 Federal Coal Mine Health and Safety Act required coal operators to invest in new equipment, modify existing machines, improve traditional practices, hire new personnel, and comply with federal record keeping.

Congress, pressured to provide compensation, invented a precedent-setting dust-control program. What emerged after a year's debate was a phased-in standard of two milligrams of coal-mine dust per cubic meter of air within three years. This standard—the lowest in the world—was predicted to produce pneumoconiosis in only 2 to 3 percent of new miners if every mine during every shift for the next thirty-five years adhered to it.[51] Operators were required to test the mine atmosphere periodically and send air samples to a federal facility for evaluation. But this dust-control system had many weaknesses. Its effectiveness depended on the inherent safeness of the two-milligram standard, which was derived from a small and comparatively short-term study of British miners using a methodology that in retrospect seems inadequate. Day-to-day compliance depended solely on the good faith of the operators. Miners had no way of knowing dust levels. Sampling was infrequent and could easily be manipulated, as indeed it was.[52] If an operator cheated on dust control, he could cheat on sampling and no one would be wiser.

The 1969 act directed the Social Security Administration (SSA) to initiate a compensation program for black-lung victims. The part of the act that described compensation was entitled "Black Lung Benefits," which suggests that Congress intended to help those suffering from the several mine-related lung diseases rather than just pneumoconiosis. The next decade was to be filled with rancorous debate and politicking over the definition of black lung; that is, how disability was to be defined, who was eligible, and who would get paid. No one knew in 1969 how many miners would qualify for SSA black-lung benefits. The individual benefit was paid at a rate equal to 50 percent of the minimum monthly payment given to a totally disabled federal worker (Grade GS-2). In 1970, this provided a miner and dependents with $200 per month. By 1984, federal cost-of-living escalators had boosted rates to $441 per month for a miner with a wife and/or child. To everyone's amazement, SSA had to evaluate 586,400 claims in the three and a half years it administered the program—more than four times the number of all working miners in 1969. More than 487,000 individuals—miners, widows, and dependents—received Social Security black lung benefits in 1974, the high-water mark of the program. SSA's beneficiary population declined to 333,000 by 1984 but total federal payments rose until 1983 because of benefit-escalation provisions. From 1970 through 1983, the federal treasury paid a total of $11.7 billion in SSA benefits, $1 billion alone in 1983. The coal industry paid not one penny of this bill.[53]

In 1972, an uneasy alliance of Boyle's UMWA, the BLA, and congressional advocates liberalized federal black-lung benefits. SSA under the Nixon administration had been interpreting the 1969 compensation provisions restrictively. The Democratic-controlled Congress extended federal black-lung payments for another year and made it easier for widows to win claims. Pressure from BLA organizers forced SSA to pay a number of BLA activists to participate in the formulation of what came to be known as the "interim standards," a more liberal set of eligibility regulations.

Congress and the Nixon administration went along with grass-roots demands for liberal regulations for several reasons. The federal black-lung program had synthesized a single-issue constituency with members in more than half of the states. Congress responded to this constituency in classic Great Society fashion: federal money eased political pressure. Second, BLA effectively organized and lobbied its issue. In this the BLA was materially aided by Appalachian-based lawyers and community organizers working mainly through federally funded organizations headquartered in Charleston. The Appalachian Research and Defense Fund (Appalred), a nonprofit organization specializing in class-action suits, supported several black-lung legal advocates. Designs for Rural Action (DRA), founded in the late 1960s by ex-Appalachian Volunteers did much coordinating and conceptualizing on black lung for the BLA in the early 1970s. The alliance between miners and radicals seemed to be a good fit at the time. The young, technically skilled, left-leaning organizers in DRA and Appalred discovered a volatile issue and a political base. Disabled miners found an organization through which to direct political pressure. DRA hired a disabled miner, Arnold Miller, to be its black-lung organizer in 1969. Miller assumed the presidency of the BLA and became its most publicized advocate. Finally, an overwhelming opposition to BLA demands did not materialize as long as the federal taxpayer, not the industry, paid the victims of private sector irresponsibility.

The Department of Labor took over black-lung compensation in 1973, but most of the claims that have been submitted to date had been filed with the SSA. Claims filed with Labor came mainly from miners who retired in the 1970s and from more marginal cases. Labor took a harder line and a more adversarial posture as fiscal reins tightened. As the rate of claims denial shot up, new anger brought the dormant black-lung chapters back to life. This pattern repeated itself several times in the 1970s. The UMWA and the BLA remnants responded with another lobbying campaign in Congress, which produced the 1977 amendments that liberalized eligibility in several areas and created an industry-financed trust fund to pay future claims. Congress tucked the tab into industry's pocket because the annual cost of black-lung claims was approaching $1 billion.[54] The new operator-funded trust fund was financed by a tonnage tax. Labor was to determine the "last responsible operator," and that company was to pay the claim. When the last operator cannot be determined, the trust fund pays the claim. Individual companies continued to contest a high percentage of all claims, and soon the trust was depleted.[55] The Labor Department's black-lung report to Congress for 1981 shows that from 1978 through 1981 the U.S. treasury financed the bulk of the program, $1.5 billion. The tonnage tax in 1981 produced only $248 million while benefit expenditures reached $618 million, forcing the U.S. taxpayer to subsidize black-lung compensation on the order of $538 million. (This subsidy is in addition to the $1 billion paid through the Social Security program.) The Trust Fund paid 95,000 benefits to miners and widows in 1981 while "last responsible operators" paid only 3,000 claims.

The UMWA decided in late 1981 to string along with a White House proposal to tighten eligibility requirements for new claimants in return for exempting 400,000 current beneficiaries from review. The Black Lung Disability Trust Fund owed the U.S. Treasury more than $1.5 billion by 1981 because of inadequate operator financing and company challenges. The Reagan administration proposed doubling the black-lung tax to $1 for each ton of deep-mined coal and $0.50 per ton of surface-mined coal for the next fourteen years to wipe out the deficit. The new eligibility rules required more evidence of disability and made it harder for widows to obtain benefits. Congress passed the bill, and the law took effect on January 1, 1982. Litigation against between 10,000 and 20,000 claimants was dropped. Survivor benefits were to be awarded only if the miner died of black lung (a black-lung sufferer who died in a car accident would no longer qualify). The law ended the prohibition against the Labor Department rereading X-rays, which, in the past, had been used to deny claims. The new law also ended the rebuttable presumption that a miner with fifteen years of service might qualify for benefits despite a negative X-ray. The Labor Department under the Reagan administration was not anticipating approval rates on new claims of more than 15 percent by 1985.

In light of the inherent and insoluble uncertainties in determining the cause of mine-related lung disease, administrative simplicity and equity—and perhaps even cost-effectiveness—should dictate restructuring the compensation program. A system based solely on disability is one alternative. Any miner with ten or more years in the mines would automatically get black-lung benefits upon demonstrating respiratory impairment. (Benefits would be geared to the degree of impairment, as they are under worker compensation.) It would be fair to reduce compensation to heavy cigarette smokers. Respiratory impairment is easier to measure than pneumoconiosis, and would shift the claims process to less subjective grounds. The controversy over X-rays and their rereading would end because X-rays don't measure impairment. A system of this sort is not an automatic pension; a claimant would have to prove his lungs are bad, after a minimum number of years in the mines. But it would switch the process of proof to simpler and more reliable tests. Moreover, it gets directly at the purpose of the federal black-lung program—compensating those who have been disabled by mining.

Another alternative might be a risk-pension system. In its simplest form, Congress would classify the health risk of all trades and industries into three groups: nonrisky, moderately risky, and very risky. Workers in the very risky category would automatically receive an employer-financed graduated pension on retirement for damage they suffered. Those in the moderately risky group would receive a percentage of the risk pension. Social Security would administer the program, and Congress would determine the dollar amount of the pension. Underground coal miners would be in the very risky category; surface miners in the moderately risky group. The assumption of this proposal is that workers in dangerous jobs perform a public service. Insofar as employers have profited from their employee's risk and damage, it is fair to ask industries to fund this pension in proportion to the number of their workers and the degree of risk. Individuals would have only to prove length of service. Unlike a national workers' compensation program, benefits would not be determined through individual adversary proceedings. If an industry or trade demonstrates measurable improvement in work-place conditions over a certain period, it could petition for reclassification. This would provide a financial incentive to invest in hazard control. A risk-pension system, however, might accelerate capital substitution and export of hazardous industries to other countries.

The most outstanding feature of the struggle for black-lung compensation in the 1970s was that marvelous, maddening collection of activists known as the Black Lung Association. Disabled miners like Woodrow Mullins, Joe Mallay, Arnold Miller, and Charles Brooks founded it. Others like Carl Clark, Willie Anderson, Earl Stafford, Martin Amburgey, Bill Worthington, Don Bryant, Walter Franklin, and Hobert Grills carried it on. Women became its most active leaders—Helen Powell and Anise Floyd, two black women from southern West Virginia, were among the most dependable and steadfast. Blacks and whites worked together in the BLA in remarkable cooperation. Young activists of many persuasions kept the network functioning and supplied with information. Some, like UMWA lawyer Gail Falk, did much of the legislative and legal work. Others worked with local chapters. In a few cases, Communists of one stripe or another influenced individual chapters. In others, Republicans ran the show. The racial, sexual, generational, and political mix was extraordinary, and often a source of organizational strength. If you saw the BLA from the inside, you saw a fragile macramé of loose threads and knotted personalities. If you saw it from Washington, it was a pesky, persistent, and deeply rooted movement of bright lawyers, dedicated organizers, and scrappy advocates. Some insiders thought the BLA to be an organizational myth—more a decade-long game of mirrors played in Washington than a functional reality; an army of generals who every so often were able to conjure up troops to march on Washington. The BLA was a network of individuals more than a mass organization, with its share of ego clashes and jealousies. The ties that bound were their work as advocates helping peers win claims and lobbying for better laws and regulations. When Arnold Miller became UMWA president in December 1973, he drew off some BLA staff to work for the union. Miller then began to see BLA militants as errant troublemakers. Black-lung issues grew more technical. Battles with legislators and administrators were increasingly fought in the jungles of science, economics, and law. While some lay advocates kept up with the technicalities, the movement, inevitably and increasingly, relied on its legal and technical supporters. The rank and file of the BLA lost interest, while some BLA leaders stopped organizing. The movement weakened at about the rate at which bogus black-lung organizations and titles appeared.

As more claimants won benefits, black-lung activism declined—the movement's very success spelled its own demise. Had BLA focused on prevention as well as compensation in later years, *working* miners might have joined the original BLA cadres. Several factors explain why the BLA did not adapt and broaden its constituency. First, the UMWA, even in the early years of Miller's first administration, would not have supported this change. Miller no less than Boyle did not welcome independent political groups within the UMWA.[56] Second, BLA's constituency and activists had left the mines and did not see prevention as their primary responsibility or interest. Third, some BLA members may have sensed that compensation and dust control were not entirely compatible political goals. If the industry were forced to spend a lot for dust control, it would fight harder against financing compensation, and vice versa. BLA's immediate interest lay in getting cash benefits. Fourth, BLA politicking around black lung soon came to define the social lives of many activists. Their sense of worth and work revolved around the BLA and the Washington connections their lobbying and lay advocacy made. The energy and resources of younger, working miners would have overwhelmed the older BLA activists and shortened their moments in the limelight. These cross-pressures presented the BLA with a no-win situation: if it ignored prevention, it would grow increasingly irrelevant; if it worked for prevention, it would lose its identity. The BLA never resolved this dilemma.

The more BLA focused on federal compensation, the weaker it became. Of all the reasons for BLA's decline, the most overlooked is that the organization chose the wrong target from the start. Black-lung victims demanded *public* compensation—first from the West Virginia legislature, then Congress, then federal agencies—but black lung was a disease for which private sector economics was to blame. BLA's demand for compensation should have been taken directly to the mine operators. Creative lawyers could have brought negligence suits against the industry. Political pressure could have been brought on legislators to allow workers to sue for damages—a discarded feature of the original West Virginia black-lung bill. Had the companies been made liable, they would probably have installed foolproof dust-control systems to reduce future obligations. Imagine the momentum rank-and-file insurgents would have accumulated had they won company compensation. Imagine the justice black-lung victims would have felt had their monthly checks been issued by their former employers. Imagine the opportunities for rank-and-file education in a campaign against management.

But would a company-focused campaign have succeeded? Possibly. The times were opportune. Boyle's UMWA might have swung into line to avoid being outflanked by its membership. A black-lung movement fighting for industry liability could not have confined itself to a single issue. It would have inevitably raised questions about profits, management priorities, investment, technological change and labor relations.

The federal black-lung program remains a remarkable testament to a handful of miners who stood on a windy street corner in Montgomery in December 1968 saying they were fed up with bad lungs and no money. The political system responded and matters have improved immeasurably. On the other hand, policymakers defused political pressure without resolving the matter completely or equitably. The Black Lung Association never managed to become an institution; and some would argue that it is better that spontaneous movements not make that transition. Had the BLA been able to broaden its scope without deadening its spirit, it might have provided rank-and-filers with a framework for other kinds of political protest. Those protests, which came in the mid-1970s, suffered for lack of such a framework.

8. The Rise of Miners for Democracy

The impulse to form unions is fundamentally democratic. Individuals who join understand that their welfare will advance at the same rate and to the same degree as that of their peers. Everyone shares organizing risks equally. Everyone works under the same terms. Despite the democracy of their origins, many American unions evolved into undemocratic institutions that denied the rank and file elementary rights of participation and representation. Many theories have been developed to explain why this process has occurred so frequently. Some argue that it is an inevitable cost of the "need" to centralize authority and create a service bureaucracy. Others argue that pressures within and without force unions to "mature" into nondemocratic, conservative oligarchies. The UMWA represents a case where the force of a single personality—that of John L. Lewis—transformed a democratic, decentralized, and politically pluralistic union into an autocracy that tolerated no dissidents. Had Lewis not been president of the Mine Workers from 1920 to 1960 the union might have evolved undemocratically anyway; but the experience prior to his reign does not support that likelihood. The absence of democracy—genuine elections, rank-and-file ratification of contracts, free debate, membership participation in making policy, honest accounting—led to willful abuses of the union's members on many fronts. It took five years—1968 to 1973—of struggle to oust Lewis's heavy-handed heir, Tony Boyle. And that was possible only because events and personalities came together during a time when the politics of the country nurtured a fight for union democracy. Boyle's fall from power was the product of many forces boring at him at the same time.

The seventy-eight-victim Farmington mine disaster, Boyle's handling of union affairs, and the 1969 black-lung strike prompted Joseph A. (Jock) Yablonski to challenge him for the presidency that spring. Boyle and his fellow officers, Vice-president George Titler and Secretary-treasurer John Owens, saw Yablonski's challenge as a traitorous act. Yablonski had been one of them, had dined at the same table, had shared their secrets. He knew where they buried their skeletons. For openers, Yablonski knew the officers were taking unjustifiably large sums from the union as salary and benefits. Nepotism was rampant: Boyle's daughter and brother and Owens's two sons were salted away in high-paying staff jobs. Cadillacs wheeled the officers about town. The UMWA paid for a residence at Washington's Sheraton-Carlton Hotel for Owens that cost almost $70,000 over a six-year stretch. Boyle established a special $650,000 pension program for John L. and the three officers, providing full salaries for life. Boyle and his cronies were painfully out of touch with ordinary miners; none had been in a mine for years. Bombast and a boutonniere were Tony's trademarks, along with a rainbow of polyester fabrics and mod ties. From his dyed hair to his elevator shoes, Boyle tried to project an image of power, an image borrowed from John L. and his other hero, Jimmy Hoffa. "I can do whatever I want—just like Hoffa," he crowed. "I'll hire lawyers and they'll get me off."[1] Vice-president Titler, a man who looked as if he had swallowed a beer barrel, was nimble with words. He characterized Yablonski as "a man so goddamn crooked he could hide behind a corkscrew."[2] Owens, the oldest of the three, (seventy-eight in 1969), sported an ill-fitting wig. Yablonski said of this leadership: "They've sat on their backsides so long they don't know what the coal miner's problems are anymore."[3] Suzanne Richards, a lawyer who assisted Boyle, and Edward Carey, the UMWA's counsel, kept this bizarre trio functioning from day to day.

Jock Yablonski was no radical. He had been a Lewis loyalist since being elected to the District 5 Executive Board in 1934. As a member of the International's Executive Board from 1942 to 1969 and president of District 5 from 1958 to 1965, he was one of the few UMWA officers who could claim a genuine rank-and-file base by virtue of election. Yet he rarely went against the institutional grain. As a career union official he too had done well financially. His family lived in a handsome stone house on a hill near Clarksville, Pennsylvania. His children were well educated: both sons became lawyers; his daughter, a social worker. He developed into a skilled troubleshooter who persuaded rank-and-file dissidents to retoe the official line. But over the years he had grown more liberal in his social and political ideas. He developed friendships with the left-leaning medical professionals who worked for the Fund in the Pittsburgh area. His wife and daughter contributed to his thinking about social responsibility. He declared his opposition to the Vietnam War in 1969 before breaking with the UMWA hierarchy. In the early 1960s, Yablonski recognized that Boyle was no Lewis, and had drawn a line with him several times. At the UMWA's 1964 convention, for example, he threatened a walkout over Boyle's District 19 goon squad, which had beaten delegates on the first day. He split with Boyle in 1965 over a black-lung compensation law in Pennsylvania, winning the point but losing Tony's trust.

On the whole, however, Yablonski went along with Boyle's operation. In 1966, for example, Lou Antal, a rank-and-file insurgent from Jock's district, testified that Yablonski had offered him a bribe to withdraw from his race for the district presidency.[4] In June 1968, the *UMWA Journal* quoted Yablonski as saying: "I have never seen anyone as devoted and concerned with the well-being of this Union and as determined as . . . Tony Boyle. . . . The hours are never too long for him to . . . work tirelessly up to 18 hours a day . . . in the interest of the membership."[5] Several months later at the UMWA's convention, which Yablonski keynoted, he said: "I lack the words which would enable me to describe the love, the devotion and the respect for the integrity of every member of our Union that our distinguished President has." Well into 1969, Yablonski continued his public praise for Boyle. A month before he broke with him, Yablonski told miners at a Fairmont, West Virginia rally: "This dynamic union of yours has been re-awakened . . . and we're going forward under this great leadership of President Boyle and his distinguished officers."[6]

On May 29, 1969, Yablonski announced at a by-invitation-only press conference in Washington's sedate Mayflower Hotel that he was running against Boyle. Boyle's handling of the Farmington disaster and the West Virginia black-lung strike had angered him. Ralph Nader, who was publicizing black lung, convinced him to run after assuring him of legal and financial help.

"If I do run, Ralph, they'll try to murder me," Yablonski said.

"Oh, they wouldn't dare," Nader replied.[7]

Nader introduced Yablonski to Joseph Rauh, a well-known liberal lawyer, who then represented him in legal proceedings arising from his campaign. Yablonski knew he was taking a tremendous personal and financial risk, but, he said, "It's time that somebody speaks up . . . regardless of what the sacrifice may be!"[8] At fifty-nine, Yablonski was only a year away from his pension. He told reporters: "I have been a part of this leadership . . . but with increasingly troubled conscience. . . . I can no longer tolerate the low state to which our union has fallen."[9] Yablonski asked the eighty-nine-year-old John L. for his endorsement. Lewis is supposed to have said that passing his mantle to Boyle "was the worst mistake I ever made," and that Boyle and his officials should be allowed to "drown in their own slime."[10] But thirteen days after Yablonski's press conference Lewis died without a word about Jock's candidacy.

Yablonski's campaign was loosely tied to an ad hoc rank-and-file organization. No convention nominated him, and no rank-and-file group was elected to advise him. No committee of miners drafted his platform. Most of his campaign money did not come from working miners. His candidacy was a personal statement, which was to evolve into a political movement. His platform emphasized procedural reforms and a modest liberalism. In Yablonski's view, Tony Boyle was the basic cause of UMWA shortcomings. His platform urged a stronger UMWA commitment to mine safety, restoration of autonomy in all twenty-five districts, mandatory retirement at age sixty-five for all UMWA officers, improved Fund benefits, improved grievance machinery, increased financing of local unions, establishment of a credit union, UMWA support for higher taxes on coal to improve education and social services, more aggressiveness with the coal operators, organization of nonunion mines, an end to nepotism, and creation of an ombudsman to work with miners. These ideas were bolder in their time than they sound now; far timider than those John Brophy offered in 1926 when he endorsed union democracy, nationalization, labor education, worker cooperatives, and a labor party. Yablonski did not spell out his collective-bargaining goals. He said nothing about layoffs or mechanization. He did not include rank-and-file ratification of contracts—a bedrock of union democracy—or a commitment to a right to strike. He did not link the state of UMWA affairs to the industrial relations that Lewis and Love practiced in the 1950s. He did not say how he would deal with the BCOA other than to suggest that unions should maintain an "arms-length relationship with management." Yablonski's campaign was not far in front of the political base he hoped his challenge would mobilize.

As the campaign progressed, Yablonski sensed that working miners were increasingly receptive. Boyle sensed it, too, and fought back. He fired Yablonski from his position as director of the UMWA's political action division. He denied Yablonski access to the *UMWA Journal* and sent underlings to break up some of his rallies. Yablonski was physically attacked after one speech in southern Illinois. He was castigated as the dupe of outsiders—particularly liberal lawyer Joseph Rauh, Congressman Hechler, Ralph Nader, the Communist party, and the oil cartel. Boyle wondered aloud why Yablonski had always said such nice things about him. Yablonski replied it was the only way he could keep his job. Boyle mobilized UMWA resources behind his slate. He appealed to pensioners—whose eighty-thousand or more votes constituted about 35 percent of the UMWA's electorate—by engineering a thirty-five-dollar hike in their monthly benefits through a sleight-of-hand with the Fund's trustees.[11] Boyle stepped up loans to districts where Yablonski had support. They received almost $829,000 more from UMWA headquarters in 1969 than in 1968. The thirteen districts that Yablonski neglected received $318,000 less. Boyle hired campaign workers through the UMWA's "organizing" department and as temporary lobbyists.

Yablonski was not without resources of his own. Although Nader had dropped out of Yablonski's braintrust because of what he saw as the candidate's lack of fire and financial commitment, Joseph Rauh remained a formidable legal ally. He hammered away at Nixon's labor secretary, George Schultz, to investigate intimidation of Yablonski's workers and other violations of the Landrum-Griffin Act. Yablonski waged his campaign at the local level while Boyle stayed in Washington, save for a few well-staged rallies. Miners liked Yablonski's rough-cut candor. Boyle would not debate him. Some liberal money flowed to Yablonski's small Washington campaign office. Yablonski brought together the grass-roots protests within the UMWA—the democracy advocates from northern Appalachia, the black-lung militants from West Virginia, disabled miners and widows, young miners, safety advocates, and political dissidents of several stripes.

Yablonski lost, 80,577 to 46,073, in December 1969. He did well in northern Appalachia, the anthracite region and where he had poll-watchers. He split the vote of working miners, but lost badly among pensioners and in the more remote districts where the UMWA competed with nonunion mines. He demanded that the Labor Department impound the ballots and investigate the election. A few years later, one of Yablonski's sons estimated that his father would have polled an additional 19,530 votes had the election complied with the UMWA's constitution and the Landrum-Griffin Act, which would have made Yablonski a winner by about 5,000 votes. The calculation is debatable. Many of the issues and lawsuits that tarnished Boyle so badly in the 1972 rerun had not surfaced in 1969. Older miners and pensioners—the bulk of the electorate in 1969—were not ready to dump Lewis's chosen heir. The vote would, of course, have been closer. An honest Boyle victory in 1969 would have forced the reformers to stretch their protest until 1974, the next scheduled election. That would have been hard. By rigging the 1969 voting, Boyle contributed to his own downfall.

At a postelection rally on December 14 near Beckley, West Virginia, Yablonski said he had "accomplished more in these last seven months than I did in the past 35 years in the union."[12] Indeed he had. He had pierced the curtain of self-flattery and misinformation that had shielded UMWA leaders from rank-and-file criticism. He had become the leader of many disparate dissidents and drawn outside allies into the miners' fight. Lawyers, politicians, college students, and liberals fought for union democracy. His courage appealed to cynical reporters who began to raise questions about union matters to which Boyle had no answers. He had discredited Boyle while maintaining allegiance to the UMWA.

The Nixon administration ignored every request to make the election fair and safe. Mike Trbovich, Yablonski's neighbor, campaign manager, and long-time friend, recalled:

> I sat in the office of one of the Secretary of Labor's assistants on the morning of Oct. 27 and I told him—*I told him*—that if they didn't do something to keep this election fair, that either Congressman Hechler or Jock Yablonski was going to get killed. . . . He was listening to me with a pad in front of him and he didn't write down a goddamned thing.[13]

On the last day of 1969, a few weeks after the election, three gunmen slipped into Yablonski's house and murdered him and his wife and daughter. After two trials that spanned several years, Boyle was convicted of instigating a murder conspiracy and financing it with twenty-thousand dollars of UMWA money. Between the 1969 election and the 1972 court-ordered rerun, the plot came unglued piece by piece. Although Boyle himself had not yet been indicted, the jury of coalfield opinion figured his mind had guided the hand that held the gun. A lot of miners who didn't especially care for Yablonski in 1969 cared even less for someone who would have him *and* his family murdered. Arnold Miller and the Miners for Democracy reaped that macabre windfall in 1972. Eventually, nine persons were convicted in the Yablonski murders in a six-state investigation involving more than 150 witnesses and costing $750,000.[14] Boyle attempted suicide by drug overdose in 1973 and was sentenced to life. He and the three gunmen are still jailed.

In the three years between Yablonski's death and the December 1972 victory of the Miners for Democracy, the character of the insurgency changed. The movement acquired a name, Miners for Democracy (MFD), and an organizational framework at a meeting following the Yablonski funerals on January 9, 1970. Mike Trbovich became MFD chairman. Leadership of the movement fell in large part to Joe Rauh and to Yablonski's sons, Chip and Ken. Between the two UMWA elections, much of the MFD's political initiative shifted from the coalfields to the courtrooms where Chip Yablonski, Rauh, and other lawyers brought suits on MFD's behalf. Yablonski supporters tended to defer to the dead man's sons, who sought to avenge their inestimable loss by defeating Boyle and seeing him jailed.

MFD became an organization without a genuine structure that operated as a loosely knit committee of organizers. A number of MFD participants like Harry Patrick would later describe MFD as a "myth." It was and it wasn't. A network of college-educated organizers helped to maintain its frail structure. Don Stillman, a former *Wall Street Journal* reporter, left a teaching job at West Virginia University to begin the MFD publication, *The Miner's Voice*. He and Edgar James publicized the MFD's legal work and boosted MFD momentum. MFD activists linked up with organizers at Charleston's Designs for Rural Action and the Black Lung Association. MFD organizers made a critical—perhaps inevitable—error in the interelection years. Lawyers and legal battles defined much of MFD's political activity and substituted for miner-led organizing within the rank and file. If organizing had been the MFD's principal focus, a permanent rank-and-file progressive force might have been built within the UMWA. Without an organic base in the rank and file, MFD fell victim to its own success, and its absence contributed to the Miller administration's problems in ensuing years. The MFD model, dependent as it was on nonunion resources, has not succeeded in any other union despite several attempts. It is a model that worked once in unique circumstances. A movement based on worker-led, rank-and-file organizing would have been a harder and longer road, but more enduring.

The men who gathered with the Yablonskis after the funeral could not, of course, predict the future.[15] Numb with grief and fear, they improvised. Each did what seemed best at the time. Rauh took the initiative. He pledged to press

Jock's charges of election violations. Rauh needed a mine-worker base from which to bring his suits. The miners at the meeting began to think of themselves as a group in response to this need. Rauh began to pepper the Labor Department with demands for investigation into the election, claiming that "failure of the Labor Department to investigate the illegalities in the UMWA election contributed to the death of Jock Yablonski."[16]

For the next two-and-a-half years, MFD lawyers led the challenge to the UMWA's leaders.[17] Their work levered the government into direct opposition to the Boyle administration. They brought district and international officers into court repeatedly over dozens of complex legal and financial issues. Their commotion mixed with that created by congressional inquiries and the drama of the Yablonski murder conspiracy. The Justice and Labor departments had resisted messing around in UMWA politics because of bureaucratic inertia and Nixonian hard-hat politics. They swung against Boyle only after the murders created pressure for action. In the U.S. Senate, Harrison Williams opened hearings on the 1969 election and the Labor Department's neglectful role. Within a month of Yablonski's funeral, the Justice Department arrested three men who proved to be his executioners. MFD miners picketed the Labor Department to draw attention to a neglected 1964 suit over the district trusteeships, which denied them free elections. Rauh helped to arrange for Chip Yablonski and Clarice Feldman to work full time on Boyle-related suits. One was filed under Section 501 of the Landrum-Griffin Act before Yablonski's death, charging Boyle and his officers with using UMWA money in their campaign, concealing financial dealings, and investing union money improperly. The 501 case eventually voided the 1969 election and opened the way for the MFD to run against Boyle in 1972. Another suit under the NLRB charged Boyle with an unfair labor practice for making pensioners pay union dues during retirement. The UMWA received about $1 million a year from its pensioners in this fashion.

A few legal breaks appeared in 1970. In July the judge in the election suit ordered the UMWA to bare financial records in eighteen of its twenty-five districts. A month later, a grand jury indicted Michael Budzanoski and John Seddon, president and treasurer respectively of District 5, for using money improperly in the 1969 election and other matters. In the fall, Rauh, Yablonski, and Feldman sued to intervene in the Labor Department's suit to overturn the 1969 election. They eventually won this case, *Trbovich v. United Mine Workers of America,* despite UMWA and Labor Department opposition.

In January 1971 Labor sued the UMWA to maintain financial records in accord with Landrum-Griffin guidelines. Labor charged that almost $10 million in UMWA funds had been spent between 1967 and 1969 without receipts or with only vague references to "expenses," district loans, or "organizing." The president of District 28, Ray Thornbury, went on trial in Washington, D.C., a month later for embezzling $1,800 that had been used for Boyle's candidacy. A Washington grand jury indicted Boyle and other UMWA officers in March for embezzling $5,000 for a campaign contribution in 1968.

The disabled miners and widows won the *Blankenship* suit against the Welfare and Retirement Fund in April. The court removed Boyle from the Fund and severed its ties with the UMWA and the National Bank of Washington. Thornbury was convicted of embezzling union money in June, and later that month, one of the Yablonski killers confessed, saying "a man named Tony" had ordered and paid for the assassination. In July, a Senate Labor subcommittee looked into the Labor Department's lethargic handling of its Landrum-Griffin responsibilities. A day after these hearings, Labor brought the 1964 autonomy case to trial.

At about the same time a panel of three judges ordered the law firm of Williams, Connolly and Califano to stop representing the UMWA. Rauh had sued more than a year before over whether this firm could be UMWA counsel *and* defend Boyle in the 501 corruption case without conflicting interests. In September, the government took the 501 suit to trial without allowing the MFD lawyers to participate. The MFD's intervenor case, *Trbovich,* however, had worked up to the Supreme Court, which agreed to hear it in November. In that month, a U.S. Court of Appeals ordered UMWA staff lawyers to stop representing Boyle, Titler, and Owens in the 501 case, forcing them to pay for their own legal defense.

In January 1972, Judge Gesell ordered the UMWA and the National Bank of Washington to pay $11.5 million in *Blankenship* damages to disabled miners and widows and their lawyers. A few days later, the Supreme Court allowed MFD lawyers in *Trbovich* to intervene in the election trial. Rauh and Chip Yablonski significantly strengthened the government's team. Paul E. Gilly, one of the Yablonski assassins, was convicted in March, and his wife, Annette, confessed her role as a go-between. Her father, local union president Silas Huddleston from Tennessee's District 19, was part of the chain of conspirators. She implicated UMWA officials and pointed a finger directly at Tony Boyle. At the end of the month, a jury convicted Boyle of embezzling UMWA money for illegal political contributions in 1968. On May Day, the 1969 election was overturned and the court ordered another for December under the supervision of the Justice Department. The next day, the FBI arrested Albert E. Pass, a ruthless UMWA official (known as "Little Hitler") in District 19, and accused him of funneling money to the Yablonski killers through one of his staff, William Prater. Huddleston blurted out his part in the conspiracy a day later. At the end of May, a federal judge granted autonomy to seven UMWA districts in the 1964 suit, and ordered elections there.

Between January 1970 and May 1972, MFD organized several coalfield protests in addition to the legal fights. Jeanne Rasmussen, a photo-journalist and wife of the black-lung doctor, organized a "Birthday Tribute" to Jock Yablonski on March 1, 1970 in southern West Virginia for about three-hundred miners. Later that month, more than five-hundred miners in Moundsville, West Virginia, heard Ken Yablonski urge them to sue to remove UMWA officials in Ohio's District 6. Several thousand came to Washington, Pennsylvania on April 1 for a Yablonski memorial service. Afterward, about two-hundred formally constituted themselves as Miners for Democracy. They kept Trbovich as MFD chairman and appointed Harry Patrick, Yablonski's campaign manager in northern West Virginia, and Kenneth Dawes, from District 12 in Illinois, as MFD co-chairs. A steering committee was organized, and each committeeman was to organize in his district. They decided to put out a publication, *The Miner's Voice,* which had been the name of a Yablonski-campaign publication. Don Stillman began the paper, along with Keith Dix, Robb Burlage, Rick Diehl, and others.

MFD was involved in several wildcat strikes over mine safety in the summer of 1970. One came in response to the Nixon administration's decision to stop *all* Bureau of Mines' inspections after a federal judge restrained the bureau from enforcing some regulations promulgated under the 1969 act. Nixon nominated J. Richard Lucas to direct the bureau a few weeks later. The coal operators and Boyle supported Lucas, while the MFD, because of his $200,000 worth of mining stock, did not. About twenty-thousand miners in northern Appalachia struck over the Lucas nomination and Nixon's handling of mine safety at the end of June. Lucas asked that his name be withdrawn, and the MFD claimed its first political victory. MFD soon challenged the Boyle officers in District 5. Lou Antal, a feisty long-time autonomy advocate, ran against incumbent Michael Budzanoski. Boyle and Budzanoski charged MFD miners with being dual unionists and Communists, but Antal's slate won. Then Budzanoski's election tellers threw out the votes of a number of locals, including the three largest, which had voted MFD. Antal's victory showed that rank-and-file sentiment had switched to anti-incumbent candidates, even though the Boyle machine was still formidable.

In the summer of 1971, MFD organized meetings to develop an independent rank-and-file agenda for collective bargaining, scheduled for the fall. The demands that emerged were, among others: a four-shift day (the fourth shift devoted to safety and maintenance); a higher Fund royalty; a two-year rather than three-year contract with a ten dollars-per-day wage hike each year; a new grievance procedure; the right to strike over safety disputes; federal dust samples (for black-lung prevention) to be taken by UMWA miners rather than company men; sick pay; and rank-and-file ratification of contracts. This "wish list" was a broad critique of the Boyle contracts. The BCOA awaited the bargaining with uncertainty. The operators were as concerned as Boyle when the MFD staged a

Miners at Lejunior, Kentucky, in the 1930s (Russell Lee). Courtesy Photographic Archives, University of Kentucky Libraries.

Period Ending 7-18-46 No. 11

Name Grant Hroth

DEDUCTIONS		CREDITS	
Powder	5 67	Coal at	
Mine Checks		96 " at .707	67 87
Self Rescuer		Yds. at	
Scrip	47 00	" at	
Rent	4 50	" at	
Lights		" at	
Coal	1 40	13 Hrs. at	1 56
Accts. Rec.		" at	
Bath House		" at	
Overdrafts		" at	13 37
Hospital	1 17	Machine Cutting	
		Labor Transfer	
Burial Fund	1 00	Overtime	7 97
Bus Tickets	4 00	Hrs. Trav. Time	
Doctor	80	Shift Differential	1 60
Insurance		Wage Equalization Approved by N. W. L. B.	
Union Dues	1 40		
U. S. Bonds		Total Earnings	
Cash Advance		Pinch	
Benefit Fund		Mine Checks	
Checkweighman	86	Self Rescuer	
Board			
S. S. Tax	1 00	Transfers	
Withholding Tax	3 30		
TOTAL	69 45	TOTAL	105 79
Due Company		Due Workman	35 84
Date 8-10-46		Less Pinch	
		Net Amt. Due	

KINGSTON-POCAHONTAS COAL [illegible] Inc.)

10M 11-45

Right, pay slip for a half-month period (Russell Lee). Below, commissary at Sandlick Mines, Kentucky (Jennie Darsie Collection). Courtesy Photographic Archives, University of Kentucky Libraries.

Left, a hand loader after a day's work in Scotts Run, West Virginia, 1937; above, children of miners at Scotts Run (Lewis W. Hine). Courtesy National Archives.

Above, company housing at Mohegan, West Virginia (Russell Lee). Below, guardsmen in Harlan County, Kentucky, during the "mine wars" of the early 1930s (Herndon Evans Collection). Courtesy Photographic Archives, University of Kentucky.

Right, rescue worker at a mine disaster. Below, coal miners (Cathy Stanley). Courtesy *UMWA Journal*.

Above, father and son miners at Kenvir, Kentucky; below, Welch, West Virginia, shopping and entertainment center for nearby mining camps (Russell Lee). Courtesy Photographic Archives, University of Kentucky Libraries.

Above, a West Virginia coal camp in the 1930s; below, miner bathing (Russell Lee). Courtesy Photographic Archives, University of Kentucky Libraries.

John L. Lewis, UMWA president 1920-1960. Courtesy *UMWA Journal*.

Left, W.A. (Tony) Boyle, UMWA president 1963-1972. Below, at the 1968 UMWA convention, from left: John Owens, secretary-treasurer; Tony Boyle, president; George Titler, vice president; and Joseph A. (Jock) Yablonski. Courtesy *UMWA Journal.*

Above, UMWA strikers and supporters. Courtesy *UMWA Journal.* Below, dissident UMWA miners from District 31 demand strike benefits from District President Leonard Pnakovich, 1971 (Rick Diehl/Appalachian Mountain Youth Collective).

Left, Arnold Miller, UMWA president, 1972-1979. Below, Indiana miners reject 1978 contract (Cathy Stanley). Courtesy *UMWA Journal*.

Above, black-lung protest rally at the Washington Monument, 1981. Courtesy *UMWA Journal*. Below, black-lung victim Gonzelee Sullivan of Grant Town, West Virginia, 1979 (Ted Wathen for the President's Commission on Coal). Courtesy National Archives.

Above, Eastern Kentucky miners lobby for black-lung legislation with Rep. Tim Lee Carter in Washington, 1975 (Earl Dotter/American Labor Education Center). Below, UMWA strikers in Martin County, Kentucky, face state police, 1975. Courtesy *UMWA Journal*.

Right, President Jimmy Carter in Steubenville, Ohio, September 1979. Below, West Virginia Governor John D. (Jay) Rockefeller IV and Sam Church, UMWA president. Courtesy *UMWA Journal*.

Right, Sam Church, Jr., UMWA president 1979-1982. Below, Richard Trumka, UMWA president 1982-. Courtesy *UMWA Journal*.

Right, coal miners, 1979 (Jack Corn for the President's Commission on Coal). Courtesy National Archives. Below, operating a continuous miner at U.S. Steel mine in Gary, West Virginia (Marat Moore). Courtesy *UMWA Journal.*

protest rally in front of UMWA headquarters a few months before negotiations, after one of the Yablonski killers said that someone named "Tony" had masterminded the murders.

In September 1971, MFD pulled together a meeting in St. Clairsville, Ohio, of 250 miners and supporters. The written invitation reflected the radical literacy of the college-educated outsiders who were helping MFD. Ed James, a former Columbia University student working with *The Miner's Voice,* wrote to Chip Yablonski and Clarice Feldman: "At certain points it sounds like an SDS (Students for a Democratic Society) propaganda sheet: 'Power to the Safety Committee,' 'Support Rank and File Demands,' and so forth."[18] But, he wrote, it came off as "extremely positive, emphasizing strong democratic unionism . . ., [and] it doesn't attack Boyle in a name-calling manner." The agenda included Nixon's wage-freeze directive, contract demands, better organization within the reform movement, and a "succinct summary" of the various court cases.[19] Trbovich chaired the meeting, which mainly updated local news and legal proceedings. The much-needed critical examination of the MFD—what it was, what it should do, how it should be governed—was left undone.[20]

When the UMWA contract expired on October 1, miners stopped work. Boyle called an official strike on October 11, but authorized strike benefits of one-hundred dollars for the forty-five-day strike only after MFD picketed his headquarters. The 1971 contract boosted the top miner's daily wage to fifty dollars at the end of the third year, raised the Fund royalty to $0.80 in two stages, and included a sick-pay provision. Boyle's effort bettered earlier contracts but fell far short of what the BCOA was able to afford and what miners expected. Many miners continued striking after Boyle signed, because they hadn't yet seen the actual document. Their skepticism was justified. The 1971 contract's new welfare benefits were only recommendations to the Fund, not bona fide guarantees. The new pay scale increased the differential (between the highest and lowest paid miners), which companies favored and miners opposed. Boyle's handling of this contract eroded his political support among working miners who had not sympathized with Yablonski. They felt he settled for less than he should have. His disingenuous explanation of the contract sowed doubts even among his supporters.

When the MFD's steering committee met in February 1972, the structural weakness of their organization was obvious. On the one hand, MFD had done well in keeping its anti-Boyle critique flowing through coalfield conversation. MFD lawyers had ensnared Boyle and his officials in proceedings that disrupted business-as-usual. MFD could pull a few mines out on strike and claim credit for the snowballing stoppage. But MFD leaders had failed to make it a membership organization of miners at the local level. Instead, MFD developed as a set of nesting superstructures of lawyers, college-educated advocates, and miner-activists. Trbovich was blamed for the weaknesses of the organization, because he was ineffective and rarely took initiatives. But his failures are only a partial explanation of the MFD's inability to build an organization.[21]

More to the point is that MFD could never become a genuine rank-and-file organization run by and for miners as long as it depended on the unique talents of its professional helpers. And without them, the MFD, as it came to be defined, would not have gone far. The MFD's history suggests how much of its activity was routed into lawsuits. Lawyers were the MFD's only full-time paid staff for many months. The Yablonskis became de facto leaders because of their legal skills and their anti-Boyle vendetta. Some in the MFD thought a better balance should have been struck between legal work and grass-roots organizing. The MFD, in their minds, should have been a rank-and-file movement with an agenda of political change, not simply a courtroom campaign against Boyle. Lawyers, they felt, inevitably ended up telling miners what to do, rather than vice versa. Some thought MFD lawyers did not always understand that MFD miners were fighting corrupt UMWA leaders, not the UMWA itself.[22] MFD lawyers, of course, saw themselves as working on behalf of coal miners and as victims of their own vitality and success. The antilawyer sentiment within MFD never came to a head. Critics never articulated a clear alternative and no one could deny that the MFD suits were succeeding. But the more progress made in legal actions, the more legal work substituted for MFD organizing within the rank and file.

MFD nominated its slate at a convention at Wheeling College in May 1972. With the exception of Trbovich and BLA's Arnold Miller, no MFD stalwarts represented organizations that showcased them. Organizers from Designs for Rural Action (DRA) had promoted Miller as the main spokesman of the Black Lung Association. Dozens of miners—Johnny Mendez, Levi Daniel, Walter Franklin, and many others—had worked just as hard to liberalize federal black-lung benefits, but Miller alone had the benefit of DRA's legal, promotional, and financial resources. Trbovich was linked to the Washington Project's lawyers, which did not give him the grass-roots legitimacy BLA gave Miller. Trbovich lacked a loyal following and the Yablonski brothers did not want to see him as the MFD's candidate. The Yablonskis thought Trbovich could not draw support from Anglo-Saxon and black miners in southern Appalachia.[23] They doubted his leadership ability. When Trbovich sought Joe Rauh's endorsement, Rauh declined on principle.[24] If Trbovich had been widely respected, the Yablonskis' defection might not have been fatal to his quest for the MFD's nomination. He was also saddled with whatever resentment there was about the Yablonskis and Washington lawyers.

The convention nominated five miners for president, but the contest came down to Trbovich and Miller when the other three withdrew. The Yablonskis threw their support behind Miller the night before the balloting after Tom Pysell, their first choice, failed to generate enthusiasm. They didn't know Miller well, but Charleston activists assured them he would do. A group of ten or so—the Yablonskis, Ed Monborne, and Bill Savitsky from Pennsylvania, Tom Pysell, Ken Dawes of Illinois, and others—decided that night to support Miller and balance the ticket with Trbovich, a Pennsylvanian, and Harry Patrick, an impulsive younger activist from northern West Virginia. Monborne and Savitsky, both presidential aspirants, placed their concern for MFD's success above their own ambitions, Hopkins concluded after interviewing a number of the participants.[25] The Yablonskis influenced Miller's nomination, but were probably not the decisive factor. Trbovich's political negatives defeated him more than Miller's positives or the Yablonskis' endorsement. Miller satisfied two criteria the Yablonskis and others established: the MFD's candidate had to come from one of the UMWA's large districts, and he couldn't be a "Hunky." The Yablonskis pushed the ethnic theme more than anyone else. Chip Yablonski was quoted as saying, "Those fucking hillbillies won't vote for a hunky,"[26] but their "No-Hunky" line may have been a smokescreen for personal doubts about Trbovich. (Their "No-Hunky" feelings had not stopped them from supporting their father in 1969.) The unstated corollary to "No Hunkies" was "No Blacks." Two of Miller's West Virginia coworkers in the black-lung movement were Levi Daniel, a black miner from Beckley, and Johnny Mendez, a miner of Mexican ancestry from Logan. Neither was able to win any MFD nominations, partially because of the racial criteria the Yablonskis and others accepted. Mendez left the convention in a huff. Daniel sought and lost each of the top three nominations and was not included in any of the minor positions. After Miller's victory, he was appointed to several union jobs, which he carried out with no particular distinction. Mendez returned home and sniped at Miller from a distance. Neither Daniel nor Mendez would probably have done any better than Miller as UMWA president, although each was denied that opportunity for the wrong reason.

The anti-Trbovich coalition turned to Miller because of who he was and where he lived rather than for what he thought and what he could do. Few were strongly committed to him personally. His flaws were unknown, and he came across as plainspoken and no-nonsense. Miller bore the marks of a hard life; his ear had been disfigured on the Normandy beachhead and he quit the mines in 1970 with an ounce or two of coal dust in his lungs. He dressed in inexpensive clothes and occasionally hung a thumb on the lip of his front pants pocket. Of medium height and build, he was not above squatting down to talk as he had often done underground. A cloud of cigarette smoke usually curled around his head. He had lived his entire life "up" Cabin Creek—a series of coal camps strung along the bottom of a valley about a twenty-minute drive from the Governor's mansion in Charleston. Arnold, as he was generally called, was no more or less than what he seemed to be.

The convention's lack of meaningful comparison among the candidates and debate over qualities of leadership reflected the simple fact that none of the MFD leaders had ever led much of anything. No track records existed for comparison. Miller, Trbovich, and Patrick—the MFD's slate in 1972—had been miners all their lives. None had any leadership experience within the UMWA save for their local unions. None had ever had to direct professionals or manage an organization. Although they had been active in the Black Lung Association or MFD, they were figureheads as much as leaders. Patrick and Trbovich shared many of Miller's virtues and weaknesses—good instincts, courage, undisciplined intelligence, dedication, inexperience, lack of intellectual depth, and self-doubt. Some who could have been nominated in 1972 might have done a little better. Several would have been measurably worse. Miller was an accurate reflection of the MFD's rank and file.

Delegates from two of the three UMWA districts in West Virginia voted for Trbovich, while a majority of the Pennsylvania delegates voted for Miller. The more miners knew about a candidate, the less they seemed inclined to vote for him. Trbovich and Harry Patrick easily won nominations for vice-president and secretary-treasurer, respectively. Trbovich never resolved his hard feelings for the Yablonskis. His bitterness toward Miller and his staff figured prominently in his public split with them in the mid-1970s.

The MFD platform extended Jock Yablonski's 1969 goals. Subtitled "The Miners' Bill of Rights," it stressed safety—it must "never be arbitrated"—and the individual's right to work only in safe conditions. A full-time, company-paid safety committeeman was urged. The MFD promised to help miners and widows secure federal black-lung benefits. The platform said the "conditions that cause black lung must be stamped out." The MFD promised major changes in the UMWA: shifting union headquarters to the coalfields, cutting salaries, requiring retirement at sixty-five, and ending nepotism. Major reforms in the Fund were proposed, including higher pensions and benefits to disabled miners and widows. The UMWA was to be democratized—every district was to have autonomy, and the rank and file would have the right to ratify contracts.

However, the MFD platform equivocated or remained mute on many issues. On surface mining, the MFD came down on the side of "proper reclamation" rather than abolition, which Miller personally favored.[27] The MFD did not want to offend surface miners. The platform also failed to set forth legislative or political agendas. The MFD did not say how it would deal with management or redefine the industry's obligations to the communities in which it operated. The platform was neither well conceived nor broadly discussed, and after it passed unanimously, journalist Tom N. Bethell, who was openly sympathetic to the MFD, wrote: "When it was read from the podium, hardly anyone listened and there was no floor debate on any of the supposedly controversial issues—such as strip-mining."[28] While the platform may have been hammered together with great care in committee, the absence of floor debate reflected a submissive confidence among the MFD rank and file and the fact that it represented a minimum consensus of the MFD. A frank discussion of strip-mining, the right to strike, equalization of pensions (between the anthracite and bituminous miners), and racism might well have split the fragile insurgency and alienated some voters. On the whole, the MFD's platform was a serviceable and in some instances bold assertion of the movement's sentiments.

The MFD's campaign against Boyle was more aggressive and better organized than Yablonski's. It had more money, for one thing. For another, almost two dozen college-educated staff worked in or around the campaign—organizing events, handling media, preparing material and researching issues, and so on. Much work was done in the MFD's Charleston headquarters by Ed James, campaign coordinator; Don Stillman, press secretary; Rick Bank, a lawyer who coordinated the candidates' activities; and others. The MFD candidates took their campaign directly to the rank and file; Miller estimated that he visited four-hundred bathhouses, where the miners change to and from their work clothes and talked with sixty-thousand men.[29] Bathhouse-to-bathhouse campaigning established MFD credibility and gave miners a chance to tell the candidates what they thought. Boyle's slate, in contrast, held a few coalfield rallies, but avoided the give-and-take of informal meetings.

Boyle had been backed into legal corners but was still fighting. He dumped Titler and Owens from his ticket, substituting two younger, untainted union politicians. His main line of attack was that the MFD was the manufactured conspiracy of student radicals, communists, Washington lawyers, Congressman Hechler, miscreant federal officials, and Jay Rockefeller.[30] These charges were blunt—and had some substance. He called the Miller slate "The Three Stooges" and the MFD lawyers "the troika."[31] He claimed that eight foundations had given these lawyers as much as $540,000 for anti-UMWA work.[32] Foundations gave money indirectly to the MFD by supporting MFD-related staff on various projects. Liberal donors contributed the bulk of the MFD's campaign fund, although several thousand miners made modest contributions. Josephine Roche donated $1,000 to the MFD, a sign of what John L.'s position would have been. The MFD never did put Boyle's outsider-baiting to rest because, indeed, outsiders were heavily involved. Boyle, too, depended on "outside" professionals and lawyers. MFD defused Boyle's outsider/communist phobia by comparing the war records of the MFD candidates with Boyle's—each MFD candidate was a veteran; none of the Boyle slate seemed to have been. Patrick, a Korean war veteran, boasted that he had battled communism in the trenches, in contrast to Boyle, who, in the 1940s, worked as a coal-company foreman.[33] Chip Yablonski, who conceived the war-record comparison, said, "It took the commie argument and just stuck it up their behinds. It really appealed to the service consciousness of men who work in the mines."[34]

In June, a federal judge outlined the guidelines for the conduct of the election. Labor Department representatives were to be stationed at UMWA headquarters and in every district and subdistrict office. MFD observers—each to be paid fifty dollars per day from the UMWA—were empowered to accompany the federal fair-play monitors. Labor oversaw all UMWA finances during the campaign. Finally, the judge guaranteed MFD equal access to the *UMWA Journal* and an address list of the membership. Later that month, Boyle was sentenced to a five-year jail term and fined $130,000 for illegally contributing UMWA funds to political candidates. Although Boyle won the September nomination of 863 locals to the MFD's 410, many of his were "bogus locals" made up of pensioners. The local-by-local vote showed pensioners supporting Boyle, and working miners—many of whom were young, recent entrants to the industry—favoring the MFD. Miller pounded at Boyle's record on safety, financial mismanagement, and Fund policies with slogans—"It's Time for a Change" and "Let's Make the UMWA Great Again." Even though the federal government supervised this election, fear persisted. Trbovich was shot at. In Whitesburg, Kentucky, armed men cornered the MFD candidates in their motel. A shootout was avoided when a quick-thinking clerk told the gunmen the MFDers weren't in. The goons departed after kicking and scratching the MFD car.

> Trbovich recalled campaigning for the first time in eastern Kentucky. To get there he drove up a steep winding road from Virginia through a pass, then down a similar stretch into Kentucky. It is a strip of road, perhaps 12 miles long, full of twists, sheer drops, no guard rails, no houses, no gas stations. Trbovich had been warned that if he were to be murdered it would be on this stretch of road. And he recalled wincing at every turn, searching for hidden assailants behind every rock, waiting always for the blast of a shotgun.[35]

With almost 127,000 miners voting in early December, Miller won more than 70,400 to Boyle's 56,300. The MFD carried eight of the union's ten largest districts and took 83 percent of the anthracite vote. The MFD carried the pensioner-heavy anthracite by a 7,400-vote margin because of its promise to equalize pension benefits, which at the time differed by $120 per month. The anthracite sweep gave MFD more than half of its victory margin.

The MFD generally did best in the northern coalfields and worst in the smaller districts in central and southern Appalachia. Boyle won about fifteen-thousand fewer votes in his winning districts in 1972 than in 1969. Working miners generally favored the MFD.

MFD won where Yablonski lost for many reasons, but three stand out. First, the MFD campaign was stronger than Yablonski's. It was better funded and better staffed. The use of the MFD media was sophisticated and skillful. Its network of activists was broader and deeper. Press coverage was more favorable[36]; the 1972 race was more tied to issues than a reformer's personality. Second, Boyle was a much weaker incumbent, besmirched by the Yablonski murder cases, harassed by a series of suits and convictions. He had handled the 1971 contract badly, and provided lame replies to MFD charges of nepotism, greed, and financial abuse. Finally, the federal judiciary and the Labor Department made sure that the election was conducted fairly and honestly. The courts took away Boyle's edge.

Yet more than fifty-six thousand miners—45 percent of the voting electorate—backed Boyle, despite evidence of corruption and disrepute. Why? First, the UMWA's political culture has always fostered organizational loyalty and unity. Only by working together and sticking together under UMWA leaders did miners rescue themselves from the semislavery that American corporations created in the first part of the twentieth century. John L.'s legacy to Boyle was membership loyalty.

Second was the fear factor. Boyle's UMWA was a bad enemy to have if you were a UMWA miner. If you criticized union brass, they might get you fired or sit on their hands when your grievance was filed. The Fund could cut your pension and health benefits. "Loyal" miners would give you a hard time. More miners probably voted for Boyle out of loyalty than fear.

A third factor was suspicion. Miners wondered who the hell Arnold Miller was and what he knew about running the UMWA. What was all this stuff about foundations and outsiders? Maybe Boyle was crooked, but what about Miller and his boys; what were they after? Miners who recalled federal injunctions and intervention in their strikes had reason to doubt the intentions of the Labor Department and federal judges. Had Joe Rauh, Chip Yablonski, and Ken Hechler put the fix in with the feds? Wouldn't the Nixon administration want to get rid of a strong union president like Tony Boyle?

Fourth, some opposed the MFD platform. If Miller were really to strike over safety, a lot of UMWA miners would lose pay. Strip miners doubted Miller had changed his mind on abolishing their jobs. Some bituminous miners didn't want to subsidize equal pensions with the anthracite. Some black miners may have seen the MFD as no better than Boyle.

Finally, some thought that Boyle hadn't done such a bad job. His contracts had doubled wages since the early 1960s and doubled the Fund's royalty in 1971. The 1969 Coal Act was a good law and the UMWA claimed credit for federal black-lung benefits. Miller had no collective-bargaining history at all; he was a gamble.

The MFD victory—despite these feelings—was as much a rejection of Boyle as an affirmation of Miller, which was why he had so much trouble consolidating his victory and spreading the MFD perspective deeply through the ranks. What motivated many MFDers was a *conservative* feeling that the union should be run properly, honestly, and fairly—the way things were supposed to be. Only in the context of UMWA history and politics were they reformers. Yet many in the MFD were no more liberal in their political ideas than the mainstream of organized labor or Boyle himself. The MFD slanted toward democratic *procedures* rather than a philosophy of political reform. It underpinned its 1972 campaign with the notion of "rank-and-file unionism," which endorsed rank-and-file participation and governance throughout the union.

It's hard to say how much thoughtful analysis lay behind this philosophical principle. MFD leaders intuitively supported the idea of miners like themselves running the union. "Rank-and-file unionism," as the MFD spelled it out, defined a new role for UMWA miners—membership ratification of contracts, free elections, local governance, membership participation in collective bargaining, and democratic conventions. But it failed to say much of anything about the role of political leaders in a rank-and-file union. MFD leaders had few models on which to draw, and, as a result, never developed clear notions about how to lead. Trbovich drifted into authoritarianism. Miller, at first, provided little leadership of any kind. Later, as events began to overwhelm him, he issued threats and denunciations with mechanical regularity.

9. Union Reform and Industry Turmoil

When Arnold Miller and the MFD insurgents swept into UMWA headquarters in December 1972, they felt they had won the war, that good had triumphed over evil. Within a few years, the sweetness of victory would turn sour, the spirit of reform would vanish, its momentum gone. Yet the bitter disillusion of those later years should not overshadow the substantial improvements the newcomers made despite their troubles. Nor should it fix the blame for Miller's fall on his own weaknesses and failures, though they were many. Their experience with the BLA and MFD had not prepared the MFD leaders to think for themselves with insight and confidence. They never had to master enough law, economics, or management to be able to direct union staff with competence. John L. Lewis, for all his faults, was flawlessly knowledgeable on issues affecting coal miners. MFD leaders never grew into their new roles. This, too, was partly John L.'s legacy. He had eliminated or made lap dogs out of the genuinely talented rank-and-file leaders. Once the insurgents lost Yablonski, they could not tap a single experienced union official for their cause. Once in office, MFD politicians came to resent their own inadequacies and dependency. Resentment shifted to distrust and finally ruptured the bonds between MFD leaders and the staff whose years of work had done much to elect them.

But it is too easy—and fundamentally in error—to explain the failures of union reform solely in terms of the personalities of the reformers and their political struggles within the union, although both factors contributed, as Tom Bethell, Miller's research director, wrote after his departure:

> The old hands who had weathered the Boyle years knew a lot about political survival. It was not long before they effectively controlled the board, which shares constitutional power with the president . . . They converted [their grudge against the reformers] into a series of skirmishes with Miller and his staff over virtually every action and policy that the UMW undertook.[1] The beginning of Miller's term coincided with structural changes among coal suppliers that circumscribed what the new union leadership could do. These changes—shifts in mining methods, location, markets and ownership—created forces within the industry that even the best and the brightest union leaders could not bend to their will—and Miller was far from either.

New forces were at work in the coal industry that complicated Miller's tenure. National coal production was growing after twenty years of stagnation, but BCOA tonnage was declining. While bituminous production grew from 603 million tons in 1970 to 824 million in 1980, almost all the gain occurred in non-UMWA, non-BCOA mines. BCOA tonnage slipped from 424 million tons in 1970 to 365 in 1980 (table 1). UMWA members under BCOA contract mined 70 percent of U.S. production in 1970 and about 44 percent in 1980.

Most of the new tonnage came from strip mines, particularly in the West. By 1980, surface mines produced more than 60 percent of all U.S. coal, compared with 44 percent a decade earlier. The shift from deep mines hurt the UMWA in two ways. First, surface mines are generally more productive, so that for every million tons of surface-mined coal, there are 200 to 300 fewer jobs for deep miners. Second, the UMWA has had considerably less success organizing strip miners than deep miners. The UMWA was able to organize only about a dozen western surface mines in the 1970s. Western companies hired young workers who seemed less concerned about pensions, job security, and job rights.[2] One supervisor explained his hiring policy in these terms: "We just love to have ranchers working for us. Those boys still know how to work. They are money-oriented. They want that four-wheel-drive truck and all those guns hanging in the back window. Maybe that is why, of course, they can't buy homes, too."[3] Oil companies, utilities, and mineral conglomerates opened many of the new western strip mines and undercut Miller's organizers by paying above the UMWA scale. If they were unionized at all, they preferred to deal with the Teamsters, Operating Engineers and Progressive Mine Workers, all of which edged into western mining during Miller's years.

Miller's seven-year presidency was also shaped by coal prices, which bore directly on BCOA's ability to satisfy UMWA bargaining demands. When the 1971 contract began, the average price of a ton of bituminous coal was $7.07, compared with $15.75 at the end of 1974, when it expired. The OPEC boycott drove up the price of all fuels, coal less than oil and gas. Even though BCOA mined fewer tons in 1974—the year of Miller's first contract—operators made much more money than in 1971. Coal price (in current dollars) continued to rise after the mid-1970s, but at a slower rate. (In constant dollars, however, price was relatively static from 1974 through 1981). Slipping tonnage and price stagnation scaled down BCOA's profits at a time when the expectations of UMWA members were rising. They demanded more from Miller in his second round of negotiations (1977–1978) than in 1974, but got less because of BCOA's economic problems. Both BCOA and the UMWA mistakenly assumed that the artificial price inflation of the mid-1970s signaled the start of a long-term boom. BCOA companies invested in new mines and equipment only to discover they had no market for their coal. High prices allowed the operators to tolerate

Table 1. Trends in Coal Production, 1970–1983

	Production in Millions of Tons			Percentage	
	NBCWA[a]	Other[b]	Total	NBCWA[a]	Other[b]
1970	423.7	179.2	602.9	70.3	29.7
1971	362.3	189.9	552.2	65.6	34.4
1972	406.0	189.4	595.4	68.2	31.8
1973	392.3	199.4	591.7	66.3	33.7
1974	367.4	236.0	603.4	60.9	39.1
1975	381.4	267.0	648.4	58.8	41.2
1976	369.5	309.2	678.7	54.4	45.6
1977	330.8	360.5	691.3	47.8	52.2
1978	273.1	380.7	653.8	41.8	58.2
1979	356.9	413.1	770.0	46.4	53.6
1980	364.9	458.7	823.6	44.3	55.7
1981	316.7	485.3	802.0	39.5	60.5
1982	352	477	829	42.4	57.6
1983[c]	313	472	785	39.9	60.1

Source: BCOA analysis of UMWA Health and Retirement Funds and Department of Energy data. March 1984.

[a] National Bituminous Coal Wage Agreements. Includes all UMWA tonnage, a majority of which is BCOA.

[b] Nonunion mines, mines affiliated with other unions, and UMWA Western Surface Agreement mines.

[c] Preliminary.

inefficiencies of many kinds in the mid-1970s. Productivity fell. Industry-wide employment rose from about 144,000 in 1970 to about 241,000 in 1978, and then dropped as demand shrank and productivity rose. The mid-1970s combination of higher prices and increasing employment fostered militancy among the UMWA's rank and file, leading to a dozen major wildcat strikes. BCOA could absorb higher wages and benefits but not interruptions in supply, which played havoc with customers. Wildcat strikes, absenteeism, and work-place confrontations led BCOA to demand "labor stability" from the UMWA in 1977 and 1981, which was the major issue in the long strikes in those years.

Miller's years were also shaped by the changing ownership structure of the industry itself. Big oil began buying BCOA companies in 1963 when Gulf took over Pittsburg & Midway. Continental Oil then bought Consolidation Coal; Occidental Petroleum, Island Creek Coal; SOHIO, Old Ben; Ashland Oil, Arch Mineral; Exxon, Monterey Coal; and so on. Other oil companies—including Arco, Mobil, Sun, Texaco, and Shell—started coal subsidiaries for western strip mining. By the late 1970s, oil companies mined more than one-third of America's coal. The Congressional Office of Technology Assessment estimated that energy companies would be producing almost half of all coal consumed in the United States in 1985 and about 70 percent of the coal that utilities did not mine for themselves.[4] Utilities and conglomerates also bought coal companies during Miller's years. A five-partner holding company took over Peabody Coal Co., the industry's largest producer, from Kennecott Copper, while General Dynamics and AMAX bought others. DuPont bought Continental Oil in 1981. Utilities were opening "captive" mines, which by the late 1970s produced about 15 percent of the coal they burned. Their captive tonnage was expected to triple by 1985. The implications of these changes for the UMWA in the 1970s were profound. BCOA was a less important factor in the industry as a whole, but its leading companies were strengthened by the enormous capital resources of their parent companies. Subsidiaries were better able to withstand long strikes than independents. It was no coincidence that UMWA-BCOA contract strikes lengthened with the takeover of the BCOA by noncoal interests. As in nineteenth-century China, outside capitalists had carved coal into distinct spheres of influence and ran it as a colony.

Arnold Miller's story is a modern tragedy. Within three years he had rocketed from the mines to the presidency of the UMWA. But his rapid assent concealed two flaws. First, he didn't learn how to use political power adroitly. Second, he lacked the inclination or the ability to grow from campaigner to president. The qualities that had brought him to 900 Fifteenth Street, N.W., the union's headquarters, didn't help him to be a smart politician and administrator. Miller was a beneficiary and a victim of circumstances, never their master. Fate and a few radical community organizers had plucked him from obscurity, mounted him on the black-lung issue, and helped him hang in the saddle for three years. The MFD nomination had fallen into his lap and his election was a backlash victory. A price boom for coal in 1974 aided his first negotiations, but the expectations of his youthful membership rose faster than his ability to satisfy them. Wildcat strikes, which had once boosted his career, plagued his presidency.

Miller's first two years were his best. His administration did well those things that tapped his staff's organizing and promotional abilities. Organizing the December 1973 convention; orchestrating a successful year-long strike against Duke Power Company's Brookside mine in Harlan County; developing publications and research; and negotiating with the BCOA in the 1974 negotiations succeeded in part because the reformers faced either a common enemy or a divided opposition. Where they had neither, they fared worse. They failed, for example, to consolidate a political base in the districts, to turn the militancy of wildcat strikes into a disciplined force, and to develop an internal education program.[5] Miller's lack of political skills undercut his programs. He could neither work with nor control the union's executive board. He buckled under the pressure of political fights and shunned them as much as possible. One of Miller's top aides told Ward Sinclair, a sympathetic reporter with the *Louisville Courier-Journal:* " 'Arnold just doesn't want controversy. He wants for it to go away. He's after the absence of abrasiveness—but the more he sees, the more he withdraws into himself.' "[6] Despite his belief in rank-and-file unionism, indecisiveness always prevented Miller from becoming a rank-and-file leader.

Much of 1973 was spent making the UMWA bureaucracy functional. From the outside, the Boyle administration had seemed teutonically effective. MFD discovered that many secretaries could neither type nor take shorthand. Files were scattered in boxes, if they were kept at all. With one or two exceptions, department heads were either incompetent or incorrigibly hostile to the newcomers. Bernard Aronson, Miller's new press aide, described it this way: "It was like the Wizard of Oz. There was this screen and a lot of smoke and noise and light coming from up above. When we took the screen away we discovered the real secret: nothing was going on up there at all, just a bunch of guys drawing huge salaries."[7] Miller told the 1973 convention in his *President's Report:* "We underestimated the extent to which the central administrative machinery of the UMWA had just plain stopped functioning. It had the appearance of functioning, but when we blew the dust off and turned the switch, the whole apparatus fell apart at our feet. . . . Simply issuing an order would seldom produce any results. It was like trying to turn an ocean liner around in the Ohio River.[8] Some departments, like publications, press relations, and research, developed quickly, while others—organizing, safety, legislative, and administrative—floundered. Those departments that did something on behalf of members had the easiest time. Others, like organizing and political action, which had to get members to do something, had tougher going. Tom Bethell, research director, described the reformers' problems:

> The transfer of power from Boyle to Miller had been bitter and chaotic; reform leaders with no previous administrative experience were already stuck fast in the molasses of bureaucracies old and new, discovering with horror that it was hard enough to get the mail answered and the dues processed, let alone launch new programs for an uncertain constituency.
>
> . . . The UMWA's outward image of shiny reform masked a troubled interior—whole departments, such as safety and organizing were in turmoil; policies to the extent they existed at all, were subject to change without notice.[9]

The fledgling bureaucrats had all manner of hurdles to jump. Holdover Boyle staffers fought for their jobs. Stress produced division within Miller's camp. Mike Trbovich, who was in charge of safety, organizing, research, and anthracite, drifted away from Miller and Secretary-treasurer Harry Patrick. He had never resolved his anger over losing the MFD's nomination to Miller. The troika model of decision making the MFD candidates used during the campaign disintegrated. By 1976–77, the three had broken publicly after months of stilted perfunctory communication.

A further complication was the role and perception of Miller's young, talented staff. All unions employ professionals in important jobs. Miller drew many of his from outsiders who had staffed the MFD. Chip Yablonski became general counsel; Rick Bank, Miller's administrative assistant; Ed James, Trbovich's assistant; Don Stillman, publications director; Bernie Aronson, press director; Tom Bethell, research director; Robert Nelson, legislative director; Davitt McAteer, counsel to the safety division. Miller appointed Harry Huge, who had argued the *Blankenship* pension case, as the UMWA's trustee on the Welfare and Retirement Fund. Conflicts broke out within the staff, between the staff and officers, and between the staff and anti-Miller union politicians. The staff was not monolithic. Their political philosophies ranged from mainstream Democrat to democratic socialist. During the MFD's campaign, conflict had surfaced between the West Virginia-based outsiders (who were more inclined toward rank-and-file politics) and the Washington-based outsiders (who were lawyers and thought mainly in lawyerly ways). Insiders adapted the nomenclature of local baseball teams to dub the sides the "Charleston Charlies" and the "Washington Senators."

Miller's opponents attacked his staff to discredit him and drive a wedge between him and his aides. Arriving en masse from the MFD campaign, the staff could claim no political neutrality in their new "professional" positions. The staff, it was charged, had an independent political agenda involving socialism (which was never defined). They were "outsiders" and came from a different political culture. Most were urban, college-educated sons of upper-middle-class professionals. Moreover, they were smart, impatient, abrasive, and outspoken. Miller and his staff never staked out positions that were very far to the left of their constituency. On the right-to-strike issue, the rank and file was more radical than the "radical" staff, most of whom argued against the wildcats. Generally, rank-and-filers cared more about results and less about who allegedly believed what. Trbovich and Lee Roy Patterson (a presidential candidate in 1977) red-baited Miller's staff because they understood that it alone stood between Miller and chaos. Miller and UMWA officials eventually gave in to right-wing attacks and took a position that miners would do all staff work. Some miners suddenly found themselves assigned jobs requiring skills they hadn't acquired, increasing Miller's troubles.

The staff issue came to a head in January and February of 1975 when Patterson, a Boyle supporter, charged publicly that "house Reds" and "radicals" were running the UMWA.[10] The controversy surfaced when Ben A.

Franklin of the *New York Times* wrote a poorly conceived story without mentioning Patterson's own political ambitions. Neither Patterson nor Franklin offered any proof to substantiate the charge. Franklin provided thumbnail sketches of seven Miller aides without quoting any of them on their political beliefs. About Bernard Aronson and Rick Bank, Franklin wrote:

> Mr. Bank and Mr. Aronson are Jews.
>
> "There are probably three coal miners in the United States with Jewish-sounding names, and they are really Poles," says one of their fellow staffers. "So there is probably some country anti-Semitism, perhaps not even consciously, in the heat that Rick and Bernie are getting."[11]

Franklin's clumsy insertion of Bank and Aronson's religion and his use of a quote to substantiate an assertion that was not mentioned before in his story muddled the conflict and created new problems for the "accused." Two weeks later, Patterson failed to prove his charges at an executive board meeting when challenged to do so. Within the next two years, Miller lost most of his staff, and Franklin's ill-conceived story contributed to the exodus.

Despite these strains, the Miller administration made substantial progress in its first two years. The *UMWA Journal*, under Stillman and later Matt Witt, quickly developed into an award-winning magazine. Staff salaries were cut, and some staff cut their paychecks a second time. Nepotism ended. Harry Patrick auctioned off Boyle's Cadillac limousines. (A few years later, Miller bought a new one, so he wouldn't feel inferior, he said, when riding around Washington or meeting with corporate brass.) Miller funded the Field Service Office (FSO) in his hometown of Charleston, West Virginia, to be the union's advocate for miners having problems with black-lung claims.[12]

Some issues unified the factionalized union. Mine safety was one. "Coal will be mined safely or not at all," the reformers said: "Safety—or Else." The anti-Miller people fell into line. The *Journal* began publicizing work-place safety and occupational health. From one full-time person, the UMWA's safety department mushroomed to forty-two in Miller's first year. Occupational health drew more interest and support.

The UMWA began its first genuine organizing effort in more than a generation. It spent more than $2 million on the Brookside strike, and opened a Denver office to coordinate a run at the big western strip mines. These campaigns met with limited success—some 120 mines were organized in the first year or two—but the rise in UMWA membership from 100,000 in 1969 to 160,000 in 1980 came principally from expansion of the work force at *existing* BCOA operations.[13] Miller set up a credit union. He began to clean up the National Bank of Washington. Tom Bethell set up a competent research department, which paid off in the 1974 negotiations when UMWA bargainers were better prepared on technical and economic issues than their industry counterparts.

While his administration made headway on improving union programs, Miller and his allies made two costly political errors. First, they dismantled the MFD framework. This left rank-and-file activists without an independent vehicle to coordinate their concerns and maintain influence on UMWA decision making. Miller and his supporters saw no need to pump new blood into the MFD, nor did anyone see where the new blood would come from. Miller folded MFD because he wanted to deprive right-wing critics of a point of attack and left-wing critics of a base of attack. But his action merely acknowledged reality. Without its constellation of outside organizers, MFD could not stand by itself. Had the MFD been more deeply seated within the membership, it might have served as a proreform caucus during Miller's years. But the political dangers of that relationship were judged to outweigh the benefits. After all, Miller and his staff saw themselves as having claimed all the white-hat territory, which left little room for constructive criticism no matter what its source.

Miller's second error was to retreat from union politics. Lewis and Boyle had personally appointed almost every district and international official. Miller replaced all twenty Boyle-appointed members on the twenty-four-member executive board with MFD loyalists as interim officers, but then did little to help them in upcoming elections. Miller believed union members should choose their own officials without any pressure from above, a simplistic notion of democratic leadership. The ousted Boyle appointees were often shrewder politicians than the interim reform incumbents. Only four of Miller's interim appointees won election, although proreform miners won eight of the fifteen elections held in 1973. Miller's laissez-faire approach to internal politics left him with a fragile base among the politicians with whom he had to work. Former Boyle supporters regained about half the executive board and districts, and those aligned with Miller were not indebted to him for their seats. Boyle supporters on the board led the anti-Miller forces, particularly Lee Roy Patterson from western Kentucky and Andrew Morris from northern West Virginia. Former Miller allies—Ohio's Karl Kafton and Alabama's Frank Clements—broke with him too. Anti-Miller coalitions controlled the board on most important questions. Trbovich lined up with the anti-Miller forces in the mid-1970s. Miller was not able to reach directly to the ranks when the board opposed him, and because the reform politicians had won without his help, they felt no obligation to help him. Miller's lack of a political base on the board and in the districts tripped him repeatedly. Increasingly, he had only his staff, but turning to them amplified his political isolation. Moreover, his staff had its own doubts about his leadership. By 1976–1977, Miller counted few dependable political supporters in the union. Many of his best staff and political allies had left. As old supporters peeled away, he replaced them with subordinates whose main job qualification was fawning loyalty.

Before these cracks split his administration, Miller's team succeeded in writing democratic reforms into the UMWA's constitution and negotiating an exceptional contract in 1974. These frameworks survived Miller and can be considered his legacy. But they cost him dearly because both episodes revealed his chronic inability to lead, which moved his disillusioned staff ever closer to bailing out.

The UMWA held its forty-sixth Constitutional Convention in Pittsburgh in December 1973. This convention was not like the old days. A single band provided the flourishes, in contrast to Boyle's four. Gone were the $100,000 worth of trinkets adorned with the president's picture. No thugs patrolled the microphones. Staff "porkchoppers" did not vote with the delegates. Critics were given a chance to say their piece. Boyle stalwarts tried to embarrass Miller with slick parliamentary moves and sharp points of order. Miller, to his credit, parried these with courtesy and made a good impression. Miller took his role as chairman of a democratic convention seriously. He called a number of close votes against his own committees and never rammed through motions or silenced speakers.[14]

While Miller presided fairly, he and his allies did little to organize politically for the floor debates. Many assumed that democratic procedures would necessarily lead to good policy. Miller didn't understand that opponents could use democratic rules for nondemocratic and illiberal ends. Consequently, the convention swung this way and that, voting one way and then reversing earlier votes.[15] In the end, though, Miller won much more than he lost. Major constitutional changes were approved. The right of rank-and-file ratification of contracts was established. Political rights were guaranteed—convention rules were democratized: free and fair elections promised;[16] districts assured autonomy;[17] and procedures established for both members and officers charged with violating union rules. Delegates passed a dues increase and authorized Miller to move the headquarters back to the coalfields. Miller saw the latter as geophysical insurance that union officials would be responsive—perhaps even hostage—to nearby members. (Miller was right about this, but his impolitic desire to build the new headquarters near Charleston antagonized a geographic cross-section of the board to the point where a coalition emerged and killed the whole idea.[18])

Behind Miller's victories was the terrible frustration his allies and staff felt with his leadershp. Miller pleaded with his staff to "tell me what to do," recalls one who did. His performance persuaded Chip Yablonski and Don Stillman to leave and convinced most of Miller's aides that he was "impossibly weak."[19]

The shaky coalition of Miller, his staff and supporters stuck together through his first contract negotiation. The UMWA's victory in the Brookside strike in August 1974 helped to paper over the dissension prior to bargaining that fall.[20] Miller made two important changes in the Lewis-Boyle collective-bargaining practices. First, the reformers democratized the bargaining process. Bargaining demands were framed at the 1973 convention and in a series of district-level, rank-and-file meetings across the coalfields. A two-step ratification process was set up. Once Miller and BCOA negotiators agreed on a contract offer, a UMWA bargaining council—district presidents, executive board members, and the three officers—had to approve it before submission to the working membership. No longer could a UMWA president accept a contract on behalf of his members, as Lewis and Boyle had done. In the three contracts since 1973, this procedure lengthened strikes but also produced better contracts. Second, Miller ended the partnership that had prevailed since 1950. Miller spoke bluntly to the delegates of the possibility of a six-month strike. When negotiations began in September, the UMWA laid more than two-hundred demands before the BCOA. The UMWA caught BCOA with coal prices edging toward twenty dollars per ton, an all-time high. The price bubble inflated profits. Pittston Co., for example, claimed an 868 percent increase in the first half of 1974.[21] UMWA bargainers felt rank-and-file heat—much of which they themselves had stoked—to catch up for the past contracts. In response to the UMWA's September proposals, BCOA proposed a few cautious and minor changes in the 1971 contract.

They reached a tentative agreement on November 13, a day after the old contract expired and 120,000 UMWA miners stayed home. The bargaining council turned down the proposed agreement on November 20. The council's action had a lot to do with Miller's unresolved troubles with rival politicians, but some of the criticism of the proposed contract was valid. A new grievance procedure raised doubts. A new two-tier pension system allowed post-1975 retirees to have vested and graduated pensions at two to three times the fixed levels for current pensioners. The plan violated miners' long tradition of equal treatment, but Miller's advocates argued that inequity was necessary to initiate an improved system for the future. It was not lost on anyone that those who were to vote on the contract would be its chief beneficiaries, while pensioners could not vote.

A second offer appeared about six days later. It included a 1 percent boost in the three-year wage-and-benefits package (from a 53 percent gain to 54 percent) and a few other changes. The council again rejected the offer, but reversed itself on the same day, 22 to 15, after hard lobbying by Miller's allies and a White House threat to invoke Taft-Hartley. Members of the council and union staff fanned out through the coalfields to explain the offer before the ratification vote. The *UMWA Journal* cranked out a special issue that included the complete contract with explanations of all changes. In the first week of December, 56 percent of the almost 80,000 miners who voted accepted. The wage increase—37 percent over three years—was the biggest ever. A cost-of-living escalator was added for the first time. BCOA agreed to increase Fund finances using new formulas based on tonnage and working hours.

The 1974 contract split the Fund into four legally separate but administratively related entities, commonly called the Funds. Separate pension plans were set up for working miners and current retirees to conform to the requirements of the new federal Employee Retirement Income Security Act (ERISA). The employers' contribution roughly doubled, and the negotiators projected Funds income of almost $2 billion over the thirty-six month contract. Other changes—paid sick days, the right of an individual to withdraw from imminently dangerous conditions, helpers on mobile machinery, increase in paid days off, and a different grievance procedure—were selling points. Opponents, however, ripped into the discrimination against current pensioners, the lack of a right-to-strike clause, and what they saw as inadequacies in the wage package (the inflation escalator, for example, was capped). If the strike had lasted longer than three weeks, a slightly better contract might have been forthcoming, but possible gains would have to be tallied against the cost of each day miners spent without income and, possibly, health benefits. A second offer would have had to add $6,875,00 in three-year wage gains to break even for each additional strike day.

Although the contract was overwhelmingly better than what it replaced, Miller gained little politically. The council's two rejections embarrassed him. The inability of some of his allies to explain the contract to the ranks mystified his supporters. Miller's opponents, particularly Lee Roy Patterson, picked at every hangnail on the contract. The contract was a good one, all things considered, but Miller was victimized by the high hopes he himself had raised. He had signaled his members at the 1973 convention that a six-month strike might be needed. Rank-and-file participation in the formulation of contract demands had raised expectations. The *Journal's* drumbeat of statistics about bloated profits may have backfired: a 54 percent contract gain over three years didn't look that good when matched against gains in profits of several hundred percent. Ideally, a contract establishes a framework that provides mechanisms for settling labor-management issues for its duration. Miller's 1974 contract did not. Its terms provoked controversy. Its dispute-resolving mechanisms didn't work.

The politics of bargaining and ratification put more pressure on the precarious balance of power inside the union. Four distinct factions appeared by early 1975: anti-Miller critics with roots in the Boyle wing; a left-of-Miller group that criticized Miller and the 1974 contract for lacking a right-to-strike clause; an increasingly silent number of miners who went along with Miller but without enthusiasm; and those—a diminishing number—who were loyal to Miller personally. On the other side, BCOA closed ranks during the next three years as prices leveled, demand for BCOA coal grew more slowly than expected, nonunion coal gained, and wildcat strikes crippled production schedules. The 1974 contract sewed the BCOA together and unraveled the UMWA.

Wildcat strikes defined the three years following the 1974 contract, logging more than 5.6 million lost workdays, or almost 6 percent of all workdays in this period.[22] Two kinds of wildcat strikes occurred. The most frequent erupted over local issues at a single mine. Such strikes punctuate the history of the industry despite operator and union opposition.[23] UMWA contracts and constitutions routinely included prohibitions and penalties against unauthorized striking. Typical contracts in the 1920s fined strikers $1 or $2 for each idle day.[24] Issues precipitating wildcats were local, but not limited to wages and working conditions. Mine operators were not above rigging wildcats when stockpiles were high or sales weak; having supervisors do union work guaranteed a strike. Most postwar wildcats never spilled past the mine where they started.[25]

The 1969 black-lung strike in West Virginia had introduced a second kind of wildcat. These began at one mine, but spread because the underlying issue touched miners across the coalfields. MFD had organized at least two multistate wildcats. These kinds of wildcat strikes erupted in the mid-1970s. The issues arose from the work place and political conditions in the coalfields, including a West Virginia gasoline-rationing plan, controversial textbooks in Kanawha County (W. Va.), shift rotation, job bidding, injunctions against wildcat strikes, black-lung legislation, and benefit cutbacks in the UMWA's Funds. A few such strikes involved more than half of all coal miners. What distinguishes them are the breadth of the issues and how quickly they gathered support across county and state lines. One, for example, had nothing to do with mining: miners in southern West Virginia in 1974 honored picket lines when white conservative Christians objected to "secular humanist" textbooks in their county schools.[26] The Revolutionary Communist party, through its Miners' Right to Strike Committee, played an important role in shaping several wildcats between 1974 and 1976. The committee's influence waned as its members were fired and its own shortcomings multiplied. An insensitive judge in Charleston, West Virginia, transformed one small shutdown into a massive outburst against federal judges who were enjoining local unions and individual miners.

Many theories have been set forth about why miners strike more than other industrial workers and why so many wildcat strikes broke out in the mid-1970s. Underlying the rash of wildcats was the rising level of rank-and-file expectations that industry profits and UMWA rhetoric fed. The age composition of the work force declined dramatically after 1970 as older miners retired and net employment increased by more than half in the space of a few years. The median age of coal miners was 31 in 1977 compared with 46 a decade before. Younger miners had not been traumatized by the layoffs of the 1950s and 1960s. They were less patient with and less accustomed to the partnership contracts of the past. A few other factors are noteworthy in these strikes. First, miners throughout the BCOA sector of the American coal industry experienced more or less similar conditions. They often reacted as if they worked for a single coal company. Many companies were not communicating well with their new work force even by their historically deaf-and-dumb standard, and thereby helped to create strike issues. One local union president, Steve Shapiro, described this problem for Feb. 3, 1978 "Op-Ed" readers of the *New York Times:*

> Many coal miners have come to see the wildcat strike as the only means available to them to enforce their contractual job rights. . . . In November, 1975, a joint industry-union hearing revealed, as every miner knows, that most wildcat strikes are caused by deliberate company provocation, company refusals to settle grievances at the mine sites, frequent resorting to arbitration and the courts, harassment of grievance committeemen by threats and layoffs, and safety discrimination by which miners who complain about unsafe conditions are assigned undesirable work. These findings were and have been consistently ignored by both the Association [BCOA] and U.M.W. The mine operators launched a successful news-media campaign to blame uncontrolled miners for the wildcat strikes in order to cover up management excesses.

Second, wildcat strikes spread because miners tend to honor picket lines. This is ingrained in their political culture, "a way of life" starting in the old days when solidarity was essential to organizing. Any political reflex can be abused and manipulated; honoring picket lines was. Third, most of the issues were serious to most of the strikers, otherwise they would not have drawn sustained support. Fourth, wildcat strikes are acts of frustration that reflect breakdowns in the routine of political control and change. That so many strikes occurred in so few years suggests that the system lagged in adapting to new political forces. Fifth, certain coal companies in southern West Virginia believed in hard-nosed labor relations. These companies refused to settle disputes quickly at the mine site, favoring protracted arbitrations instead.[27] Their workers lost confidence in the grievance machinery's ability to achieve justice. A grievance over a firing, for example, might take two years to resolve, leaving the miner jobless. With regular channels blocked, disputes burst into strikes—the other familiar grievance procedure. Finally, most major walkouts began in southern West Virginia in UMWA districts 17 and 29. Why there? These counties are the most heavily unionized in the country; coal politics dominates them. When work-place disputes occur, they polarize entire communities, and mediating influences don't exist. Miners have a clear sense of their political strength owing to their numbers. Because Arnold Miller was one of their own, southern West Virginians may have been less inclined to give him the benefit of the doubt.

Court decisions that allowed employers to get injunctions against strikers complicated resolving the wildcat stoppages. The Norris-LaGuardia Act of 1932 had prohibited injunctions against unions, but the Taft-Hartley opened the door against wildcatters. The Supreme Court ruled in *Boys Market v. Retail Clerks International* 398 U.S. 235 (1970) that federal courts could enjoin a strike that breached a no-strike obligation, which need not be written into the contract. *Boys Market* allowed employers to use the law to bludgeon workers with restraining orders and injunctions. If the strike persisted, it pitted miners against the government with the courts as policemen. As wildcat strikes increased each year from 1970 through 1976, coal companies obtained injunctions against hundreds of UMWA locals and individuals. Smith found a tremendous increase in the number of local unions restrained by the U.S. District Court in Charleston: 2 locals restrained, 1970; 20, 1971; 69, 1972; 127, 1973; 267, 1974; 295, 1975; and 188, 1976.[28] Knee-jerk injunctions sparked two of the largest wildcat strikes during the Miller years—strikes for which miners were assigned all the blame. Coal operators backed away from injunctions after 1977 because the court orders themselves were causing strikes and the 1976 ruling said a sympathy strike could not be enjoined unless the issue was arbitrable under the contract.

The era of wildcat strikes ended in 1978, following a 111-day contract strike against the BCOA, for several reasons. First, miners were exhausted from striking. Each unanticipated strike depleted their financial reserves. Even the most radical miners recognized that the well was running dry. Second, demand for West Virginia metallurgical and high-sulfur steam coal weakened considerably in the late 1970s. Layoffs and short workweeks made it harder and riskier for miners to sacrifice income. Third, wildcats worked sometimes, but not always. Wildcats did not make the UMWA Funds rescind cuts in 1977, nor win prominer black-lung legislation in 1976. On the other hand, they did cool industry willingness to use injunctions, and successfully pressured West Virginia politicians on gasoline rationing. Fewer issues arose after 1978 that lent themselves to mass protest. Fourth, the bonds of social control tightened after the 1978 strike. Companies used layoffs to weed out militants, and the UMWA toughened its policy against strikes. Miller's successor, Sam Church, who moved into the presidency in December 1979, was as opposed to wildcat strikes as John L. Lewis.[29]

The wildcat strikes in the mid-1970s arose in a unique context—increasing profits, rising expectations, and changing political ground rules—that won't be reproduced for some time. Only two strikes—the 1969 black-lung protest and the anti-injunction strike of 1976—planted the miners' flag on new political ground. Most of the others failed or ended inconclusively. All are now recalled with mixed emotions.[30] There remains the spirit of fighting back. The emotional and financial disruption these strikes inflicted lingers as well. Some miners remember being forced to strike against their will. Others believe the strikes made sense at the time. Communists in the Miners' Right to Strike Committee wearied the rank and file of striking and helped to isolate themselves.[31] Multimine wildcats seem to succeed only in special circumstances—a clear grievance that draws public sympathy, broad rank-and-file perception that regular channels don't work, sufficient job and income security to permit a long strike, and a clearly defined political objective.

The wildcat years represent lost opportunities. The UMWA's rank and file had not been so ready to fight since the 1920s. Yet no UMWA official was able to focus this anger, and union labor was often at cross-purposes with itself. Had Miller been able to shape the surge he might have created a new force in American unionism. To do that, however, he had to be a respected leader. Instead, union politicians used the wildcat turmoil to batter Miller's leadership, blaming him personally for the chaos. Fourteen of twenty-four board members voted to oust him from office in October 1975 after 80,000 miners wildcatted against injunctions and the UMWA was fined $700,000 and sued for more than $10 million. The board's vote was unconstitutional and played to mixed reviews among the rank and file.[32] Lacking institutional leadership, the wildcat movement needed to spawn indigenous leaders, as Poland's Solidarity did, to negotiate on its behalf. That did not happen. Most wildcats left no residual leadership, organizational network, or common goals.[33]

The wildcat strikes destabilized the relationship between BCOA suppliers and some customers. They created doubts about the reliability of coal deliveries. Some customers began diversifying sources of supply. But those who had been around the coal business stockpiled coal for insurance against temporary disruptions. Coal supply is sufficiently elastic and consumers sufficiently savvy to compensate for strikes. C. L. Christenson, a noted coal economist at Indiana University, and William Miernyk of West Virginia University have pointed to what Christenson called "the offset factor": coal stockpiles are built up before strikes, and production is stepped up after they end. Over a year's time, "the same amount of coal would have been produced with or without the strikes," Miernyk claims.[34] While the BCOA wailed about the unexpected disruption caused by wildcat strikes, it is debatable whether BCOA would have sold any more coal over a three-year period had there been none at all. BCOA tonnage declined in years when there were few wildcats as well as when there were many. Overall, the strikes created some temporary loss for the BCOA sector, but much of that was made up when work resumed. BCOA also argued that strikes contributed to declining productivity, and justified antistrike proposals in the 1977–78 bargaining on that ground. BCOA knew better. Allen Pack, president of Cannelton Coal, saw little causal relationship between the two: "Wildcat strikes do not affect productivity that much. Productivity is measured basically from how much you mine a day with how many people. So if you are not mining any coal, of course, you have no productivity."[35] Wildcat strikes affected the politics and psychology of the coal operators more than their pocketbook. They grew paranoid about "labor stability" and made it the centerpiece in the 111-day strike in 1977–78. Miller accepted the industry's logic that miners were to blame for what should properly be thought of as labor-*management* instability, and wound up fighting his own membership.

When the UMWA held its forty-seventh constitutional convention in September 1976, its political pot was at a rolling boil. The split between Miller and the executive board had disfigured union politics for more than two years. The presidential election and contract negotiations loomed in 1977. Trbovich and Miller no longer talked to each other. Patrick and Miller were about to part. Miller's "radical" staff was leaving, and the rank and file was seething. Compared with the 1973 convention, only a few constitutional changes were made. Mostly the 1976 convention is memorable for its sandbox style. Almost 1,900 delegates—800 more than in 1973—squeezed into a room about one-third too small for their number. Convention planners had bungled hotel arrangements, a constant irritation. Microphones came alive or died according to some electromagnetic force having nothing to do with conventional circuitry. The proceedings were infused with intraunion politicking. Trbovich informed the delegates at one point that "Socialistic, Revolutionary, and Communistic elements" had infiltrated the Miller staff and threatened to destroy the union.[36] He painted a simple picture of reds and whites:

> Out of VISTA came a group of people who were going to save the coal miners from the free enterprise system and the democratic government. They came to Appalachia; they worked up until 1972, and lo and behold they got into the reform movement of the United Mine Workers, and when we won the election they moved into the United Mine Workers.
>
> . . . These elements, these people [staff] who do not believe in the philosophy of a democratic government, have disrupted and have undermined the leadership of this organization.
>
> . . . These people who have implemented their philosophy into this Union, who have ill-advised you, have practically destroyed this Union.[37]

Trbovich said the *Journal* had "crucified" the executive board, and criticized union staff for confiding to the media their opinion that some board members were "greedy men, functional morons." Trbovich claimed the Miller administration had organized only one western mine and $29 million of the UMWA's reported $80 million in assets was uncollectable. Left-wing reporters were run out of the hall, and a few rank-and-file delegates fled, fearing for their safety. Miller's patient attention to parliamentary fairness allowed him to navigate this rough water, but not without serious damage.

The convention gave the Miller administration a somewhat resigned endorsement despite the histrionics. The delegates acknowledged both his failures and successes. Dues collection was one issue where delegates scolded the administration. Miller and Patrick had computerized the central dues-collection system but had not worked out its bugs. The mix-ups that resulted were pervasive. The convention voted to have districts collect dues for redistribution. Yet Miller fended off a similar effort to decentralize authority when he defeated a move to shift organizing and safety to the districts.[38] The delegates worked out guidelines for the upcoming negotiations. Local unions submitted almost 6,000 collective bargaining resolutions, 4,800 more than in 1973. The convention proposed comprehensive changes in health and safety—a full-time health and safety committee at each mine; a full-time dust weighman to monitor federal dust sampling; automatic continuous dust sampling to replace sampling the individual worker; increased health and safety training; and expanded responsibility for the miner-elected local safety committees. On collective bargaining, the convention supported a "substantial wage increase," a six-hour day, abolition of compulsory overtime, local approval of all shift-rotation arrangements, increased personal and sick-leave days, maternity leave, uncapped cost-of-living escalator, unemployment benefits, seniority based on length of service alone, changes in grievance and arbitration machinery, and dozens of other changes. The convention endorsed the collective bargaining committee's position "that the time has come . . . to negotiate into our contracts the right of Local Unions to strike over over local issues threatening the safety, health, working conditions, job security and other fundamental contract rights of our members." This promised to be the toughest demand to win.

The convention also set the stage for the union's 1977 election, which matched Miller against Lee Roy Patterson and Harry Patrick. Why Miller ran for a second term—he had promised not to in 1972—is a mystery. For three years, his job had brought him little pleasure, respect, or power. His health had deteriorated. He lost most of his former friends, allies, and staff. His accomplishments had come during his first years, when the momentum of victory and a bright staff compensated for his inexperience. From afar or at his knee, people considered him inoffensive man tragically out of his depth. Miller could not persuade a single member of the executive board or district president to join his slate. Instead, he chose Sam Church, Jr., a blocky ex-miner whose main qualification for the vice-presidency appeared to be loyalty to Miller. His secretary-treasurer was Willard Esselstyn, a likeable young Pennsylvanian who had worked for a short time in COMPAC, the UMWA's political action arm, after stints as an air force cook and miner. Miller was again running with candidates who had no administrative experience, let alone accomplishment.

Miller's principal opponent was Lee Roy Patterson, the board member from West Kentucky who led the Boyle remnants and disaffected MFDers like Mike Trbovich. Patterson had done little as a board member except promote himself, but he had the support of most of the executive board and district officials. A strip miner—the first ever to run for the UMWA presidency—his politics were not fired in the kiln of danger and struggle of underground mining. Patterson promised to return dignity and order to union affairs. He capitalized on the feeling among miners of all political persuasions that more Miller meant more embarrassment. Understandably, he never invoked Tony Boyle; but it was apparent that he had consolidated much of the Boyle bloc.

Harry Patrick, the third candidate, was a simple, right-and-wrong kind of guy. A stubby, friendly, somewhat insecure miner from northern West Virginia, he gained prominence in MFD after decking a Boyle henchman on national television. He was not a natural politician like Patterson and had been carried past his depth by his quick rise from miner to UMWA secretary-treasurer, a job for which he had no qualification. Patrick has said he spent much of his first two years in Washington developing a taste for quiche and good booze. But as Miller's leadership disintegrated, Patrick began taking union affairs more seriously. Some of Miller's staff—Rick Bank, Tom Bethell, Ed Burke, Tom Woodruff, Martha Spence, among others—drifted into his camp even though they doubted his ability to beat Miller or Patterson. Patrick moved to the left of the old MFD agenda. He slated capable but unknown candidates for the other two international offices, leaving Patterson with the most experienced and geographically balanced ticket. Patrick's campaign was the last hurrah for the MFD political model. Like Miller he depended on staff to develop ideas and political strategy. Though he shared some of Miller's insecurities, they did not paralyze him. Had Patrick won, he would have promoted a social-democratic unionism absent Miller's idiosyncrasies. Patrick started his campaign late, however, and could not make up ground fast enough. Few thought he could win. He ran to keep the anti-Miller majority from swinging behind Patterson. Patrick blasted Patterson for his ill-considered remark that "if elected, he would be willing to consider affiliation with the Steelworkers," who had given his campaign several thousand dollars and a public relations specialist.[39] That slip may have cost Patterson the June 1977 election.

Patrick and Patterson combined beat Miller, but Miller beat each of them, with more than 55,000 votes to Patterson's 49,000 and Patrick's 34,500.[40] Miller won the big districts in Pennsylvania and West Virginia, and lost almost everywhere else.[41] Seasoned observers looking at the local-by-local results figured that Miller won the pensioner vote and lost the working miners. He won 40 percent of the votes cast and about 20 percent of the electorate. No endorsement could be read into these results. His plurality can probably be laid to a feeling that neither Patterson nor Patrick promised to be that much different. Reelection left Miller with two questions: what to do, and with whom. He had been reelected without a platform and the staff that had been responsible for his first-term accomplishments. The 1974 contract was to expire on December 6, 1977, and he faced negotiations without the resources he had depended on three years before. The answers to these questions were postponed when, four days after the election, the UMWA Welfare Fund announced major cuts in medical benefits, precipitating the last big wildcat strike of the mid-1970s. The Fund trustees claimed that wildcat strikes had caused a cash shortage. Like most affairs involving money, the lack of it was more complicated than the trustees claimed.

When the MFD took over the UMWA, the Fund needed an overhaul. Lewis and Boyle had never negotiated enough royalty income to sustain the Fund's original design or provide adequate pensions. The Fund's staff clung stubbornly to worthy principles. They screened health-care providers for quality and controlled costs. They insisted that beneficiary coverage start with the first dollar of costs.[42] Decentralized administration allowed local Fund staff to learn firsthand about quality and cost problems. Unlike a company-financed insurance plan, the Fund provided broad medical support services for its beneficiaries—central drug purchasing, rehabilitation, benefits counseling, and technical assistance to clinics. It spawned a network of some fifty consumer-controlled, primary-care clinics, with salaried staff, throughout the coalfields. The Fund paid for these group practices through negotiated retainers, which allowed the clinics to provide many services, on the one hand, but subsidized nonminer clinic users on the other. The Fund assumed central importance in the lives of UMWA members because of its comprehensiveness. UMWA miners over the years saw the Fund as "theirs," as a union institution.[43] While miners trusted Fund doctors, they had good reason to wonder about the trustees, who kept severing beneficiaries to keep the Fund solvent.

The UMWA and BCOA contract reorganized the Fund in 1974, dividing the medical and pensions programs into four trusts with a common board of trustees. Miller appointed Harry Huge to be UMWA trustee on the Funds, as they were renamed. The plight of beneficiaries genuinely moved Huge. Lacking any administrative or health-care experience, he fell back on legal friends and computer technology to solve problems. Huge hired a veteran from the Law Enforcement Assistance Administration, Martin Danziger, to direct the Funds. Danziger had no experience in the coal mining industry, in pension administration, or with health care organizations. His principal qualification for the job was his "considerable administrative experience," as one *Annual Report* put it. Huge and Danziger tended to equate efficiency of service with quality care. They concentrated on improving administrative services, and as a result, health providers instead of health consumers became their constituency. Where the Fund of the 1950s was willing to fight medical dogma, Huge and Danziger picked no bones with state medical societies and the AMA, whose medical philosophy served as an omnipresent restraint. Beneficiary participation continued to be totally absent in Funds policy making. Huge and Danziger ignored the good in the Funds' past. Efficiency, to them, meant scrapping existing administrative procedures and denigrating veteran staff, many of whom had served since the early 1950s with demonstrated competence. Quality of care had not generally been an issue in the past, rather, it was access to care. Lacking the wisdom to retain the good while discarding the bad, Huge and Danziger chucked both. They rejected the activist principles of the early Fund after a brief flirtation with talk about "substantial changes in focus and attitude."

Dr. C. Arden Miller, president of the American Public Health Association, was commissioned to do an in-depth critique. He recommended leaning "toward a health care policy that promotes prevention of disease . . . [and] in the interest of the good health of beneficiaries, [the Fund] should . . . advocate . . . social and governmental change." He felt that "all types of Fund beneficiaries should be included on the governing bodies of all agencies which do business with the Fund."[44] Miller's ideas dovetailed with the Fund's original philosophy, but his notions never got very far in Danziger's computers. The social advocacy rhetoric in the Fund's 1974 *Annual Report* was notably absent from later editions.

The final twist was that Huge and Danziger did not even manage skillfully. Suzanne Jaworski Rhodenbaugh, a former health-service specialist with the Johnstown Pennsylvania Fund office, charged in 1977:

> Technocrats . . . have made clear that neither people nor programs rank in importance to their introduction of a centralized, computerized method of paying medical bills and pension checks.
>
> Yet they have failed miserably at managing. Many cost and quality controls in the health program have been lost. Medical bills are paid late (if not lost); duplicate claims are paid; pension checks to retired miners are delayed; eligibility controls are often out of control. Virtually all experienced top-level staffers have been retired, fired, or have quit in disgust. In their place have come dozens of would-be technocrats who know nothing of labor, health *or* pension programs, or management. These technocrats don't stay long, however, and the incredible turnover fuels the problem.
>
> So much of the Funds' program has been gutted while it was "modernized." And direct health expenditures and administrative costs have risen dramatically.[45]

Under Huge and Danziger, administrative costs rose 90 percent between 1974 and 1976; medical care jumped 73 percent. Medical-cost inflation, of course, explained much of the latter, but Huge and Danziger's proprovider reforms and computer wizardry did not, as promised, contain costs well. Danziger deemphasized regional offices in favor of centralized accounting. This "reform" terminated an effective method of checking the reasonableness of doctor bills. Freed from restraint, doctors predictably overcharged. In many cases, the Funds paid these charges rather than antagonize the providers.

UMWA and BCOA negotiators refinanced the Funds in 1974 and funded them separately. Considered together, the Funds were solvent in the early summer of 1977. However, the 1950 Pension and Benefits Trusts, which provided pensions and medical care to 82,000 pre-1976 pensioners, were bankrupt. No single factor explained their insolvency. The BCOA and the UMWA laid all the blame on wildcat strikes. But George Getschow asked in the December 1, 1977 *Wall Street Journal* whether the Funds "would be solvent today" if there hadn't been so many wildcat strikes. The actual income loss from these strikes, he found, "represents only about 5 percent of the funds' total projected income of nearly $2 billion over three years. That " 'wasn't enough in and of itself to cause a problem,' " said Paul Jackson, a consultant to the Wyatt Company, a Washington actuarial firm. While the rank and file bore responsibility for most of the wildcats, two of the largest—against injunctions in 1975 and 1976—were reactions to industry provocation. Miller, Huge, and BCOA didn't mention that the number of beneficiaries—800,000—greatly exceeded the projections of the 1974 negotiators. Royalty income fell short of their expectations by many millions more than $100 million lost to wildcat strikes. BCOA output failed to grow as much as the companies had predicted, for reasons having nothing to do with the wildcat strikes. Severe winter weather cut production in 1976-1977. The rate of medical inflation was higher than the negotiators had assumed,[46] and the Funds' administrative costs rose more than anyone expected.[47] From 1973 to 1977, the Funds' *Annual Reports* show that 37 percent of the increased expenditures ($260 million) were due to new beneficiaries, 33 percent ($232 million) to pension increases arising from the 1974 contract, 26 percent ($183 million) to increased health care costs, and 4 percent ($27 million) to additional administrative costs. Without a contractual guarantee of benefits, health coverage for more than 800,000 people had become the stakes in a crapshoot of macroeconomic forces and labor-management politics.

The Funds' trustees, with the consent of the UMWA and BCOA, had twice in 1976 reallocated money among the four trusts to meet cash shortfalls. Although reallocation cost the BCOA nothing, the operators made it clear they would not help a third time. On May 16, 1977, the trustees faced a third shortfall. They might have tried to scale down payments to health providers. Instead, they chose to cut benefits without ever asking the UMWA and BCOA to consider a third reallocation. As Funds' chairman, Huge controlled the timing of the cutback and, hence, Arnold Miller's political fate. Huge was supporting Miller's reelection and had given him a thousand-dollar campaign contribution. Insiders at the Funds leaked information to both Patterson and Patrick that Huge was delaying scheduled cuts until after the election. Both went to the press, but most reporters played their warnings as campaign rhetoric. One who didn't was *Coal Patrol*'s Tom Bethell, a Patrick supporter, who wrote: "The trustees decided in May to make cuts so that the news would not be released until the election was safely out of the way."[48] The cutback letters had been set in type before the June 14 election. When they arrived a few days after Miller's thread-thin reelection, eighty-thousand miners across the coalfields stopped work for twelve weeks in protest. The UMWA's executive board voted 22–0 for Huge's resignation.

More damaging to miners than the benefit cutbacks was the underlying concept of Huge's new copayment scheme, which required miners to pay 40 percent of the cost of medical services and the first $250 of hospital expenses (with a $500 limit). Huge chose copayments instead of a number of other fixes—a renegotiated bailout, a joint company-miner assessment, a short-term restriction on certain services, a cap on certain physician and hospitalization payments, or some combination of these and other schemes. Copayments made beneficiaries bear all the burden of a mess only partly of their own making. Huge's copayment plan was originally billed as an emergency stopgap to address an immediate shortage of cash. However, when BCOA began negotiating a new contract with Miller a few months later, the operators insisted on retaining copayments as part of a drastic restructuring of the Funds. Huge advised Miller to accept BCOA's plan to shift medical coverage to company insurance during the bargaining.

The trustees also scotched retainers for the Funds-dependent clinics, which reduced services, staff, or standards. Total doctor-office visits fell 31 percent at five clinics the West Virginia University Department of Community Medicine surveyed after copayments began.[49] An Appalachian Regional Commission study found that physician staffing declined 42 percent from July 1977 through September 1978; nonphysician staff fell 25 percent.[50] A number of clinics skittered along the edge of financial ruin. Fewer patients came for clinic services, and accounts receivable rose. Many clinics never recovered from the sudden change in their funding.

When long-lived authoritarian regimes topple, it takes time and talent to recreate stability in a pluralistic framework. The UMWA after Lewis and Boyle was no different from France after Napoleon, Spain after Franco, or China after Mao. Democratic unionism took root and spawned pluralistic politics and rank-and-file activism that created new and unexpected pressures. The excesses and failures that dogged the UMWA's effort to reform itself must be seen in the light of its uniqueness: no other union in modern American labor history has even tried to do it. Democracy accelerated change and politicized union affairs. Factions formed and a spectrum of political ideas—the broadest since the early 1920s—emerged. Ambitious politicians built new ladders and began to climb them. Democracy drew from the ranks a new generation of union leaders, miners in their 20s and 30s who would soon stand in for Miller and older miners who had been the backbone of the MFD. Without the reforms Miller and his allies wrote into the UMWA constitution in 1974, an aggressive young miner like Richard Trumka would have had little chance of unseating a conservative incumbent like Sam Church in 1982.

Many observers argue that democratic unionism in the UMWA has been disastrous. Those who blame rank-and-file democracy for the Miller administration's shortcomings confuse cause and effect. With a couple of exceptions, the rank and file carried out its new responsibilities reasonably well. The model didn't fail. Rather, democratic unionism found itself after a few grace years in forbidding economic straits without savvy leaders. If one has to choose between blaming the model and blaming the leaders for the failures of UMWA reform, the latter bear the greater burden. Democratic reform did not impede effective union leadership. However, it did make it more difficult and exposed every mistake to rank-and-file scrutiny. The argument that "too much democracy" explained the Miller administration's failures—a line that hooked experienced reporters like Ben A. Franklin of the *New York Times*—is not supported by the facts. Sometimes miners and their elected representatives made wise political decisions; at other times, foolish ones. But the fact that it was ordinary miners making these decisions on their own behalf was not at fault.

10. Wrong War, Wrong Place: The 1977-78 Strike

A few weeks after the Funds' strike folded, the Miller administration limped into negotiations with a December 6 expiration. Miller was exhausted, occasionally irrational, and frequently absent. He was sick. His membership was surly and openly rebellious, his staff decimated, his board contemptuous, his name the butt of sour jokes. Although the 1976 convention had worked up a comprehensive list of bargaining objectives—including strengthening the Funds, a right-to-strike clause, equalization of pensions, and safety and health improvements—UMWA negotiators came to the table with few concrete proposals and almost no backup research.[1] Staff departures left the union with little technical support. Neither of Miller's new officers, Vice-president Church and Secretary-treasurer Esselstyn, had negotiating experience. Miller had almost no support among executive board members and district presidents who formed the bargaining council, which had to approve any offer before submission to a rank-and-file vote. Further, the Funds' cutbacks—coming only days after his reelection—embittered even the minority of members who had voted for him.

The BCOA, despite internal divisions, began with closed ranks. The industry had used the OPEC embargo to boost prices from $8.53 in 1973 to $19.82 per ton in 1977. But coal flooded the market, and prices tailed off. BCOA's 130 companies lost sales to nonunion operators in both East and West.[2] Soft markets, stagnant prices, and nonunion competition drew BCOA members together for their battle with the UMWA. They discarded the 1974 strategy of labor accommodation, which assumed that a contract "victory" would enable Miller to control his membership and BCOA production and profits would climb without interruption. Both assumptions failed under pressure. The contract raised rather than dampened rank-and-file expectations. Wildcat strikes increased, and industry hard-liners said miners struck more because it took fewer days to make what they needed. BCOA feared that *its* coal—most of which originated in the East—would be cut out of the industry's future growth because of disruptive union miners. BCOA's negotiations team was quarterbacked by the organization's least flexible members, Consolidation Coal and U.S. Steel. Interlocked through directors, bankers, and social ties, they led the never-give-an-inch bloc that planned to rebalance power with the UMWA and snap labor back into line.

"Labor stability" was the phrase BCOA coined to summarize what it wanted from the UMWA. It has a nice, positive ring. Reporters made "labor stability" synonymous with the public interest, a value in the American trilogy: life, liberty, and the pursuit of stability. A more appropriate goal for this industry would have been "labor-management stability." The emphasis on *labor* stability made it seem as if a mysterious working-class pathology—or perhaps the Marxist children of Long Island stockbrokers—caused miners to jerk about in fits of industrial epilepsy. BCOA's demand for labor stability was really part of a half-articulated reaction against the noose of change operators felt looped around their necks—women in the mines,[3] federal regulations, increased property taxes, citizen activists, strip-mine abolitionists, black-lung payments, ERISA, long hair and back talk. Their anger focused on labor, a familiar enemy. To the BCOA, stability meant UMWA leaders would control their members and management would control what happened in the mines. When a BCOA spokesman told the *Wall Street Journal* in November 1977 that "the industry has always maintained that it wants a strong union," he was not talking about rank-and-file assertiveness, but about Lewis of the 1950s.

Early in the negotiations, BCOA presented its "labor stability" demands, from which it did not deviate until mid-February 1978, more than two months after the strike began. This package of penalties, the likes of which had not been seen in the coal industry since the 1920s, would have given employers the unappealable right to discharge a miner for wildcatting. Verbal encouragement of a strike would be sufficient grounds for firing. Any who honored a strike could be summarily suspended without pay, and arbitrators could not reduce a penalty were a miner found guilty. Striking miners would have their pay cut 40 percent on their return to work for each day they wildcatted. The Funds were to get the fines.[4] Cash penalties of up to two-hundred dollars a month were to be imposed against chronic absentees and strikers. Operators were to be allowed to establish production-incentive plans—in effect, speedups. The powers of the UMWA safety committees were to be compromised. The BCOA's theory of retribution held that miners were to be penalized individually for collective protests. As Allen S. Pack, president of Cannelton Coal Company, said: "I believe that the only way we'll get labor stability is to penalize the individual, whether it's for the man to pay back the welfare fund or something else. I don't think penalizing the local or the district or international will work."[5]

Miners saw the issue of stability less simplistically. They too wanted stability; strikes penalized them at least as much as their employers. But they saw stability in a broader context and in a longer perspective. Stability to the ordinary miner included job security and equity; stability, in other words, on management's part. Management, of course, would not guarantee employment and chose not to adopt practices that promoted job tenure. Where stability meant inequity, miners saw no virtue in it. What management called instability was from the miner's view nothing more than pressure for equity. Miners also held management responsible for some of the wildcat strikes that occurred during the 1974 contract. Management's policy of dragging grievances through lengthy arbitration predisposed miners to strike to force immediate change. Miller expressed a common perception: "What generally precipitates a strike is that our members have been wronged to start with. Most of them could have been prevented if the operator would have just lived within the contract."[6] Miners also knew that coal companies sometimes welcomed wildcat strikes when orders are slow or stockpiles high. Many felt, for example, that BCOA companies welcomed the wildcat over the Funds' cutback in the summer of 1977 because markets were soft and they wanted to sap rank-and-file resources before negotiating a new contract. (That strike was one of the few in which the operators did not paper the coalfields with court injunctions.)

UMWA members sought to control wildcat strikes through negotiating a right-to-strike clause in their new contract. They reasoned that such a clause would reduce the level of striking by establishing a democratic authorization procedure, limit strikes to the mines where they began, and remove the federal courts from private sector disputes. The 1976 UMWA convention proposed going to arbitration or striking over safety, health, working conditions, job security, and other contract rights.[7] The delegates proposed a system of checks and delays that would have inevitably led to fewer strikes. An issue, for instance, had to affect a large number of local members to qualify for consideration. Local officers were to select which issues qualified and then schedule a vote for seventy-two hours later. Although the proposal was not clear on this point, it seemed as if a strike would be authorized only if a majority of miners in the local union approved.[8]

The main problem with this idea was that Miller didn't care much for it, even though he told reporters he would insist on winning "a limited right to strike over local issues at each mine by majority vote."[9] Knowledgeable observers—indeed, anyone who had mastered fourth-grade arithmetic—could see that Miller's press statements did not add up to a genuine commitment. Wildcat strikes had tormented him and subjected him to ridicule. Until the 1976 convention endorsed a local right to strike, Miller opposed it. Even during negotiations he said: "If we don't bring these work stoppages under control there's not going to be anything for anybody to fight over. The union is going to be busted. I want the membership to understand that that is not a figure of speech."[10] Miller regularly criticized strikers, red-baited them, and even expelled some of the leaders of the Miners' Right to Strike Committee from the UMWA. By mid-December, BCOA negotiators revealed that Miller had never presented a serious right-to-strike proposal. This suggests that Miller and other UMWA negotiators never disagreed with the BCOA over labor stability or the penalties devised to achieve it. Rank-and-file rebelliousness had wounded UMWA officials as much as the BCOA. Both saw the rank and file as a kind of Third World peasantry in need of pacification.

The other major BCOA proposal involved the UMWA Funds. BCOA viewed its pension obligations to pre-1976 retirees (under the 1950 Pension Trust) as an expensive burden. Providing them medical benefits was added insult. BCOA toyed with palming off these pensions on the federal Pension Benefit Guarantee Corporation. The UMWA felt an obligation to the older retirees, but union negotiators understood how money spent on the old-timers meant less for working miners and future pensioners. This also handicapped UMWA organizing campaigns insofar as it reduced the wages and benefits the UMWA could offer the unorganized. Neither the UMWA nor the BCOA took seriously the rank and file's interest in equalizing pensions between the old-timers and post-1975 retirees. BCOA would agree to only token increases in pensions in 1977.

Restructuring the UMWA Funds' medical plans was the BCOA's primary target. Under the 1974 formula, health care cost a lot, and the UMWA, through trustee Harry Huge, continued to control Funds policy. BCOA wanted health care to cost less, and to terminate union influence. BCOA proposed to institutionalize the cuts the Funds made in June 1977 to cope with the temporary cash shortfall. The Funds' health-care system was to be dismantled, except for the old-timers. All working miners and future pensioners would be converted to employer-managed health insurance. A sliding scale of copayments—first begun the summer before—would become permanent. These were drastic, unpopular changes. The shift to company insurance would end a unique, thirty-year-old experience in labor-oriented health care. If the BCOA succeeded in winning copayments (deductibles) and company-insurance plans, many of the Funds' best features—first-dollar coverage, preventive medicine, community health services, seed money for clinics, and UMWA influence over health care—would vanish. Miller and Huge conceded everything in return for a BCOA promise that health benefits be "guaranteed."[11] In that stroke, they gave away the UMWA's most effective draw for organizing nonunion miners—the promise of cradle-to-grave health care.

Although miners struck when their contract expired on December 6, what actually occurred was an employer lockout. BCOA was willing to keep its mines closed through February because of poor market conditions. Demand for coking coal was off, both abroad and at home. U.S. Steel's hard-line vice-president, J. Bruce Johnston, a BCOA bargaining leader, was certainly aware of that. Anticipating a strike, a *New York Times* editorial of December 7 said, "Companies have built up three- and four-month coal stockpiles." A long strike would draw down stockpiles; demand would tighten; prices would rise. Wall Street understood the situation. Jack Kawa of Wheat First Securities told one meeting of industry analysts that fall that BCOA's best strategy would be to "prolong the strike as long as possible . . . before the government intervenes." He concluded BCOA "probably wouldn't mind a six-month strike, and with some justification." The Carter administration realized a long stubborn shutdown was likely. Labor Secretary Ray Marshall said after the strike was settled that the administration "from the outset . . . [understood] that both sides in the coal dispute were prepared for a 60-day strike."[12]

The Carter administration added its own volatile chemistry to the negotiations. The administration was not of one mind on the coal talks. Marshall favored the miners, but he didn't have Carter's ear. The White House staff, led by domestic policy director Stuart E. Eizenstadt, opposed concessions to strikers and advocated intervention to resolve the strike. Energy Secretary James Schlesinger and trade adviser Robert Strauss carried the BCOA standard in White House debates. The White House conduit into the negotiations was Harry Huge, one-time college roommate of Carter's labor liaison Landon Butler, and Miller's closest adviser.[13] In many respects, Huge replaced Miller as the UMWA's chief negotiator, and was largely responsible for the two proposed contracts that miners pitched into history's wastebasket.[14] Although Carter officials worked different angles on the negotiations, they shared a single interest in getting miners back to work. As the strike wore on, Carter's embarrassment mounted. He seemed floundering and indecisive. The White House cared little about equity in the contract offers, only that the strike end and the president exit, if not as a great statesman, at least not as a fool. The White House wanted labor stability as much as management, but Carter didn't want a long, bloody free-for-all to achieve it. He knew the UMWA had to win something in return. Therefore, the White House took steps even before negotiations opened to bolster Miller's bargaining ability. To that end, Wayne L. Horvitz, new director of the Federal Mediation and Conciliation Service (FMCS), "maneuvered behind the scenes to strengthen the internal organization" of the UMWA by recruiting "support staff and consultants for the union on the ground that a well-organized union was a prerequisite for successful negotiations."[15] Horvitz tapped Marvin Friedman of Ruttenberg, Friedman, Kilgallon, Gutchess and Associates of Washington, D.C., to act as a consultant to the union. Friedman, however, had had no previous experience or background in the coal industry.

The White House strategy to break the BCOA-UMWA impasse was to get one company to negotiate separately with the UMWA and use that contract as a template for the BCOA hard-liners. Months before the negotiations, William Hobgood, who worked with Horvitz at FMCS, identified Pittsburg and Midway (P&M), a subsidiary of Gulf Oil, as the wedge company. P&M is a midwestern surface-mining company that traditionally followed BCOA contract terms but was not a member. Hobgood hoped P&M would cut a deal with the UMWA before BCOA did to force the P&M terms industrywide. The UMWA was ready to negotiate P&M terms with any BCOA company individually. Hobgood's gambit carried with it the threat of breaking up the BCOA, which some in the BCOA favored but which Consolidation Coal opposed. Hobgood and Huge worked the P&M angle in December and January while BCOA sat on its demands like a catatonic hen.

A powerful factor in this confrontation was the news media. The longer the deadlock, the more important it was for the press to explain it to the public. As long as reporters simply wrote the news—that is, what differed from one day to the next—the bizarre pantomime that was passing for negotiations could continue. Had reporters looked behind the self-serving statements of the various parties, public opinion might have created pressure on the BCOA to make a reasonable offer. The battle for the hearts and minds of the press was not waged between well-matched adversaries. In any strike, at least two parties are involved in the dispute. In this case, there were four—miners, UMWA negotiators, BCOA operators, and the White House. Each tried to shape reporting to benefit its own interest. That's normal; smart reporters expect it. But the competition for favorable coverage was, in effect, rigged against the miners. Their story was being reported for the most part in simplistic stereotypes, or not at all.

BCOA had finally learned the fundamentals of public relations by 1977. In 1974, the UMWA had buried the owners—and reporters—with position papers and the data to back them up. In 1977 the numbers were in the other camp. BCOA had gone so far as to hire a vice-president for public relations—practically a revolutionary act in the coal industry—named Morris L. Feibusch, an affable, easygoing man who often dined with reporters. In subtle ways, Feibusch defined the terms of the negotiations for the press. He was plausible; he would readily admit that some coal operators came on like our simian ancestors, but he conveyed the idea that they were in the minority, that the industry was firmly in the grip of people so positive in their thinking that no responsible reporter could write of them otherwise. And few reporters did. The UMWA, by comparison, was often silent or put out confusing signals, providing little guidance for reporters looking for contrasting views. It was a bit like trying to judge a debate between Howard Cosell and Harpo Marx.

In defining the news for their own purposes, the operators had clear advantages over the miners. The case of Ralph Bailey vs. The Communists is an interesting illustration.

In mid-1977 the new business editor of the *New York Times* held a series of private luncheons with various industrialists. One of these was Ralph Bailey, chairman of the board of Consolidation Coal. The meeting was so successful that Bailey hosted a rematch at a private dining room in the headquarters of Consol's parent company, Continental Oil. To this second meeting, reporter Ben A. Franklin was invited. Responding to questions about the coal industry's labor relations, Bailey said the trouble could be traced to a handful of communist miners affiliated with the Revolutionary Communist Party (RCP) who had organized the Miners' Right to Strike Committee. The business editor thought this was a good story and told Franklin to get on it. Franklin had covered the coalfields for more than ten years out of Washington. He had written about miners, the union, and the industry off and on throughout that time. He knew that for Bailey to blame a few communists for the problems of a generally backward industry was, as he reportedly told friends, "complete bullshit." Nevertheless he wrote the story that Bailey and the business editor wanted, which ran in the *Times* on November 25, just two weeks before the old contract expired. Under a Madison, West Virginia, dateline, Franklin wrote that red-baiting had arrived in the coalfields "on the strength of the discovery of a handful of Maoists in the mines." Franklin did not argue that the RCP miners were the sole source of the industry's problems. But the timing of his story could only help to create confusion over whether miners really had grievances or were simply allowing themselves to be manipulated. The presence of a few avowed revolutionaries in the mines was not news in the coalfields. RCP members had never concealed their ideology from either the press or their fellow workers. The *Times* had had three years to write about them, but didn't get around to it until Bailey suggested they were news. The *Times* story followed within a few days similar stories in the Marshall University student newspaper, the *Huntington Herald-Dispatch*, and the *Charleston Gazette*. Franklin noted that Bailey's company "has compiled informal files on the activities of the Miners' Right to Strike Committee," but made no mention of how its chairman had engineered those "files" into a *New York Times* article. Obviously it was in Bailey's interest to divide the rank and file by planting the impression that all would be well if only a few Commies could be liquidated. Anyone who believed that would be less likely to ask what might be wrong with management or with the terms of a new contract.

In theory, union negotiators and the rank and file are on the same wavelength and have the same story to tell. But Miller and his membership were not. The rank and file had no "official" voice. Miners had no press secretaries to explain their views to puzzled reporters. Each miner became, at least briefly, the spokesperson for all the rest—a situation that didn't work well at all. The most important point that many miners wanted to make was that they were rejecting the proposed contracts in their entirety. So they burned them. That was a quick, telling way to make a point, especially to TV reporters, who knew that they would never get sufficient air time to attempt a more thorough analysis anyway.

Miners quickly learned that fire attracts TV crews, so they would light up at the first sign of a network camera. Soon it became a ritual that trapped both miners and reporters. Miners would shout generalities about the allegedly worthless document they were torching, and reporters would jot down something about the ugly mood. The specifics went up in smoke. On the news that night, the spectacle could only communicate to the public that miners were violent, irrational arsonists.

That's not to say that the UMWA negotiators, who were nominally equipped with press spokesmen did much better. The press staff that came to the UMWA with Miller in 1972 were long gone. In 1977 reporters soon learned that the official press representative, Paul Fortney, knew very little about what was going on in the negotiations and was authorized to say even less. The union's negotiators were generally unavailable to reporters. Those few reporters—such as John Hoerr of *Business Week*—who had direct access to union negotiators (access based on years of keeping in touch with these sources) found that the negotiators themselves seemed to be at sea: confused, bewildered, often bitter about Miller, Huge, or their own window-dressing roles. These scattered, off-the-record revelations contributed to the tone of the coverage that characterized the union representatives as beyond hope. Notably, not one industry negotiator was as candid with reporters about the internal disarray on his side.

On February 6, 1978, BCOA and UMWA negotiators announced a tentative agreement following a two-month shutdown. It had not come easily. Miller was a chronic absentee from the bargaining table. Sometimes, nobody knew where he was or what he was doing. Huge had eased into control. BCOA suffered strains of its own. "Three different types of companies [oil, steel, and coal] comprised the BCOA," FMC's Horvitz wrote afterward, "and their varying corporate objectives led to distinct, and often conflicting, labor relations policies and priorities."[16] Roderick Hills of Peabody Coal, the BCOA's biggest producer, was a newcomer to coal politics. He favored decentralizing the industry-wide bargaining structure that Consol dominated.[17] On Christmas Day, Hills organized a secret meeting with the UMWA and Horvitz to end-run the official BCOA neogtiators. He failed, and by January 1, U.S. Steel and Consolidation Coal had isolated him and were still calling the shots.[18] Some smaller companies and independents wanted to stop the lockout and sign a contract. They fumed and fidgeted, but couldn't budge the big boys. Consol's Bobby Brown and U.S. Steel's Johnston negotiated the February 6 offer, which seemed to reward their dogmatic rigidity. Johnston, a former steel negotiator and vice-president for employee relations, had nothing but contempt for union miners and only a little less for coal leaders who had allowed labor to gain the upper hand. He is reported to have told a BCOA executive meeting: "It is our fault we haven't trained this union [the UMWA]."[19] In the February 6 offer, the

UMWA agreed to every one of BCOA's major demands—penalties for striking and absenteeism, bonus-for-production incentive plans, medical copayments, company health insurance, no right to strike, no pension equality, and no safety improvements. The "sweetener" was a 37 percent increase in the UMWA's economic package. Miller put his political credibility on the line by supporting the offer without reservation. He said "I want the membership to understand we were damn persistent. And . . . my own decision is that there ain't no more to get."[20] This proposal was so bad that the UMWA bargaining council rejected it on a straw vote, 33–3, the next day, and formally rejected it on February 12, 30–6. The bargaining council—all elected officials—had no trouble hearing the chorus of "Nos" from their constituents.

At that point, Jimmy Carter threatened everyone with some unnamed "drastic action," including everything from arbitration and seizure to Taft-Hartley injunction. On the 20th, Bill Hobgood's gambit unfolded when the UMWA and P&M announced a tentative contract. Hobgood had worked out the language with Merlin Breaux, director of industrial relations for Gulf Oil. Under intense White House pressure, the UMWA's bargaining council voted 25–13 to offer the P&M terms to the BCOA or any individual company.[21] The P&M offer was only slightly better than the February 6 proposal. It brought the average hourly wage to $10.20 at the end of the third year. Employers could discharge strike leaders and picketers but not miners who simply stayed home. Health care was converted to company insurance with a $300 to $700 deductible. Although the P&M offer included much of what BCOA wanted, BCOA hard-liners felt it did not measure up to their craftsmanship of February 6. Consol and U.S. Steel saw the P&M ploy as a step toward decentralizing the BCOA, which directly threatened their influence over industry labor relations. Tails will not wag this dog, Consol barked; BCOA rejected the P&M terms. Carter then invited top officials of U.S. Steel, Continental Oil, Bethlehem Steel, National Steel, and Pittston to the White House. He threatened federal seizure if they didn't accept the P&M framework that day. BCOA tried to work out a last-minute change to allow local-by-local votes on incentive plans, but the UMWA declined. BCOA reluctantly accepted the P&M terms.

The P&M maneuver had bigger stakes than getting another troublesome UMWA- BCOA contract out of the way. It tapped the desire of some in industry and government to replace "industrywide" bargaining with either company-by-company agreements or "pattern bargaining" in which a couple of companies cut a template for the rest. Coal producers could be regrouped in several different ways: underground from surface, metallurgical from steam coal, or UMWA district from district. The oil industry used pattern bargaining and wanted its coal subsidiaries to follow suit. Roderick Hills of Peabody Coal supported the P&M initiative and its logic because he felt the strike was "to a very large degree the necessary result of a union and an industry that have tried to maintain a monolithic, heavily structured collective bargaining relationship that simply does not fit today's industry."[22] Hills saw three sectors, underground metallurgical coal, underground steam coal, and surface steam coal: "What this means . . . is that we must have some separation of the industry into three parts for a great many purposes—including separate bargaining over the problems that have exacerbated the poor bargaining relationships that now exist." Hills was right that today's coal industry is composed of noncompetitive parts, but the principal motive behind pattern bargaining was that it placed labor at a disadvantage. Decentralized bargaining in the 1920s contributed to the UMWA's collapse. Pattern bargaining weakened the unions of American oil workers. Logic suggested it would not work any differently in the coal industry.

Harry Huge clearly understood the implications of the P&M decentralization strategy. Huge wanted to protect his standing with the Carter administration, so he did Hobgood's bidding even though he knew the P&M gambit was bad for the UMWA. He read the operators' intent correctly five weeks before the contract expired when he told a reporter: "The operators want to abandon national bargaining because their position would be stronger in negotiating local or regional contracts where the union would have less clout."[23] Without Huge and Miller's vice-president, Sam Church, Hobgood's strategy would probably have died in preliminary discussions. Had any of the parties to the P&M deal bothered to sound mine-worker opinion it would have been clear that what floated at the White House would sink in the coalfields. None did. Seven hundred UMWA miners at P&M voted down the offer. Then on March 5, the UMWA rank and file clobbered it, 2–1.

The Washington big shots were stunned and enraged. Who were these mule-headed hillbillies anyway? Why, indeed, did miners reject the March 5 offer, which, after all, had taken some of America's better minds at least six months to arrange? The simple answer is the right one: the P&M offer wasn't very good for miners. The 31 percent wage boost amounted to almost no gain in real income when inflation, deductibles, and tax-bracket creep were figured.[24] Management could still fire a wildcat picketer (but not a miner who stayed home—a slight moderation of BCOA's original demand). BCOA retained the deductibles and company insurance, which were the object of much grass-roots anger. Most of the UMWA convention's recommendations were absent. Important ones like the right to strike, pension equality, inflation protection, improvement in the grievance machinery, and safety were either excluded or included without substance. The offer seemed an arrogant affront to miners who had not worked for three months. A sellout.

The negotiators and the press reported the rejection another way. UMWA miners were "in a state of virtual anarchy and open rebellion against [their] president," said the March 6 *Washington Post*. Another reason, many reporters said, had to do with the miners themselves. Ben Franklin of the *New York Times* on March 5 probed the "Celtic ethnicity . . . the Anglo-Saxon, not to say Druid, heritage of southern mountain coal towns" for clues to their behavior. Rejection was explained in terms of miners being "inbred" (*Newsweek*, March 20, 1978). As *Time* put it, young miners are "independent, outspoken, and not addicted to regular work." *Newsweek* of March 6 wove these scraps of muddleheaded sociology into one crazy quilt that covered everyone: miners were "a breed apart from the rest of the populace—clannish and fatalistic, wary and independent, hell-raising and violent, promiscuous and enduring.[25] This bizarre analysis helped to persuade the public that miners alone were to blame for the impasse and drew attention away from BCOA's lockout strategy, which was, at the very least, partially responsible. Although Miller's idiosyncrasies slowed the negotiations, BCOA's own gear-grinding also impeded forward motion. Hard-liners had so torqued up industry hysteria that few in the BCOA were willing to counsel moderation for fear of being seen as "soft" or traitorous. The more reasonable companies had little influence on negotiations. A handful of top coal executives—not the BCOA negotiating team—retained final authority to approve contract terms. UMWA negotiators felt they were dealing with the BCOA's second string and reasoned they might get better terms if the chief executives were forced to the table. (That eventually happened in mid-March.)

The March 5 rejection forced Jimmy Carter to take the "drastic action" he had threatened two weeks earlier. His choices were: (1) do nothing, on the assumption that even coal miners and coal operators would eventually make a deal; (2) seize the mines and negotiate; (3) seek a Taft-Hartley injunction. Doing nothing would have made him appear weak and at the mercy of events. Seizure would have settled the strike, the soonest, and *probably* the most fairly. Assuming Congress authorized seizure, the White House would have had to make a reasonable offer directly to UMWA miners; otherwise, they would continue to strike and humiliate Carter. Miners were in no mood to swallow a BCOA pill in a White House capsule. A Taft-Hartley injunction was supposed to force strikers back to work under threat of fines and imprisonment.

Within the administration, opinion divided over seizure or injunction. Robert Strauss, a veteran Democratic fund raiser and Mr. Fixit, Energy Secretary Schlesinger and Eizenstadt favored Taft-Hartley. A small delegation of coal-strike observers met with Eizenstadt on March 7 to discuss President Carter's options following the contract's defeat. The strike was fast becoming the President's "equivalent of the Vietnam war," I wrote in a memo to Eizenstadt. "The Administration finds itself about to fight the wrong war in the wrong place on behalf of the wrong side. The president has three options: get tough, get entangled, or get out." The delegation lobbied for federalization leading to an equitable contract, a get-tough position with the BCOA and a get-reasonable position with the UMWA. Eizenstadt said miners had no claim to sympathy in the White House and thought an injunction was not inappropriate. Labor Secretary Ray Marshall, on the other hand, leaned toward seizure but later changed his mind.[26] Few in the White House believed miners would end their strike because of a Taft-Hartley injunction. Carter merely hoped the injunction would show that he was acting boldly.

Carter invoked the emergency provisions of the Taft-Hartley Act on March 6, 1978, a day after the UMWA rejected the P&M terms of February 24. The administration received a temporary restraining order (TRO) without any trouble. Carter had to prove that the strike was producing a national emergency for a permanent injunction to be issued. That gave the White House pause; Carter officials knew the strike was actually having little impact in the economy. Nonetheless, they were determined to effect a settlement through a Taft-Hartley framework, so they set about fabricating a justification. This involved stampeding public opinion with predictions of economic catastrophe. Indeed, the government's TRO memorandum argued not that the strike had created an emergency as of March 6 but that one would result if the strike dragged on. The Carter White House used the media as a megaphone to whip up a panic against the strike.

Back in October and November 1977, Washington had been confident that the coming strike would not affect the nation's economy. "It looks like the UMW could strike forever and there would be a lot of coal," a government official told the *Wall Street Journal*.[27] Many reasons were given: the dwindling percentage of the nation's coal mined by UMWA members; tremendous stockpiles; plans to transport non-UMWA coal into strike areas; and the capability of "wheeling" electricity from one region to another. Helen Dewar in the *Washington Post* quoted Labor Secretary Marshall as saying that a UMWA strike "wouldn't be a national emergency." She concluded that "government and industry officials are saying that it would be about three months before the nation's energy supplies would be seriously affected," although some states might feel the pinch sooner.[28] All through December and January bland confidence prevailed in Washington and was duly reported. Even a full two months into the strike, John F. O'Leary, deputy energy secretary, testified before the House that "the coal shortage caused by the strike . . . has not reached critical proportions yet, and there is no need for President Carter to declare a national emergency."[29]

Just as the UMWA's bargaining council was rejecting the first tentative agreement, these headlines appeared: "Coal Supply Falls, Foes of Pact Grow";[30] "Coal Walkout's Impact Mounts on Wide Front";[31] "Carter Orders Plans to Send Coal to Areas Hit by Miners' Strike."[32] Through the beginning of March, reporters wrote that the nation's economy was "on the brink of a serious setback . . . if the government doesn't get coal moving again."[33] "Cities Turn Off the Lights as They Run Out of Coal," the *Times* claimed.[34] *Time* reached for the sky: "Entering the Doomsday Era: Shutdowns and Blackouts Loom as the Coal Strike Rumbles On."[35] The only hard news in these stories was that some high-volume consumers of electricity were voluntarily conserving power and that a few states were thinking about imposing mandatory power cutbacks.

The White House, industry, and coal-state politicians encouraged the press to extrapolate layoffs at a few plants and cutbacks at individual generating stations into a national pattern. The day after the UMWA bargaining council formally rejected the first contract, the *Post* sketched a picture of miners throwing a monkey wrench into the great engine of democracy: "Federal energy officials say there is little they can do to alleviate the effects of the coal strike, which now threatens to throw hundreds of thousands of people out of work in America's industrial heartland in the next few weeks."[36] The source of this scenario turned out to be Senator John Glenn, (D., Ohio), who predicted that nearly three-quarters of a million workers might be idled by the strike in Ohio alone. *Post* reporter J.P. Smith did not question the reliability of Glenn's numbers and quoted them as established fact. Glenn's projections were flapdoodle, but within a few days others were outshouting him. On February 15 the *Post* quoted Ohio officials as predicting 1.3 million layoffs in Ohio by the end of February. On February 20 *Newsweek* quoted unnamed administration officials who promised that if the strike did not end by March 1 "power shortages could idle at least 4.5 million workers." The *New York Time*'s top business reporter, Leonard Silk, described the effects of the strike as "alarming" and "immediate" on March 9. He quoted William Nordhaus, a member of Carter's Council of Economic Advisers, that unemployment would rise sharply in March "with 3 to 4 million unemployed above existing unemployment of about 6 million by mid- to late April—and that would be only in the mid-Central region." The cold breath of calamity seemed to be sweeping across the land. Well into March, the *Wall Street Journal* was sticking with Secretary Schlesinger's numbers (a million laid off by the end of March, 3.5 million by the end of April) without attempting any analysis of how those numbers were derived.

But what of reality? The day before Silk and Nordhaus weighed in, the *Times* carried a story reporting that only 22,000 layoffs could be attributed to the strike. On March 20, *Time* reported that "so far, coal-related layoffs number 50,000—far below the hundreds of thousands some had predicted." Even in West Virginia, where the strike shut off more than 90 percent of coal production and a 30 percent mandatory power cutback was ordered, William Robbins reported that "most companies were coping without major dislocations" and quoted the plant manager of a glass company: "The immediate impact on us is none."[37] The actual high mark appears to have been about twenty-five thousand strike-related layoffs—a significant figure, but hardly Armageddon. The layoffs appear to have been mostly of short duration.

A little-heralded report issued after the strike from the General Accounting

Office—*Improved Energy Contingency Planning Is Needed to Manage Future Energy Shortages More Effectively*[38]—revealed that the White House knew that neither power shortages nor extensive unemployment were occurring even as its spokesmen peddled false forecasts to the press. GAO found that "there were no power blackouts or shortages attributable to the coal strike . . . , [that] all power needs were successfully met by the [utility] system; in fact, generally more power was available than was expected." Although some states, "imposed mandatory curtailments for short periods of time . . . , only one state [Indiana] actually enforced the curtailments . . . and the economic effect on consumers was minimal." Even the steel industry, which is heavily dependent on UMWA-mined coal, traipsed through the strike with little discomfort. Spokesmen for U.S. Steel, Armco, Bethlehem, and others said that bad winter weather rather than the coal strike restricted steel shipments.[39]

How did the administration concoct a prediction of 3.5 million unemployed when it knew utilities were providing all essential power to industrial consumers? The White House had two ways of monitoring strike-caused unemployment. The first involved a weekly telephone survey of 1,000 national companies with more than 1,000 employees each in the eleven states most dependent on coal. The Bureau of Labor Statistics (BLS) did the surveying for the Department of Energy. BLS reported to the White House, according to the GAO, "that there was very little unemployment directly related to the coal strike . . . [and that] severe weather conditions were more often blamed for unemployment problems than the . . . strike."[40] At no time did BLS tell DOE that more than 25,500 workers were off work because of the strike. Moreover, the facts as the Energy Department knew them were quite clear: coal production was not even cut in half as of March 8; weekly coal production had climbed slowly but steadily since late January; by the middle of March, about 15 percent of the electricity consumed in the eleven states most dependent on coal was being wheeled from adjacent power grids; and, finally, western coal was moving east of the Mississippi to offset any shortages.[41] The White House also used a computer forecasting model that projected unemployment based on best-case and worst-case assumptions. Best-case assumptions predict what may actually happen and reflect actual circumstances during the strike. A worst-case scenario projects the most unfortunate consequences that can be foreseen. Schlesinger plucked the 3.5 million number from the worst-case hypothetical extreme. Administration officials never explained to reporters that their forecast of apocalypse was based on "what they knew to be a relatively unreliable assessment method."[42] GAO found that "the hard data available to the Administration showed minimal unemployment attributable to the strike, the continuing upward trend in coal deliveries, no major increase in strike-related violence, and continued availability of electric power."[43] Throughout January and February the press broadcast Schlesinger's disinformation. No reporter ever asked BLS for the results of its telephone survey. Others who knew the facts—the UMWA, BCOA, utility associations, state agencies, and Marshall's Bureau of Labor Statistics—kept quiet. The public interest, it seemed, required the public to be misled. Carter obtained the Taft-Hartley TRO only by showing federal judge Aubrey W. Robinson the cynical, baseless computer forecasts of James Schlesinger.

Finally, by the end of February and early March, some reporters began to detect the White House con. Their stories began to juxtapose conflicting predictions. George Getschow of the *Wall Street Journal* recalled:

> The White House wanted the media to play up that false image, that a national calamity was imminent, and the press accepted their statements because of our tendency to seek out the sensational in a story. When the White House began issuing press releases about the national emergency, we'd call utilities around the country to make sure. Most didn't think it was there and said so. However, some, like Duquesne Light, were crying wolf on their stockpile figures to give the impression of an emergency.[44]

After the strike, many publications informed their readers that they had not been drawn to the brink after all. In an editorial, the *Washington Post* noted, "The startling thing about the strike, as it wore on, was the minimal disruption that it caused."[45] The *Post*'s Art Pine argued on April 1 that the "political side" of the White House had portrayed the strike "in the worst terms—in part to maintain pressure on the two sides," despite the caveats of economists that the projections were "grossly exaggerated." *Fortune* got half the post mortem right: "The picture painted in newspaper headlines of a nation brought to the verge of massive power curtailments and layoffs was greatly exaggerated."[46] But *Fortune*'s Faltermayer did not tell his readers about the phony government estimates that fed the headlines. The *Wall Street Journal* explained the doomsday estimates in a different way. Urban C. Lehner reasoned that Carter really believed Energy Secretary Schlesinger's prediction of 3.5 million unemployed and "intervened too quickly."[47]

As predicted, miners ignored the temporary restraining order. The coal operators didn't. They knew the injunction would fail, which would push Carter toward seizure and perhaps a contract like the one in 1947 that gave miners a fair deal. Political considerations—not economic ones—might force Carter to federalize the mines.This prospect was worse than settling the strike with the union's rank and file. The rejection of the March 5 contract had undermined support for Consol and U.S. Steel's rigid posture, and BCOA "moderates" asserted themselves. Two old boys—Stonie Barker, Jr., president of Island Creek Coal (a subsidiary of Occidental Petroleum) and Nicholas T. Camicia, chairman and chief executive officer of the Pittston Co.—took over negotiations from Consol's oil man, Brown, and U.S. Steel's Johnston. Stonie and Nick were walking in downtown Washington a few days after the March 5 vote when Stonie turned to Nick. " 'Hell, Nick,' " the tall, round-faced Mr. Barker said to his companion, " 'let's go over and see the union boys and talk this over.' " It had taken three months for the operators to stop giving orders and begin talking. Nick and Stonie phoned Harrison Combs, the union's general counsel, who had been consistently receptive to BCOA's demands for "labor stability." The union people insisted that Camicia and Barker bench the junta types, Johnston and Brown. Another "moderate," William V. Hartman of Peabody Coal, was added. The old- boy team did not insist that Miller absent himself from the negotiations. When the final talks took place in mid-March, Miller had long since ceased being anybody's stumbling block except his own.

Negotiations accelerated when Judge Robinson denied Carter a permanent injunction because he found no evidence of a national emergency. Camicia and Barker shelved BCOA's most antediluvian "labor stability" demands in the proposed contract they announced on March 14. This offer—a 38 percent increase in total compensation—boosted the wage gain a nickel to $2.40 over three years. Highest-paid miners—a relatively small number—would get $73.32 per day in the first year instead of 1977's $55.68. The offer eliminated the inflation escalator in the first year and imposed a $0.30 per hour cap in the second and third years.[49] When inflation, taxes, and medical deductibles were counted, real income might rise 5 percent to 9 percent if markets firmed enough to provide full-time employment.[50] (While editorial writers painted this as inflationary, Consol's Bobby Brown later said that "management rights won in the agreement will produce offsetting productivity increases" to the UMWA's 38 percent economic package.[51]) For pre-1976 retirees on fixed pensions, monthly payments went up $25 to $275 in the first year instead of taking three as under the February 6 offer. The new proposal cut the number of annual hours needed to qualify for pensions from 1,450 to 1,000.

The March 14 contract no longer allowed management to fire miners who honored a wildcat strike or to fine workers and cut off medical benefits. Penalties against chronic absentees were dropped. However, appended to the contract was a Memorandum of Understanding that said that Ruling 108 of the Arbitration Board would be in force during the contract. This fine-print section gave employers the right to discharge or discipline miners who advocated, promoted, or participated in a wildcat.[52] ARB 108 had been handed down just seven weeks before the December 6 expiration, so miners did not know how it would be used. BCOA took some of the sting out of the 108 appendix by allowing foremen to settle disputes within twenty-four hours with the understanding that no precedents would be set. This commonsense change promised to reduce grievances and arbitration cases, the latter having cost each side about $4 million since 1974.

BCOA kept the clause that allowed mine-by-mine plans awarding cash for extra tonnage. The UMWA had opposed "incentive" plans over the years because they encourage risky shortcuts and make it harder to negotiate normal wage increases. The bonus plans were tied to increases in either production or productivity. The new language required a majority vote among those voting to install a plan. Miners could not rescind approval once it was given. The plans were not allowed "to lessen safety standards as established by applicable law and regulations." However, this language only required the mine operators to obey the law: he was supposed to do that anyway. Some plans would be linked to fixed goals, allowing a certain number of disabling injuries before the bonus was threatened. The pressure on miners and management to set a high cutoff number was evident. The numbers game also creates incentives for both sides not to report injuries. Even more ominous were the health implications of the bonus plans. No penalties were triggered if dust levels increased. Because health impacts "are hard to measure on a day-to-day basis, they are not to be considered at all. . . . Miners will be encouraged—and will encourage each other—to mine without proper ventilation and water sprays. Twenty years from the day they pocket a couple of extra fifties, they'll get their real bonus in the form of . . . pneumoconiosis, bronchitis and emphysema."[53] A few dozen mines authorized bonus plans after 1978. Markets for BCOA coal were so soft that management had little need for additional tonnage. The consequences of cash-for-coal bonuses are likely to appear when BCOA sales pick up in the late 1980s. If the U.K. experience and common sense are reliable guides, the bait of extra cash will produce tonnage at the expense of miners' health and safety.

While BCOA planed down some of its roughest edges, Barker and Camicia would not reverse the shift to company medical plans for all 160,000 working miners and 6,000 post-1975 retirees. The UMWA Funds continued to cover only the 82,000 pre-1976 pensioners. When they die, the last scrap of UMWA influence over coalfield health care will vanish. Private insurance has little interest in building a coalfield medical infrastructure, raising the quality of health care, or maintaining clinics and hospitals organized over the years by the Fund. One hospital in southern West Virginia shut down entirely, and at least sixty clinic doctors left, in the two years following the Funds' original cutback in June 1977. Camicia and Barker promised to "guarantee" health benefits, but they did not spell out what benefits were guaranteed. The contract disingenuously referred the curious to "trust documents" on file at the UMWA Funds in Washington, D.C. The principle of "guaranteed" benefits was a good one for UMWA negotiators to have won. It promised no-cut health benefits regardless of BCOA output and profits, but no miner knew whether the guarantee would hold up.[54]

The March 14 proposal retained the deductibles the Funds had initiated in June 1977, which the UMWA conceded in two earlier offers. But the maximum annual deductible for working miners was reduced from $700 to $200, with a $100 deductible for pensioners. Copayments promised to save BCOA between $70 and $75 million a year. The miner-paid copayments applied only to physician care and medication. Health analysts believe this discouraged people from seeking preventive physician care, resulting in delayed treatment and, ultimately, higher costs. Copayment reduced demand for physician services in other systems—Stanford University Group Health plan and the Saskatchewan plan—when it was imposed.[55]

Miners read the March 14 offer with their bellies rubbing their backbones. Three months out had won them less defeat, not more victory. The loss of the Funds could not be made up in other areas. BCOA had cleared the bargaining table of all the rank and file's priorities—the right to strike, pension equality, refinancing the Funds, and safety. Miners could not continue to battle their own union leaders, the White House, the news media, and employers.[56] Their strike could not create enough political or economic pressure to split the coalition that opposed them. Utilities were getting nonunion coal and "wheeled" power; most stockpiles still had forty to sixty days. Continued striking would cost miners more than they could gain. Had Miller been capable of rising to their defense, miners might have backed him even then. But as Joe Taylor, a local union official in Ohio said; "They realize Miller is so stupid they can't get anything better."[57] And Frank Brunk, an eighty-nine-year-old pensioner from Charleston, said, "The miners just don't have confidence in Miller. Lewis would have raised more hell than a jackass in a tin barn."[58] This barn held its share of donkeys but none was raising hell.

So on March 24, miners accepted the contract, 58,802 to 44,457. Each time they had said "No!" they followed constitutional procedures.[59] Yet observers said contract rejections constituted anarchy and too much democracy.[60] That logic falsely argues that the obligation of democratic citizens is to obey their leaders whether or not they represent their wishes. The equivalent of "America—Love It or Leave It" is "The Contract—Take It and Shut Up!" Had UMWA officials been better representatives; had the White House refrained from using deception to engineer antiminer contracts; had BCOA bargained in good faith; had the press done a better job of handling a complex story—the 1977–78 strike would have been shorter. Union democracy would not have been blamed for industry intransigence. And the American public would have a clearer idea about why the coal industry's historical narrative is so punctuated by exclamation points.

The 1977–78 strike was a watershed in coal politics that recast the balance of power between labor and management. It helped to maneuver Miller into retirement. It created a political vacuum at UMWA headquarters, which Miller's vice-president, Sam Church, filled by making peace with the operators as John L. had done in 1950. It precipitated changes in the BCOA, which adopted a form of pattern bargaining in which a contract acceptable to three companies was made to fit all members. The strike revealed the amateurishness of the Carter administration. Finally, the strike showed miners their weakness despite their solidarity. With less than half of the U.S. production under UMWA contract, a strike no longer created a crisis. Even a 111-day strike did not draw down supply to the level of demand. One investment observer noted the benefits to suppliers of the strike: "It's remarkable that after [this] strike the market is as soft as it is. Thank goodness we had such a long strike, otherwise we'd be awash in coal."[61] Coal consumers worked around a UMWA strike by building extra stockpiles, wheeling electricity, and buying nonunion coal. The oil and steel companies who dominated BCOA were quite happy to allow miners to strike themselves into submission. Oil companies were perfectly willing to sell as much oil at $35 or $40 a barrel while miners struck, and steel wanted to bring down labor costs in coal to make its products more competitive. Some, like Wayne Horvitz of FMCS, interpreted the BCOA's intransigence as an expression of the coal industry's backwardness.[62] But intransigence was simply a reasonable business posture.

The new balance of power put the UMWA in a terribly weak position after 1978 and produced another inequitable contract, a contract so bad that another insurgent was able to oust a conservative union president on a platform of reform and justice in 1982.

11. The Interwar Years: A Phony Peace

The coal operators got "labor peace" after 1978, but it took a recession to do it. BCOA tonnage slumped, and coal prices fell in constant dollars. Demand for metallurgical coal dropped sharply because of world-wide cutbacks in steel production. Lagging electricity output cut BCOA sales to utilities. Power plants in the Midwest shifted about 50 million tons of their annual buy to low-sulfur non-BCOA coal from the West to comply with air-pollution limits.[1] Falling oil prices limited the market for BCOA steam-coal abroad. The erosion of BCOA markets left between sixty to seventy thousand coal miners without work in the mid 1980s. Thousands of others worked irregularly. Exhausted, broke, and in many cases unemployed, UMWA miners laid low after 1978.

Industry and union leaders began to speak of the need to cooperate, to settle differences peaceably and carve out a bigger share of the U.S. energy pie for coal. They realized that profits and jobs depended on the ability of BCOA-UMWA mines to deliver coal on schedule. Arnold Miller, the UMWA's distracted president, allowed his vice-president, Sam Church, to take the lead in working out a new relationship with the BCOA. In a joint interview in Westmoreland Coal Co.'s magazine, *Triangle,* Church and BCOA's Joseph Brennan sounded the need for labor peace. Church said the labor-management "atmosphere is now problem-solving."[2] Brennan agreed. "Avoiding a strike [in March 1981] is our number one objective . . . but it is part of our overall objective—which is to promote the use of coal over the next 25 years. . . . Where there are things we do fight about, we have to do that in a professional manner geared to making sure our growth is not hindered."

Labor peace was certainly what the Carter administration wanted. The president had wrestled with energy policy since Day One of his term. Coal was the centerpiece of his effort to limit America's consumption of oil and slow down nuclear power. It was cheap, abundant and capable of being burned directly for steam or synthesized into a gas or liquid. But coal also brought substantial environmental and human damage, safety and health costs, and UMWA miners who created havoc with orderly supply schedules. Carter tried to balance his desire to boost America's coal consumption with the need to protect the country's environment and miners' safety. This balancing act was struck on dozens of coal issues, each of which tilted one way or the other depending on which interest-group coalition prevailed. Taken as a whole, Carter's policy zigzagged toward a coal-based energy future. (Critics mistakenly explain this in terms of his vacillation rather than as the inevitable product of competing interests.) In labor relations, on the other hand, the president saw no reason to balance the need for coal with equity for the very workers who had embarrassed him. The administration looked for ways to facilitate a status quo truce between the BCOA and its workers.

During the 1977–78 strike, Carter announced the formation of the Commission on Coal, whose purpose was to look into matters

> pertaining to productivity, capital investment, and the general economic health of the industry, collective bargaining, grievance procedures, and any other such aspects of labor management relations as the Commission deems appropriate. Health, safety and living conditions in the nation's coal fields. The development and application of new technologies to the industry. The impact on the coal industry of federal regulations and such other matters as the Commission deems appropriate.[3]

The idea of a commission arose in February 1978 when Carter was fumbling for any light in the dark tunnel of the coal negotiations. Originally, the commission was to undo the knots that tied coal to industrial turmoil. Earlier presidential investigations produced frank, insightful reports, which, had they been followed, might have measurably improved labor-management relations in this century. But following the 1977–78 strike, it became clear that miners had stopped wildcatting. The White House realized that weak demand had brought "peace" to the coalfields. The last thing Carter wanted was a presidential commission asking hard questions and stirring up old animosities. The commission that had been set up to assess the labor-management relationship would lay on the president's desk a report that discussed everything except that.

Carter appointed West Virginia Governor John D. Rockefeller IV to chair the commission. Rockefeller was neither impartial nor uninterested in the final recommendations. Since his election in 1976, he had lobbied on behalf of the coal operators, particularly strip miners, to tilt federal policy toward coal and weaken federal surface-mining regulations. He had publicly urged UMWA miners to accept the February 6 contract offer (which the bargaining council rejected out of hand), describing it as "trying to meet those human needs that have faced our miners across the years—this agreement does meet those human needs and meets them well." Rockefeller wanted industry and UMWA leaders to end wildcat strikes, which he saw as coal's main liability in securing future sales. The other members of the commission could be counted on not to disturb Rockefeller's agenda: W. Dewey Presley, a Dallas banker and board member of Continental Oil was a "neutral" commissioner; Willard Wirtz, a former secretary of labor and Washington lawyer, was the other "neutral" commissioner;[4]

Jesse Core, a former steel executive and Penn State professor represented the operators; and Marvin Friedman, whose experience in coal was limited to counseling Miller during the strike, was the UMWA appointee.[5] Both UMWA and BCOA did not want the commission meddling in their on-again, off-again relationship.

The fundamental problem in the coal industry is its chronic inability to tailor supply to fit demand. Excess capacity has destabilized supply, weakened producer finances, precipitated strikes, and encouraged inefficiencies. The commission ducked this basic issue by taking the simplistic but politically popular approach: increase coal demand as much and as fast as possible and the mismatch won't be so bad. The commission had no interest in considering questions whose answers might require its private sector constituency to change basic attitudes. Its report provided no analysis of why the industry has alternated between strikes and repression and ignored the 1977–78 strike completely. No effort was made to assess the impact of oil company ownership on coal's industrial relations. The commission suppressed at least one of its reports on collective bargaining, which recommended pattern bargaining; abolishment of the UMWA bargaining council (or replacing rank-and-file contract ratification with bargaining council ratification); appointment of a permanent mediator and contract arbitrator; and separate bargaining on health care and retirement benefits.[6]

Instead, the commission proposed timid superficialities: (1) 1981 negotiations should begin a year in advance; (2) union leadership should stop wildcat strikes; (3) labor and management should try to improve productivity; and (4) the two sides should establish a labor-management committee for "continuing action on the broad range of issues of mutual concern."[7] The commission cloaked controversial subjects in gossamer vagueness. Strong statements were made as abstractions. For example the Commission said, "Coal companies not prepared to operate safe mines and miners not prepared to observe safe mining practices have no place in the industry."[8] This is a nice but meaningless statement in the absence of a formula for determining which companies and miners are unsafe and a proposal for removing them from the industry. Similarly, the commission recommended that coal companies "make available at fair market value" land suitable for housing and that local governments give "serious consideration . . . to [the] selective use of the powers of eminent domain to acquire lands otherwise unavailable for housing development."[9] This might have been something other than rhetoric if the commission had used its resources to help local governments plan coalfield housing. When the commission really supported an idea it spelled out how to go about implementing it, as with its proposal to convert more than one-hundred oil-burning electric-generating plants to coal.

The commission laid before the American public a *Coal Data Book* and a pictorial essay, *The American Coal Miner.* The former pulled together facts and figures from government and industry sources that had apppeared elsewhere. *The American Coal Miner* did not explain why miners were angry. Its point was to demonstrate how much happier and better off they were than in 1947, when the Boone Report documented the need for across-the-board improvements in coalfield living conditions, particularly medical care.[10] *The American Coal Miner* focused on living conditions rather than on working conditions or collective bargaining. By showing how community conditions had improved—indeed, it is inconceivable that they would not have—Rockefeller produced an upbeat report for his fellow Democrat in the White House. *The Miner* said: "The real story of the American coal miner is this: even in communities with problems of poverty, the working coal miner has risen to the status of a middle class citizen, and ought not to be regarded as a member of an impoverished class."[11] Why, then, the ordinary reader might ask, had his fellow middle-class citizens in the coal mines "threatened" national security only months before? The Rockefeller Commission seemed to suggest that designer jeans and a pickup truck were now the measure of one's class position in America. Had the commission traced coal's political history since 1947 rather than compare snapshots of community life, it might have enlightened the American public.

The commission waxed specific on only one subject: pumping demand for coal. It proposed that 114 oil- or gas-burning utility plants shift to coal and certain kinds of boilers be prohibited from burning oil and gas. Federal grants were to cover the $15 billion cost of conversion and new construction. An estimated two million barrels of oil per day might be saved by 1990.[12] The Carter administration submitted an oil-backout bill in March 1980 to convert only fifty power plants at a cost of $12 billion to save one million barrels per day by 1990. Carter scratched the others from Rockefeller's fantasy list because for technical, economic, or logistical reasons they could not be switched. Most of the 15 to 40 million tons or so of new coal these converted plants were to burn would come from the eastern United States, home of the UMWA, BCOA, and Rockefeller's own coal-dependent West Virginia.[13]

Sam Church and the leaders of three coal associations said they too thought oil-to-coal conversion was vital to our national security.[14] Utilities must be switched to coal, environmental laws implemented in a "practical and sensible fashion," synthetic fuels subsidized, exports encouraged, and federal coal R&D increased, they said.[15] There's nothing surprising—or out of line—in a union president endorsing the products his members make. But union leaders in coal had rarely been able to distinguish between working with management for the benefit of their members and selling out their members for their own benefit. The National Coal Policy Conference in the late 1950s and 1960s was notorious for being a joint industry-labor lobby that promoted industry at labor's expense. UMWA leaders after 1978 should have approached a common lobbying effort with care and shrewdness. Instead, they backed every pie-in-the-sky plan the operators said might get a few more mines in operation.

The problem with the coal-for-oil strategy was that it is not necessarily the cheapest, most environmentally acceptable or efficient way to reduce oil imports. Switching some oil-burning utilities made sense. But a $12 to $15 billion conversion subsidy is no bargain compared with alternative strategies—conservation, small hydroelectric construction in New England, "unconventional" domestic sources of gas and oil, and certain biomass technologies. For $12 billion, Carter could have outfitted four million households with $3,000 solar collectors and cut the equivalent of one million barrels of oil per day within a few years. High oil prices were the driving logic behind conversion, so that when world oil prices fell in the early 1980s, the commission's rationale dissolved. Moreover, no one had ever asked the Rockefeller Commission to devise a plan for ending oil imports; indeed, the commission never evaluated or compared other strategies.[16] It offered not a scrap of cost-benefit evidence to prove that coal-switching was America's wisest strategy. The commission's panacea—MORE COAL—was a political answer, a special-interest answer. MORE COAL drew labor and management together. MORE COAL promoted West Virginia. MORE COAL advanced Rockefeller's political career. The commission's budget of more than $2 million was used to lobby the public on behalf of one segment of the private sector and the politicians who served it.[17]

The commission included a lukewarm endorsement of existing environmental legislation. It recommended that "the accelerated construction of new coal-fired utility and industrial boilers, the reconversion of coal-capable boilers, and the use of coal-oil mixtures . . . meet the provisions of the Clean Air Act," and that "the surface mining required for current and recommended expanded coal use meet the provisions of the Surface Mining Control and Reclamation Acts."[18] The industry's position was that it could live with environmental laws but not the interpretive regulations of EPA and the Office of Surface Mining. That position became the commission's: "With respect to increased coal use and protection of the environment, the differences that now exist must give way to accommodation."[19] Environmental protection—not coal expansion—needed to "give way to accommodation." Where and how the environment was to give way was not specified. The commission claimed there would be little net change in current air pollution if new clean-burning coal plants replaced existing "dirty" oil-burning plants. True enough, but this assumes current pollution costs are acceptable—that's a likely 7,500 to 120,000 annual deaths from airborne sulfur oxides alone.[20] Surely an assumption of that scope deserved some explanation; the commission provided none. UMWA vice-president Church was more forthright in attacking federal environmental regulations that constrained coal demand: "As far as we're concerned, some of these laws [surface-mining regulations] are really ridiculous. OSM [Office of

Surface Mining in the Interior Department] could actually come in and shut down 80 percent of the mines the way the laws are written."[21]

Coal blamed the air-pollution regulations of the Environmental Protection Agency for the demand slump in the late 1970s that idled eastern miners. EPA and what Church described as the "EPA lobby" (presumably messianic environmentalists in hiking boots and thin girls from the Ivy League) were convenient scapegoats. Eastern coal's slump after 1978 had more to do with *how* utilities chose to comply with EPA's regulations than with the regulations themselves. The slump was also caused by other factors—consumer conservation, economic decline in the northern industrial states, foreign competition in steel and autos, short-sighted investment policies by domestic steel companies and utilities, coal transportation bottlenecks, and the industry's own blemished record of supply interruptions. Still, coal operators had grounds to argue that federal regulations hampered coal *demand*. However, neither the operators nor the commission could demonstrate that regulations affecting the *supply* of coal—those dealing with mine safety, surface-mining reclamation, water quality, leasing, and land use—should be altered in light of the industry's historical ability to mine more than it could sell and excess production capacity of 15 to 20 percent. Yet the commission endorsed an "accommodation" of supply-side environmental rules.

Church made no effort to persuade the commission to strengthen federal mine safety and health programs. Once the UMWA-BCOA antiregulatory ball began moving, it tended to roll over all cost-increasing regulations, not just a few "excessive" ones. Safety was a hatchet Church buried with the BCOA in the months after the 1978 contract. UMWA and BCOA submitted testimony to the commission that said "Further improvements in health and safety can be accomplished only by a new and realistic approach built upon a foundation of a cooperative effort designed to accurately identify problems of safety and health and the means for their solution."[22] One key union representative said this meant: "The union is less aggressive on safety these days because everyone in the national office wants to keep things as quiet as possible in the coalfields."[23] In September 1979 BCOA and the UMWA announced that henceforth they would hammer out all differences on proposed federal health and safety regulations *before* each commented on them publicly. This undercut a basic premise of federal rule making: independent analyses of proposed regulations from all affected parties. Rather than press for regulations stronger than those government proposed, the UMWA implied it would settle for less in the name of industrial cooperation. The UMWA backed away from or settled short on several mine safety and health regulations after 1979. When mine fatalities and injuries rose in 1979, 1980, and 1981, Church blamed state and federal mining officials, not the industry, not himself.

As the Rockefeller Commission polished its prose in the winter of 1978–80, Sam Church laid plans to oust Arnold Miller, whom the 1977–78 strike had broken and discredited. A month after the strike, *Business Week* reported: "Miller has turned over the day-to-day running of the UMW to Sam Church, Jr. . . . Church, who even chairs board meetings, 'is becoming Miller's Haldeman,' says one insider. 'If you need something, you go to Sam.' "[24] Between the March 1978 contract ratification and Miller's resignation in November 1979, Church established an independent identity in the eyes of public officials and coal executives. Realizing that Miller's weakness created a vacuum that everyone wanted filled, Church projected himself as one who could provide stability and strong leadership. This sat well with anti-Miller critics from both the Boyle and MFD factions who formed Church's political base. He lined up the Patterson faction on the board. He also took control of the UMWA's now-76 percent interest in the National Bank of Washington and entrenched friends on its board. He actively encouraged the Rockefeller Commission to facilitate industry communication through him rather than Miller.

By the middle of 1979, Miller was on the brink of political and physical collapse. His board was rebellious. A recall petition drew thirty-one thousand signatures. Most of the rank and file thought he "was so sorry that the dogs wouldn't bark at him," in the words of one miner from McDowell County, West Virginia. His top aide, Frank Powers, a savvy West Virginian who handled administrative and political chores, preserved Miller's hold on office. Powers, a Charleston native in his mid-thirties, had worked off and on again with Miller since black-lung-strike days. Although of the same generation, Powers had never been an "insider" among the "outsider" staff who had worked for the MFD and the first Miller administration. Miller's erratic behavior often forced Powers to act as a surrogate president, which soon led Church and the anti-Miller forces to aim their attack at him. When Miller was hospitalized in July, Church encouraged the anti-Miller members of the board to isolate Powers, whom he saw as his main obstacle to becoming president. Board members called for Miller to fire Powers, and several days later, he resigned.[25] Miller now faced his vice-president, his board, and his members alone. He split publicly with Church about a month before he resigned. "I think it's evident [that Church wants the presidency],' " Miller said, " 'but there ain't no vacancy yet.' "[26] Miller shuttled in and out of the hospital four times in the last months of 1979. He could no longer tolerate stress and emotional pressure. Church understood Miller's vulnerability, and "turned up the jets to drive him out of office or into a grave," Powers claims. Miller was hospitalized with a heart attack in November 1979. The union's convention was less than a month away, and he was in no shape to chair what promised to be a political gullywasher. Church proposed that Miller retire as president emeritus with full salary through the end of

1982.[27] Church asked legal handyman, Harry Huge, to bear this offer to Miller's hospital room. Huge quickly switched allegiance to Church and persuaded Miller to accept the offer.[28]

On November 16, 1979, when Miller resigned after seven years, he could claim some achievement. He had led the only successful rank-and-file movement for union democracy in American history. He and his allies had thoroughly reformed the UMWA and, in the process, showed that American unions are not necessarily imprisoned by the iron laws of autocracy and conservatism. He could claim a major role in negotiating a benchmark contract in 1974. But Miller did not wear well over the years. He had not grown into the job. Reporters found him bitter and disillusioned in 1979. Union reform, Miller said, had "gone too far."[29] Ben Franklin caught the bathos in Miller's story:

> Mr. Miller's inability to lead his membership . . . is all the more poignant because he has, quite literally, set his people free. After his election in 1972 at the head of a reformist ticket, Mr. Miller managed a nearly complete restoration of democratic processes to an autocracy. But that revolution called for a dynamic leader. Mr. Miller's betrayal was that he was weak, laconic and leery of confrontation. Ultimately he became suspicious of his most loyal, if impatient friends.[30]

Miller drowned in the backwash of strikes, industry resistance and union politics that his own waves had made. He lost his original reform bearings. Yablonski, had he survived, would have handled union politics better, but the undertow of rapid change would have pulled at him too. Only the most extraordinary and farsighted leader could have survived—let alone shaped—the economic turbulence of the mid-1970s. The union needed a person who could change tactics rapidly without abandoning basic commitments to union democracy and aggressive collective bargaining. Yablonski, independent and courageous though he was, would probably have fallen short, too. Miller was often saddled with blame for events beyond his control, but one responsibility that he alone bears was picking Church as his vice-president. That choice reflected how far Miller had drifted from the ideas that had taken him from Cabin Creek in the winter of 1968–69 to fight for black-lung compensation. He had only himself to blame for Church—and reportedly did so with regularity in his retirement.

Church nudged a sick benefactor out of the UMWA presidency in the same way John L. had done fifty years before with Frank Hayes. A tobacco-chewing Virginian in his mid-forties with an icebox build and a matchbox temperament, Church reproduced Lewis's girth, but not his stature. He grew a beard after a year or so into his presidency that was reminiscent of Henry VIII's. Church was elected district field representative in 1973 with Miller's endorsement after working as a mine electrician and local union officer near Big Stone Gap, Virginia. Church never disguised his antireform roots: "My local union voted 98 to 2 for Boyle. . . . I did not support the reform platform. I was on the other side."[31] Some of Miller's "radical" staff brought him to Washington as a troubleshooter in the contract office. Church quickly identified Miller's insecurities. He once demonstrated his loyalty by punching staffer Rick Bank as he sat at his desk after accusing him of leaking information on Miller to Jack Anderson.[32] Miller soon promoted Church to be his assistant. When Miller could persuade no one else to join his reelection slate in 1977, Sam agreed. Church had little standing among the rank and file. His philosophy fit with the business unionism of most of his predecessors. He told the delegates at the 1979 convention: "Without labor unions, revolution would be the only alternative. . . . Revolution would be a certainty. Communist tyranny would be the order in America."[33]

What Church lacked in vision, he made up in ambition. Lewis looked to Shakespeare for inspiration; Church turned to Machiavelli. He boasted to reporters that he had read *The Prince* " 'three times' . . . holding up his three fingers to emphasize the point."[34] He frequently threatened union officials and staff. When interoffice memos failed to communicate his meaning, he hammered out his message with his beefy fists. At one meeting of the executive board, witnesses reported, he went after Secretary-treasurer Bill Esselstyn with a whiskey bottle.

Church met UMWA members at the union's convention in Denver a month after Miller's resignation. He gathered several windfalls. Layoffs and the 1977–78 strike had quenched much of the fire in the ranks. Miller's departure denied critics an axis of unity and smoothed the way for any successor. The delegates did not want to weigh Church down with dissension, or christen his term with a replay of the fitful factionalism of the 1976 Cincinnati convention. Clark observed that many delegates sought unity "by assuming a passive, noncombative role . . . [which] translated into the almost slavish acceptance of the chairman's rulings on votes, parliamentary questions . . . , procedural issues . . . and the delegates' endorsement of nearly all administration-backed proposals."[35] Church's ties to the Boyle-Patterson faction facilitated the transition. John Darcus, vice-president of District 31 in northern West Virginia, summarized the prevailing mood: "Sam, he's new and he hasn't had the opportunity to develop any record. . . . Since the union is in such bad shape, everyone is willing to give him the opportunity to see if he could take this union and move it forward."[36] Sam read the delegates with great skill. He looked bigger and more powerful than Miller. His podium style fell somewhere between Miller's patience with democracy and Boyle's intolerance of it. He sought no major constitutional changes such as abolishing rank-and-file contract ratification. He nibbled at the edges of Miller's reforms—winning the one-time right to appoint a vice-president, affirming the president's right to appoint committee chairmen at conventions, and preserving the president's power to appoint organizing and safety staff.[37]

Church's main goal was to bring more money into the UMWA treasury. He convinced the delegates to raise monthly dues from $12 to $26.69 per month, double the new-member initiation fee to $200, establish—if needed—a *selective* strike fund, and grant the authority to assess each member ten dollars for legal expenses.[38] The executive board could activate the strike fund through an assessment of $25 per month per member. A fund of this sort could support a strike against any company that broke from the BCOA pack.

Church left the convention a clear winner. He had not lost a single important fight. The delegates reported to their locals that he seemed to be a take-charge good ol' boy. No one questioned him about his two do-little years overseeing safety and organizing.[39] No one asked him about his role in the National Bank of Washington. No one asked him about the partnership he was promoting with the BCOA or about the $1,800 that three oil companies—Consol (Continental Oil), Exxon, and Gulf—contributed to a $2,700 "unity" dinner the Labor Department stage-managed at the convention.

Church repeatedly invoked John L.'s memory from the chair, and the press dutifully publicized the comparison. John Moody, veteran reporter for the *Pittsburgh Post-Gazette,* said Church left the convention "as the strongest UMW president since the legendary John L. Lewis."[40] *U.S. News & World Report* of December 24, 1979 rubbed its hands in anticipation of Church's reign. Sara Fritz, its reporter, wrote that Church had John L. Lewis's same "domineering style" and "he intends to rule with an iron fist." She quoted Paul F. Clark, a labor-studies instructor from Penn State: "The emphasis on having a democratic union isn't as great now as it has been in the past. . . . The miners see more serious problems facing the union, and democracy is a luxury they can't afford." Ben Franklin said Church translated the convention's mandate into "new respect for the beleaguered union from the coal industry." He quoted one unnamed coal executive: "[Church] has proved to be very aggressive on behalf of his membership and very responsive to the needs of the industry. The industry's expectation of Mr. Church's presidency are being more than met."[41] The evidence? Franklin said industry officials credited Church "with having at least momentarily avoided spontaneous walkouts [over the lack of mine safety enforcement] by publicly proclaiming his 'outrage' that two mine workers had died in the pits in the first 15 days of 1980."[42] Church had also hired several management consultants from West Virginia University to propose administrative changes at headquarters. This limited record of achievement sat well with the coal operators, but Franklin might have asked what Church was doing to serve his members.

Press coverage of Church at the convention once again revealed how easily reporters were swayed by their collective expectations and the "news" itself. Moody, Franklin, and other reporters had given Miller's reformers a free ride in the early 1970s. They accepted the white hat–black hat division the MFD drew between themselves and the incumbents. When that simplistic perspective could no longer explain Miller's problems as president, the press interpreted intraunion turmoil as an expression of Miller's personal inadequacies. Reporters were ready for a new personality at the union's helm by 1979. They wrote about Church without a context, without examining his behavior as vice president, without fitting him into the sweep of union politics. He was "news," and they were not expected to explain the news, simply report it. If Church wanted to bolster his political base by likening himself to John L. Lewis, veteran reporters were willing to help his image. The failure of the press to interpret Church and his policies within the context of the industry's politics and economics contributed to the news media's later inability to make sense of the 1981 strike and Church's maneuvers with the National Bank of Washington.

Anyone could see the new alignment of UMWA leadership with the major coal companies after Church became president. Church repeatedly stated his feelings about BCOA in his first presidential months: "I believe that rather than fighting each other over every issue, as I think we did in the past, at least until I took office as vice president . . . there are areas that we can work together in for the betterment of our people. I want to see the coal companies making a profit because I think we'll be able to get more for our people then."[43] In the December 1979 *UMWA Journal,* Joseph Brennan, BCOA president, said he "had a tremendous amount of respect for Sam. I have found him to be a very effective and professional spokesman. . . . We can disagree without having to go to war."[44] Church responded: "Joe and I have formed a good talking relationship. . . . We both have had problems, and, in a sense, we have helped each other. . . . We don't have to holler at each other to accomplish something.[45]

Ben Franklin of *The Times* perceptively compared Church's public courtship of "Big Coal" to Lewis's industrial "hug," which embraced "major coal producers, coal-hauling railroads, coal-burning electric utilities and the manufacturers of the mining machines that . . . had automated hundreds of thousands of men out of the mines." Franklin was not unaware of the eerie repetition of coal history when he wrote:

> The new industry-labor courtship is so far largely cosmetic, but nonetheless astonishing. Last month . . . Church and the major coal trade associations—the Bituminous Coal Operators' Association, the National Coal Association, and the American Mining Congress, all alumni of the earlier Lewis alliance—staged a glittering Capitol Hill pour to persuade members of Congress that coal is okay.
>
> . . . Industry and union spokesmen are unwilling to predict whether the labor-

management hand-holding will blossom into a rebirth of what Mr. Lewis christened The National Coal Policy Conference. They leave the impression that it could.[46]

The Church-BCOA rapprochement was concerned as much with image as substance. The turmoil of the 1970s gave coal the reputation of being the Daffy Duck of hydrocarbons, a fuel whose supply was unpredictable. Coal customers knew that layoffs had quieted miners and "labor peace" would probably last only as long as job insecurity. Church's posture raised the possibility that miners might once again be controlled by a strong union leader working with management. Coal operators saw "labor peace" as one part of their campaign to repaint the public's picture of coal, which they saw as a first step in selling more coal to utilities and industry. The coal industry's image of being dirty, dangerous, and backward was deeply rooted in public opinion. The American Mining Congress paid Opinion Research $75,000 in 1981 to survey public attitudes toward mining. Respondents ranked mining at the bottom of their lists of ten industries that contributed to the American economy and "quality of life." Only the chemical industry ranked below mining in health, safety, and environmental problems.[47] Don Donnell, president of Starvaggi Industries, a major strip mine operator, lamented, "We're being portrayed as rapers of the land and maligned at every turn. . . . Mount St. Helens did more damage in a couple of days than was done in the whole history of strip mining . . . and it didn't file for a permit or environmental impact statements."[48] Donnell and other managers like him did little to further their cause by suggesting that strip-mining appeared benign only in contrast to a major natural disaster.

The UMWA and several coal-operator associations addressed their image problem by forming the American Coal Foundation in 1981. Financed with $1 million in contributions, the Foundation sponsored "education" campaigns in the media to help the industry curtail rail rates, build slurry pipelines, change air-pollution laws, and loosen health standards that restricted coal combustion.[49] Much of the Foundation's work was undermined by the contract strike that year that confirmed the industry's continuing inability to achieve stability through equitable labor relations.

With all the news stories about Church following in John L.'s footsteps at the 1979 convention, no reporter bothered to ask whether Sam was also emulating the old man's penchant for banking. John L. acquired 75 percent of the National Bank of Washington (NBW) in the late 1940s and discovered bankers could do things that labor statesmen could not. A horse-country socialite, True Davis, ran the bank for Boyle with an agreeableness that made other bankers recoil. When Boyle was spinning from courtroom to courtroom in the early 1970s, NBW gave him a $179,000 unsecured line of credit to finance his defense.[50] Miller's reformers replaced Davis and his crowd with more experienced bankers, Donald D. Notman as board chairman and Walton W. (Hank) Sanderson as president. Miller took little interest in NBW affairs as Notman and Sanderson improved bank operations for a time. But by the mid-1970s bank examiners of the Comptroller of the Currency criticized the pair for rigging short-term profits through the sale of assets and making questionable loans.[51] In June of 1976 an examiner's report criticized $47 million in NBW loans "as having potential or definite weaknesses, a level twice as high as that considered reasonable by banking experts."[52] Sanderson and Notman blamed the bad loans on True Davis, but in March 1977 examiners continued to criticize $49.3 million in NBW loans. In 1978 several large loans that Notman had backed were in default. In that year as well, "the union men under Sam Church increased their grip on decision-making at the bank . . . [and] over the next two years, the bank made some of the worst loans in its history."[53] Notman and Sanderson were canned in favor of even more pliant bankers. When Sanderson refused to help Church retire campaign debts from the 1977 election, he was dismissed.

Church, according to investigative articles first in the *Baltimore Sun* and then in the *Washington Post*, parlayed Miller's lack of interest and his own seat on the NBW's board into functional control. He had friends and contributors to the 1977 Miller-Church campaign named as directors. Ronald G. Nathan, a lawyer who advised Miller during the 1977–78 negotiations, "quickly allied himself with Sam Church."[54] He was appointed to the board and in 1979 became general counsel to the bank, billing $191,000 in fees to NBW that year. Nathan, in turn, persuaded Church to appoint two friends of his: Irving B. Yoskowitz, a lawyer, and Marvin J. Goldman, owner of a chain of Washington movie theatres and contributor to the Miller-Church campaign. Church seated two cronies from Baltimore, Willie Runyon and Wilson Lau, although neither could point to a single reason why a Washington bank would need his advice.[55] Reba Huge, a "real estate associate" in a Washington firm, took a seat; she happened to be the wife of Harry Huge, who was now handling Church's legal affairs.[56] Nathan Landow, a prominent District of Columbia builder with ties to both the Carter administration and underworld gambler Joe Nesline, was also named. Joseph B. Danzansky, ex-president and director of the Giant Food chain, became board chairman. Bruce D. Lyons, a hotshot real estate developer and client of Nathan's, and Gary Callen, an aide to Church, also joined. Eleven of the NBW's twenty-one directors in the spring of 1980 had some tie to the UMWA, mostly to Church personally.

The *Sun* and *Post* reporters exposed an incredible pattern of misuse of the bank under Church's influence. In the summer of 1980, federal examiners identified $53 million in NBW loans with real or potential weaknesses, a figure nearly eight times the size of the bank's earnings and only $3 million less than its net worth of $56 million.[57] NBW directors owed the bank $13.45 million, a sum representing nearly 25 percent of NBW's stockholders' equity in 1979. Of this sum, federal examiners classified more than 60 percent as "high-risk," a NBW proxy statement revealed in 1980.[58] Almost $5.7 million of the $9 million in new loans "criticized" by federal bank examiners in 1979 went to insiders. Theater mogul Goldman, for example, received a $1.25 million loan from the NBW five months after he contributed $5,000 to the Miller-Church reelection campaign and two months after he was named to the board.[59] No other Washington bank would bail out Goldman because of the poor cash-flow and collateral complications. Ronald Nathan asked Goldman for the $5,000 contribution, which the NBW loaned to him unsecured at 8 percent. Miller never reported the $5,000 as a campaign contribution, insisting it was a loan. Goldman never asked Miller for repayment.[60] Goldman's cash flow fell short—as professional NBW officers predicted—and the bank had to "restructure the loan in June 1979, delaying payment on the principle."[61] Another director, Bruce Lyons, fell delinquent on $3.6 million in NBW loans, and may have falsified an NBW application form to obtain another $1 million.[62] Director John W. Lyon, president of Excavation Construction, Inc., had arranged loans of $5.4 million for his companies by 1980. NBW reported in a 1980 proxy statement that only $217,000 was adequately secured when Excavation Construction filed for reorganization in bankruptcy. The NBW charged off $500,000 in loans made to Lyon and his wife in July, 1980, and he resigned from the board two months later.

Ronald Nathan parlayed his friendship with Church during the 1977 campaign into a board seat and the job of general counsel. Dale L. Jernberg, the NBW career official whom Church appointed to replace Hank Sanderson as bank president, vied for Church's ear and control of the bank.[63] Nathan won that fight. Nathan, it appears, did not borrow money from the NBW on his own. He preferred to wrap himself in partnership deals. He was, for example, a partner in the Winding Brook Joint Venture, which was building townhouses in Chantilly, Va., with a $4 million line of NBW credit. He also held a 7 percent interest worth $47,500 in a Washington real-estate venture, McPherson Square Limited Partnership. NBW approved $2.1 million on February 13, 1979 and an additional $400,000 on February 20, 1980 to this group. The McPherson Square deal was put together by Nathan's friend Jeffrey N. Cohen, to whom NBW loaned $6.6 million between October 1978 and February 1979. The *Washington Post* discovered that Nathan had also "struck a secret deal" with Mickey Nocera, a Washington businessman, for a $100,000 piece of Nocera's investments in West Virginia gas wells, to be financed with $1 million in NBW money.[64] Nocera and his partners' company, Energy Resources, failed, and the NBW loan was in default. Nathan appears not to have revealed his interest in either McPherson Square or Energy Resources, as he was legally obliged to do.[65] Nathan resigned from the NBW board in August 1980, when he was being investigated.

Joseph B. Danzansky, whom Church had named chairman of the bank in late 1978, was a well-respected leader in Democratic politics, Washington business, and professional sports. Danzansky became the link between the NBW and a constellation of Washington real-estate developers, investors, and businessmen. In the year and a half before his death in November 1979, the NBW committed more than $30 million in $100,000-plus loans to businesses or deals in which Danzansky, his family, or his business associates were involved. Some loans and lines of credit were extended to corporations like Giant Food or the Potomac Electric Power Company, which were blue-chip customers. Minutes of the NBW's executive board reveal that Danzansky and his wife borrowed $750,000. During his term, the NBW executive committee approved loans to his brother-in-law's business (more than $3 million) and various businesses in which he or one or both of his sons had an interest (more than $8 million). When Danzansky died, his estate inventory filed with the Register of Wills in the District of Columbia showed more than $5 million in assets. Included in his nineteen partnerships and sixteen stock holdings were the following that the NBW executive committee approved for loans or credit lines during his chairmanship: $1,031,500 to the Newport Motel, Inc. (Danzansky owned five shares worth $105,000); $1,350,000 to Newport Associates South (5 percent worth $54,000); $5 million line of credit to Giant Food (Danzansky, a director, held a $313,000 retirement plan, $177,000 in stock options, and $1.4 million in stock in Giant); $750,000 line of credit to S.J.R. Communications, a subsidiary of the San Juan Racing Association ($929,000 worth of stock); and $4 million to Potomac Electric Power ($2,400 in stock and a seat on the board). Other large loans were approved for Danzansky's former law firm—Danzansky, Dickey, Tydings, Quint, and Gordon—and its partners (one of whom was his son Stephen) who were involved with him in various investments. The minutes show that Danzansky did not always reveal his holdings before the committee voted on the applications. It comes as something of a surprise to find Danzansky, as the chairman of a bank owned by coal miners, investing in 115 shares worth $3,800 (and sitting on the board) of the Columbia Gas System, Inc., Can West 1973 Gas Partnership ($15,800), South Tex 1973 Oil Partnership ($24,000), and two coal ventures, Maidsville Coal Mining Partnership ($30,000) and United Coal Resources ($50,900). His debts seemed to violate federal law, which requires that executive officers of national banks not borrow more than $10,000 from them. When federal bank examiners questioned Danzansky's conflict, Nathan, as bank counsel, wrote an opinion that Danzansky was not an executive officer because he did not participate in setting bank policy. (Minutes of the bank's executive committee show that this was untrue.) The issue was dropped when Danzansky died.[66]

None of these insider deals could have been swung without the consent of Sam Church, who personified the UMWA's majority ownership in the bank. Church made and broke NBW careers. He was responsible for the departures of

Notman and Sanderson. He moved Jernberg up and, after Jernberg leaked damaging information about Nathan's deal to federal prosecutors and bank examiners, had him fired. NBW loan officers were understandably reluctant to scotch a request that seemed to carry Church's imprimatur. Under Church, a NBW directorship came to mean few-questions-asked money for your own deals and those of your friends and family. It meant making money that couldn't have been made as easily without that seat. And if your loans came due and you couldn't pay, the miners' bank didn't hassle you.

What did Church get in return? No hard evidence has surfaced that he profited from any of the loans. He did, on the other hand, touch NBW directors for campaign contributions in 1977 and 1982. His largesse with NBW assets allowed him to hobnob with movers and shakers in a city known for its moving and shaking.[67] Church, like Lewis, may have felt a need to wheel and deal with successful businessmen. Though neither was rich, the bank allowed each to walk as an equal among those who were. The bank gave Sam power, entrée, and a measure of respect.

The worst loan that Church figured in was for $4.5 million to the Laurel Raceway, a harness-racing track located between Baltimore and Washington with which Church was reportedly not unfamiliar. The FBI was investigating the track's principal owner, Joseph E. Shamy, when the NBW approved his loan in August 1978 despite a prior conviction and disbarment in New Jersey for unethical behavior. No other bank would lend Shamy money needed to repay a loan and operate the track. Although NBW officers and directors knew of Shamy's troubles and the high risk of the loan, they approved it for two reasons: "First, Sam Church, Jr., wanted it made. . . . Second, Dale L. Jernberg, who hoped to replace Walton W. (Hank) Sanderson as NBW president, wanted to please Church. Sanderson opposed the loan."[68] The story of the Laurel loan shows how Sam Church misused NBW assets and besmirched its name.

When Shamy tired of having bankers shoo him away, he hired Baltimore lawyer and political wheeler-dealer Maurice Wyatt to find a lender. Wyatt had handled patronage for Maryland governor Marvin Mandel, himself a man experienced with racing interests and legal entanglements.[69] Wyatt was a friend of NBW directors Runyon and Danzansky. Wyatt looped in two of Mandel's codefendants, Harry W. Rodgers III—who began selling insurance to Shamy's track—and Irvin Kovens, with whom Shamy and others joined in a $2.7 million Atlantic City real estate venture.[70] Church's Baltimore sidekick, Willie Runyon, hired Wyatt to handle some legal work, and Joe Danzansky's Giant Food retained him as a lobbyist in Annapolis. Danzansky told NBW president Sanderson that Wyatt would call him about a loan to Shamy's track. After meeting with Shamy and Wyatt, Sanderson didn't like the looks of their proposition. NBW's senior officials voted against the loan in August 1978. At that point, Danzansky and Runyon began to peddle the loan to Jernberg, who wanted Sanderson's job. Jernberg consulted Church, and later, on a fishing trip to Ocean City, Church told Jernberg: "If there is a way to make the loan, we ought to make it."[71] By that time, NBW loan officers had prepared a detailed critique listing ten reasons why the loan should not be made, including rumors of a $250,000 kickback in the form of a "finder's fee" that Shamy was to pay an unnamed person or persons.[72] The credit department produced a negative assessment based on inadequate cash flow, Shamy's doubtful integrity, and insufficient collateral, which one bank official characterized as "absolute crap."[73] Jernberg ordered the credit department to tone down its report and took the application to a committee of senior officers. To his surprise, they voted to reject the loan. Jernberg ordered the committee's minutes destroyed and then submitted it to the executive committee of the board. Only one committee director of the ten who were present—including Church, Nathan, Runyon, and Danzansky—voted against the $4.5 million loan on August 30, 1978. Two months later, Shamy bought $10,000 worth of tickets to a fund raiser Church and Runyon organized to retire debts from the 1977 Miller-Church campaign. As NBW credit officers predicted, Shamy didn't repay the loan. The FBI's probe into Shamy's affairs led to indictments of him and his two track partners on fraud charges in May 1979. The loan was declared in default on July 31, 1979, but NBW had to lend Shamy another $325,000 to satisfy debts that if left unpaid would have jeopardized its collateral in the property.[74] The bank finally sold the track to Washington lawyer Frank J. DeFrancis for $4.8 million, of which NBW lent $4.2 million packaged with favorable terms and sweeteners. DeFrancis offered the track and $400,000 in securities as collateral. In July 1980, a federal examiner suggested that NBW write off $900,000 as a loss on the DeFrancis loans, which the bank did.[75]

When the *Sun* series appeared in May 1980, Church was spun into his first scandal. The NBW story amazed even cynical veterans of Watergate, but mineworkers saw only badly condensed versions of the articles in their coalfield papers. Church arranged a coalfield tour, during which he denied everything. Local reporters didn't know the right questions to ask and had only a once-through understanding of the facts. Church claimed to have no control over bank hiring and firing and never to have influenced directly the approval of any loans.[76] Church told one coalfield rally he was "not a dictator" and that press "attacks" were "to be expected."[77] Two days later, however, he told a rally in Welch, West Virginia: "We own the bank. We can do what we please with the bank. It is one of our greatest assets."[78]

Church faced a stiffer audience in Washington, where a federal grand jury subpoenaed NBW records in July 1980 and federal examiners pored over the debris. About three dozen bank employees departed. NBW reported a 42 percent drop in earnings in 1980 and a 35 percent drop in 1981. Earnings per share were $5 in 1979, $2.77 in 1980, and $1.90 in 1981. Had NBW earnings in 1980 and

1981—the first two years of Church's UMWA presidency—simply remained at the 1979 level, NBW would have recorded almost $4 million more in earnings.

Church fired Jernberg and two other officers, giving the impression that Jernberg was to blame for the bad loans. Several directors resigned. The American Federation of State, County and Municipal Employees withdrew $26 million from the bank. The NBW hired a law firm—Gibson, Dunn & Crutcher—to investigate the charges against itself. They found "some procedural problems with a few loan transactions" but "no improper conduct" by UMWA officials.[79] Church then appointed himself in September to head a special committee to clean up the bank's operations!

In late October after the *Post* series, federal banking officials threatened Church with a "cease and desist" order that would have forced him to stay away from the bank.[80] Harry Huge, whose wife was an NBW director, represented Church in negotiating less severe "articles of agreement" by which the UMWA was allowed to name only eight of the NBW's twenty-five directors. Federal officials named Luther Hodges, Jr., the NBW's new chief executive and insisted on a veto over directors and issuing stock dividends.

It is not wrong for a labor union to own a bank or other financial institution. Indeed, far-sighted unions should be using their capital assets to benefit their members either directly through services or indirectly as investments. The story of the UMWA's ownership of the NBW, however, is a walk through a series of warped reflections of these purposes.

Apart from the financial losses Church caused UMWA miners, the real danger in the NBW's slide down this particular hill was that organized crime may have been waiting at the bottom. Underworld capitalists are always interested in easy loans and financial laundromats, and they have proved frighteningly capable of infiltrating unions. Nathan Landow, a Church-appointed NBW director and major Democratic Party fundraiser, has had a long and multifaceted relationship with Washington gambling figure Joe Nesline and through him with others who were associated with Meyer Lansky and the Carlo Gambino Mafia family.[81] The tortured history of the Mob and the Teamsters' pension funds is all too familiar to the American labor movement. A bank would serve organized crime even better than a pension fund. When Church ran for election in 1982, he retained Paul Locigno, director of Research and Governmental Affairs of the Ohio Conference of Teamsters, the power base of Jackie Presser, a then Cleveland-based union boss with reported links to organized crime who was selected Teamster president in 1983. Church also turned for campaign research to a paranoid Detroit-based right-wing group, Intellico, which had done anti-insurgent research for Presser and Detroit Teamster leader Rolland McMaster, who was once imprisoned on thirty-two counts of labor extortion and is an associate of Florida mobster Santo Trafficante. Church was also observed meeting during the campaign with Eugene R. Boffa, a convicted bank swindler who operated Country Wide Personnel, Inc., a labor leasing company in Jersey City, New Jersey. Boffa provided Teamster labor at below cost to employers who signed agreements with Country Wide, which received a fee of 8 to 10 percent over costs. Teamster officials approved Boffa's sweetheart deals and he had "cordial relationships" with a number of Mob-related Teamster officials, including Tony Provenzano, Russell Bufalino, William Bufalino, and Salvatore "Sally Bugs" Briguglio.[82] Had Church won election in December 1982, it is not unreasonable to suppose that Mob elements would have made a move to infiltrate the National Bank of Washington even after federal banking officials made an effort to end insider abuses there.

How quickly the UMWA seemed to revert to industrial subservience when coal demand fell and surplus capacity cheapened labor! The Rockefeller Commission created a public framework for this realignment. BCOA was receptive. For all the democratic reforms, the UMWA's rank and file could not prevent its president from abusing union assets and redirecting union policies. Church, like Lewis in the 1920s, used market softness to solidify his position within the union, though less successfully. As the 1980s began, the pressures within the union and within the industry for collaborative partnership appeared stronger than the vision of an independent union with an independent agenda.

12. The 1981 Strike and the UMWA's Struggle for Survival

Coal demand has shaped the industry's labor relations for the past one-hundred years. Labor gained when sales were increasing; labor lost when they were not. Strikes and lockouts occurred in times of growth as well as slump, however. A strike could be for a greater share of a growing economic pie, or against cuts proposed during slack demand. Similarly, lockouts have been used in both circumstances. The 1980s are a unique period in the history of the American coal industry because coal demand is reasonably strong nationally but declining in the sector under UMWA contract. Market forces threaten the union and the organized part of the industry with a kind of economic rationalization that promises long-term inefficiencies in exchange for short-term gains.

The first consequences of this process were visible in the BCOA's demand for concessions in the 1977–78 and 1981 strikes. The UMWA under Church's leadership cast its lot with management, not for lack of alternatives but for lack of vision. It became ever clearer as the new decade began that traditional labor tactics were ineffective and often counterproductive. BCOA's labor policies were as much the product of forces over which its members had little control as they were the result of events within the organized sector. The 1980s mark a time of profound change, a time that requires the UMWA to adapt if it is to survive.

In the early 1980s BCOA found itself the victim of deep changes in the pattern of U.S. coal production. Total output had increased from 603 million tons in 1970 to 830 million tons in 1980. But BCOA production from 1971 through 1980 had never equaled 1970's total of 424 million tons. BCOA mined 365 million tons in 1980. BCOA negotiators blamed disappointing sales on wildcat strikes. But after the 1977–78 strike, wildcat strikes fell by 90 percent, productivity rose, and Church provided cooperative leadership—and BCOA sales continued to decline. BCOA tonnage fell to only a quarter of national production in 1983. BCOA's problems were linked to general economic trends, the special troubles of its main customers, and its own weaknesses. Economic recession undermined BCOA's traditional markets in the North and East. Falling oil prices in the 1980s removed the economic rationale for two likely growth areas—coal-based synthetics that would replace natural gas and oil and the conversion of power plants from oil to coal. America's shift away from electricity-using manufacturing and basic industries trimmed BCOA's sales of steam-coal to utilities, as did consumer conservation. BCOA metallurgical coal sales shriveled as American steel mills fell to operating at a fraction of their capacity in the early 1980s. BCOA's exports of metallurgical coal dropped because of the worldwide slump in steel output. BCOA also had to contend with structural handicaps in its mining operations. Much of its tonnage came from relatively high-cost underground mines rather than lower-cost open-pit strip mines. BCOA mines tended to be older and less efficient than many of the new, nonunion mines that opened in the 1970s. BCOA had to contend (as did the rest of the industry) with excess production capacity, which was estimated nationally to be between 100 and 150 million tons of coal that could be mined but not sold.[1] The penalties of excess capacity fell heaviest on underground mines, where the cost of maintaining idle capacity was greatest.

As the post-1978 recession deepened, a central battleground for the UMWA and BCOA was productivity—the tons of coal each miner produced per shift. Productivity increases were essential to remaining price competitive and surviving demand slumps. As the 1978 contract drew to a close in the first months of 1981, BCOA had reason to be pleased. While productivity had fallen from 19.9 tons per worker shift in 1969 to 14.8 in 1977, following two decades of steady gain, it had risen again after 1978. The gain in the 1950s and 1960s was the product of a unique combination of factors—among others, the widespread application of more efficient mining machines and the absence of reclamation and work-place safety rules. When coal's externalized costs began to be figured in the 1970s, productivity fell to more realistic levels. Coal operators blamed the decline on federal safety and environmental regulations, expensive contracts, and strike-happy workers. They rarely mentioned how their own inefficient mining practices and poor judgment (they expanded capacity in the mid-1970s, when sales were stagnant and prices climbing) contributed to the slide. The downswing bottomed out in the late 1970s, when "the things that made productivity go down had washed through the system," Forrest E. Hill, coauthor of DOE's *Analysis of the Labor Productivity Decline,* told *Business Week.*[2] The operators learned to "live with" the safety and environmental regulations.[3] Inefficient suppliers closed as prices leveled. Companies boosted productivity through capital investment, layoffs, and getting the survivors—through bonus plans or intimidation—to mine as much coal as before.[4] Island Creek raised productivity from 2,159 tons per worker per year in 1977 to 3,256 in 1980. The company's production rose from 17.6 to 20.9 million tons, but employment fell from 8,151 to 6,412 miners. Consolidation Coal produced 43.7 million tons in 1977 with 18,855 miners, or 2,318 tons per worker per year. By

Table 2. Stock Prices for Five Coal Companies, 1979–1980

Company	17 January 1979	17 January 1980	% Gain
Pittston	19⅜	28½	47
Eastern Gas & Fuel	16⅛	24¾	54
MAPCO	29⅞	41⅞	40
North American	20¼	46¼	128
Westmoreland	12½	32½	160

Source: *Washington Post*

1980, its production was up to 45.5 million tons, but employment was down to 15,326. Consol's productivity was 2,969 tons per worker per year—a gain of 28 percent. Ralph Bailey, Conoco's chairman, explained the need to boost productivity when demand is weak: "At a time when the market is soft—as it has been for the past couple years—that's the time to move into your mines and make them more efficient. . . . At Consolidation, we have improved our productivity a little over 10 percent in each of the last two years."[5] Consol had learned that labor behavior was not the basic cause of productivity decline in the 1970s, and the key to improvement lay in managing better within the prevailing context of regulations.

The productivity gains that Consol and other companies had achieved since 1978—estimated in 1981 to be improving at about 10.5 percent annually—enabled the industry to remain profitable. Consol's Bobby Brown linked productivity directly to future sales: "Productivity determines our ability to compete in world markets."[6] Productivity improvement allowed BCOA companies to maintain profits despite lagging production. Frank Young, Conoco's director of economics, explained to reporters that in 1970 Consolidation, the second largest coal company, produced 57 million tons and made $21 million; in 1979 the company produced 47 million tons on which it made $124 million. "Oil is more profitable, but more risky," he said. "When you build a coal mine, you know you will get coal."[7] Stock prices reflected coal's solid performance (table 2). Between January 1979 and January 1980, the price of five top coal stocks rose by at least 40 percent.[8] Stockbrokers at the end of 1980 saw gains in earnings for four of these stocks (table 3).[9]

Coal's ability to turn a profit without price increases or output gains was an added incentive for oil companies to deepen their coal holdings in the late 1970s and early 1980s. The first round of oil investment in coal began when Gulf Oil bought Pittsburg & Midway in 1963. Continental Oil (Conoco) took over Consolidation Coal in 1967, and Occidental Petroleum acquired Island Creek in

Table 3. Earnings per Share for Five Coal Companies, 1979–1981

Company	1979	1980	1981
Pittston	$1.45	$2.10	$2.30
Eastern Gas & Fuel	2.27	2.40	2.65
MAPCO	3.83	4.20	4.65
North American	3.84	3.95	3.60
Westmoreland	0.26	0.55	1.50

Source: Company annual reports.

1968. Oil bought well-established operating companies, most of whose mines were east of the Mississippi and mainly UMWA. The second round occurred in the mid-1970s, when such oil companies as Texaco, Arco, Kerr-McGee, Shell, and Sun started their own coal operations, principally western surface mines. Round three came at the end of the 1970s. By that time, oil and gas companies produced more than 25 percent of national coal output, and as much as 35 percent of all steam-coal not mined by utilities themselves. (The Office of Technology Assessment estimated that energy companies would probably mine almost half of all coal used for energy purposes by 1985.[10] In that year energy companies would mine about 70 percent of the coal used by utilities—coal's biggest market—excluding that which they themselves produced.[11]) By 1982, only one of the top fifteen coal-producing groups in the United States was an independent company whose principal business was mining coal. Oil companies, conglomerates, and utilities owned the major coal suppliers. Energy companies were also major holders of coal reserves—the key to future production.[12] Electric utilities, energy companies, natural gas pipeline companies, conglomerates, and smaller oil and gas firms increased the absolute and relative amounts of federal coal they leased in the 1970s (table 4).[13] Companies that have integrated coal either horizontally or vertically now dominate the industry.

Oil ownership of both nonunion and union mines bore immense consequences for UMWA negotiators.[14] Cash-rich oil companies could pay higher than union scale to keep the UMWA out of their new western surface mines. They carried no obligations to pensioners, which allowed them to offer unusually big pay and benefit packages to their miners. They preferred to bargain separately from the BCOA. Where oil-owned companies negotiated with the UMWA, the union found itself disadvantaged at many levels. First, the union could not assemble as much financial information on subsidiaries as on independent producers. Securities and Exchange Commission rules do not require a

Table 4. Number of Acres and Percentage of Federal Leased Land Held by Business Activity Category

Category	1950		1955		1960		1965		1970		1975		1980	
	Acres	%	Acres	%	Acres	%	Acres	%	Acres	%	Acres	%	Acres	%
Electric utilities	0	0	2,000	3	8,263	6	45,363	15	132,038	18	142,077	19	163,259	21
Energy companies	0	0	0	0	0	0	9,491	3	132,274	18	138,409	18	155,024	20
Peabody Coal Co.[a]	0	0	0	0	0	0	(6,251)	(2)	(59,121)	(8)	(68,923)	(9)	62,009	8
Steel	4,993	12	14,817	19	19,888	14	34,158	11	46,114	6	49,448	6	60,015	8
Independent coal	14,584	35	25,022	33	41,557	29	77,273	25	78,297	11	58,837	8	55,410	7
Oil and gas (minor)	0	0	0	0	0	0	2,080	1	26,911	4	42,193	6	45,926	6
Unincorporated persons	11,129	27	17,618	23	25,678	18	41,475	13	78,995	11	66,515	9	43,215	5
Natural gas pipeline	0	0	0	0	0	0	0	0	0	0	32,522	4	36,317	5
Nonresource-related diversified	0	0	0	0	0	0	4,610	1	10,015	1	12,580	2	35,675	5
Kemmerer Coal Co.[b]	475	1	1,752	2	6,849	5	18,504	6	33,793	5	33,988	4	32,191	4
Metals and mining	5,009	12	5,009	7	9,266	6	17,708	6	107,504	15	118,300	15	17,620	2
Landholding	1,360	3	4,576	6	11,504	8	13,411	4	43,581	6	26,225	3	4,661	⟨1
"Other" lessees	2,907	7	3,240	4	18,288	13	39,134	13	41,153	6	37,051	5	77,861	10

Source: Office of Technology Assessment, *Development and Production Potential of Federal Coal Leases*, p. 373.

Note: Numbers may not add to 100 percent because of the holdings of uncategorized lessees, who held less than 2 percent of land under lease at any analysis date.

[a] Numbers in ( ) tabulated in metals and mining category (1970 and 1975) or independent coal company category (1965).

[b] Kemmerer Coal Co. was purchased in March 1981 by Gulf Oil Corp.

parent corporation to provide detailed data on every subsidiary. Second, a strike exerts less pressure on subsidiaries than on independent suppliers. Oil companies can carry a coal subsidiary through a strike if necessary. Third, intransigence in bargaining squeezes smaller, less capitalized suppliers and in this fashion ripens them for takeover. Independent companies in the heart of UMWA territory—Carbon Fuel, Cannelton, Amherst Coal, among others—were prepped for acquisition by long strikes. Fourth, oil management brought a unique set of bargaining priorities to UMWA negotiations in 1977 and 1981. It was the oil companies that insisted on converting health coverage to company-managed insurance and spearheaded the effort to decentralize collective bargaining.[15] Fifth, and finally, there was Conoco's Bobby Brown.

Bobby Brown, president of Consolidation Coal and lead BCOA negotiator in the 1981 talks, found himself up to his neck in coal politics after climbing the corporate ladder in the oil business. He was Conoco's man at Consolidation Coal. Brown is a pin-neat executive in his fifties who favors dark suits and thick black-rimmed glasses. He parts his still dark hair sternly just left of center, leaving a crest of waves dipping across his forehead. If a banner were to display his labor philosophy, it would read: "Never give an inch, except to gain a foot." Though lacking practical experience in coal, he had a plenitude of clever strategies to put the UMWA in its place. Brown steered Consol on a love-me-or-leave-me course after the 1978 strike. Unlike other operators, Consol made only token efforts to improve its labor relations or promote the "labor peace" BCOA sought after 1978. Two *Wall Street Journal* reporters wrote that "many union and company officials in the coal industry . . . [believe] that Consolidation's labor-relations stance these days is so tough that the company is raising temperatures throughout the coalfields at a time when the rest of the industry is making its most earnest effort ever to cool labor strife."[16] Consol's policy was to arbitrate grievances rather than settle disputes at the mine site. Most BCOA companies dismissed suits against wildcat locals after the 1977–78 strike; Consol didn't. Consol focused on watering down the "prior practices and customs" its mines had with local unions concerning working conditions. Each time a Consol supervisor tried to clip some customary practice a dispute ensued, and Consol went to arbitration. One journalist concluded that Consol was "attempting a radical restructuring of social control within the unionized sector of the coal industry . . . [intended to] break down the traditional identification of workers with their unions so as to replace it with identification with the corporations."[17]

In February 1980 Consol fired a local union president and two mine committeemen at its Four States mine in northern West Virginia who had protested the company's refusal to give an older miner a temporary light-duty job. In similar circumstances in the past, Consol had made an assignment of this sort more or less routinely. Four States was an old, inefficient mine that sold to customers that were stockpiled to their eaves. One miner figured Consol set up a wildcat: "Everybody was on edge. . . . There'd been a lot of pressure. Everyone knew they [Consol officials] were trying to instigate something."[18] Six thousand Consol miners in the area walked out to protest the firings. Four States Local 4060 had never been troublesome, and this was the first time it ever sent pickets to shut down other mines. Church appointed a two-man UMWA commission to negotiate an end to the strike. "The UMW leader said he repeatedly talked with B. R. Brown . . . suggesting the pickets and Zemonick [Michael P. Zemonick, the local union president] be disciplined but not fired."[19] Instead of backing his fired members, Church asked Consol to transfer or reinstate them as new employees at other mines.[20] Consol turned him down; an unmistakable signal that "Consol is going into the next round of contract talks with a tough, hard-line position."[21] An arbitrator upheld one of the firings and then ten others for picketing. Brown, Church, and a federal judge finally settled the strike in mid-April. Contempt of court fines were dropped and Consol promised not to fire any more miners. The eleven who had been discharged were left dangling.[22]

Consol's behavior stunned Sam Church. The *UMWA Journal* ran a front-page story with the lead: "Consolidation Coal Company's obstinate, inflexible attitude turned a minor job dispute into one of the largest unauthorized work stoppages in the history of northern West Virginia.[23] The *Journal* confessed it couldn't figure out "why the company chose to allow a minor incident to escalate into a major confrontation."[24] Eldon Callen, a UMWA spokesman, laid the blame directly at Consol's door: "President Church had them back to work once and they worked for a week, but Consol was so intent on creating the disturbance that they went out again."[25] The Four States strike seems to have been a social experiment in which Consol was testing miners' ability to sustain a multimine wildcat strike and Church's willingness to back his members.

Consol was also laying down new rules for its fellow operators. Brown had been humiliated in the 1977–78 negotiations. U.S. Steel, Consol's traditional ally, had had to lead the bargaining much of the time. Roderick Hills of Peabody Coal, a lawyer who knew little about the coal business from a company that played to mixed reviews within the industry, had even challenged Consol's premier role. Brown's visceral antiunion hatred so stalled negotiations that his fellow operators finally benched him. As a result, Consol left BCOA, announcing that it intended to negotiate separately with the UMWA. Consol may have believed that other companies would disaffiliate, thus ending industry-wide bargaining. U.S. Steel used Consol's grand exit to bully BCOA into allowing the old guard to recapture leadership.[26] The big companies, in consultation with Consol, then voted changes in the BCOA bylaws to "lure Consol back into the industry's collective bargaining fold."[27] One anonymous trade source observed:

"Consol wants to run the show, pure and simple, and they cut a deal to do it."[28] Another noted with some amazement that Consolidation helped "restructure an organization it didn't even belong to."[29] Once BCOA approved the new bylaws, Consol reaffiliated. The change empowered three big companies—Consol, U.S. Steel, and Peabody—to negotiate on behalf of the BCOA in 1981. The three would coordinate their work with a new eleven-member CEO committee, nine of whom automatically represented BCOA's nine largest producers, with two at-large elected members. The bylaws say the CEOs must be the heads of *parent* companies, not their coal subsidiaries. This meant only two of the eleven CEOs represented companies in the 1981 talks whose primary interest was producing coal.[30] The new rules authorized this committee to accept a contract in the name of the entire BCOA membership. Companies representing about 40 percent of BCOA tonnage were excluded from participating in collective bargaining.[31] Consol's power play worked. The smaller companies were snapped back into line.[32] The impulse to decentralize industry-wide bargaining was blocked, at least for a time. Negotiations in 1981 were back in the hands of those with the least to lose from a long shutdown.

BCOA-UMWA talks opened in the fall of 1980 against an economic backdrop of inflation, high interest rates, unemployment, and cutbacks in federal domestic spending. Across the northern industrial heartland, executives in many basic industries talked of "labor concessions." Brown was confident that Church would go along with cuts and get the rank and file to approve them. He and Church kept up a drumroll of statements about the necessity of a strike-free contract. Jay Rockefeller had promised that the 1981 negotiations would be a "different Kabuki dance."[33] The negotiators would get a proposal to the miners close to the March 27, 1981 expiration date, and "labor peace" meant they should ratify it. The enigmatic aimlessness of Arnold Miller was part of the past, reporters said; Sam Church, a pragmatist, would demonstrate how miners would rally behind a forceful leader modeled after John L. Lewis. More than twenty years after his retirement, John L.'s persona still defined UMWA leaders and their political behavior. Like Lewis, Church tried to be a strongman within the UMWA while accommodating BCOA, a combination he believed would ensure his winning the UMWA's next election in November 1982.

Brown's plan for the 1981 negotiations was to agree to nothing that jeopardized the productivity gains management was squeezing out ton by ton. This dictated no give on work rules, safety, union rights, and job security. Brown hoped to use soft markets and another agreeable UMWA president to erode UMWA power in the 1980s. BCOA devised language to encourage unionized companies to market nonunion coal instead of that which they mined themselves. The easiest way of matching nonunion coal was simply for BCOA companies to sell it as their own. The "labor peace" that BCOA and Church had worked on so diligently would now be brought to bear against UMWA miners.

The union president's drive for a strike-free contract—on which he pinned his election strategy—would maneuver the rank and file behind Brown's agenda. Church even raised the possibility of UMWA miners working without a contract, although he never pressed that very touchy point. Brown assumed Church to be a strong-arm leader who could bull a take-away contract through a ratification vote.

On March 23—four days before the contract ran out—the negotiators worked out an offer which the UMWA bargaining council ratified, 21–14, after debating for eight and one-half hours the next day. Miners struck on the 27th—"no contract, no work"—and a ratification vote was scheduled for March 31. Of the 100,000 miners who voted, nearly 67,000 voted "No." The BCOA choked with rage; Church was stunned. BCOA had not sent him to the coalfields empty-handed. Wages would have increased 32 percent—and the economic package 36 percent—over three years. Pensions for those who retired before 1976 (under the 1950 Trust) were increased from $275 to $315 per month in the third year. Pensions for post-1975 retirees (under the 1974 Trust) were raised. (These pensions, which were figured according to age and years of service, continued to be roughly twice as high as those covered by the 1950 Trust.) Church made much of a new hundred-dollar monthly pension he had won for widows of the 1950 Trust pensioners and a new dental plan for working miners. In return for a two-dollar monthly payment, up to $750 in dental services would be allowed annually for each miner and dependents. Church had stopped BCOA from converting pensions to company-managed plans, as was done with health coverage for working miners in 1978. He also held off the operators' demand to be allowed to schedule continuous twenty-four-hour, seven-day operation. Finally, BCOA agreed to dissolve the Arbitration Review Board. Reporters, who had been proclaiming Church the messenger of labor peace—and if not peace, pacification—and predicting a strike-free contract, read the contract for what it gave the miners, not for what it took away. This analytical error led them to misconstrue how miners would perceive the contract despite their stories that described the proposal as "overflowing with concessions," containing "every major improvement the miners had demanded," and "so full of goodies . . . that any tradeoffs with industry seem small by comparison."[34]

Two-thirds of the miners read this offer in a different light. They saw Bobby Brown's union-weakening strategy written in boldface in three basic changes. The first had to do with the scope of the agreement. Previous contracts required that UMWA members perform all mine-site subcontracting work—construction, hauling, repair, and maintenance. AMAX challenged this clause, and the U.S. Supreme Court in 1980 upheld a Third Circuit Court decision that it constituted an unfair labor practice. On the advice of UMWA counsel Harrison Combs, Church agreed to delete this protection even though other lawyers said the decision held only in the third circuit and the UMWA could negotiate around the ruling by, for example, prohibiting all subcontracting or adding a union-standards clause.[35]

A second, related change in scope involved subleasing. BCOA companies leased some of their reserves and mines to low-cost independents that were not signatories to the UMWA-BCOA contract. The BCOA company would buy its lessee's output at a cost lower than its own. The proposed agreement would have allowed BCOA companies to continue subleasing with no protection for UMWA workers or UMWA recognition at the lessee's operations.

The third threat was termination of the requirement that BCOA companies pay a $1.90 tonnage royalty to the 1950 Trusts for health and pensions on all non-BCOA coal they bought and resold. On BCOA-mined coal the royalty was only $1.38, so BCOA operators would save $0.52 on every non-BCOA ton they handled in the future. Church claimed he traded the royalty for the one-hundred-dollar monthly widows' pension. He told reporters the new pension would cost BCOA $0.50 per hour, versus $0.045 per hour for the royalty. Pensions for the widows would cost about $48 million. A $0.50-per-hour rate would generate about $123 million to cover a $48 million expense. The $.045 cost of the royalty would generate only about $11 million. Church's numbers suggested he had somehow persuaded the BCOA to commit more than twice the cost of the new widows' pension, a sum more than eleven times greater than the cost of the old royalty. Miners doubted that their employers had suddenly lost the ability to add and multiply; they wondered whether Church was giving them the real facts, and even whether he knew them himself.[36] The actual spread in costs to the operators between the old royalty and the new pension was probably closer to $0.10 per hour and $0.25 per hour respectively in the first year or two. Even if UMWA members didn't question their president's arithmetic, they could easily see how trends affected each side of the trade. Widows were a declining population. Each year, BCOA would pay them, as a group, less and less, while presumably marketing more and more non-BCOA coal. Energy and Environmental Analysis, a consulting firm in Arlington, Virginia, calculated that BCOA companies could save as much as $6 per ton or between $40 and $80 million a year by processing nonunion coal through their facilities.[37] Coupled with the new subcontracting and leasing language, miners saw the loss of the "non-signatory" royalty as an irresistible incentive to circumvent the UMWA contract. These three changes, they felt, would cost many of them their jobs and their union. *Business Week* quoted one top, unnamed industry executive that the agreement "makes up for the excesses of the past."[38] BCOA expected that "the new pact will enhance last year's 8% productivity increase in UMW mines . . . [and] gains of even half this size would largely offset the new pact's cost."[39]

Rank-and-file miners criticized other features of the offer as well. Church abandoned the cost-of-living escalator started in 1974.[40] BCOA added a forty-five-day probationary period for new workers, which would probably be used to weed out likely union activists. Although the Arbitration Review Board was eliminated, all of its rulings were continued. Absenteeism—the control of which was central to maintaining productivity—was dealt with harshly: two consecutive unexcused absences and the culprit could be canned. The proposed contract made only one minor improvement in health and safety, although the 1979 convention had endorsed about three dozen recommendations.

Coming on the heels of two years of "labor peace," the rank and file's sweeping rejection of the BCOA's offer was a slap in the industry's face. BCOA negotiators were disgusted by the prospect of bargaining with someone who appeared no more capable than Arnold Miller of selling his members a bad offer. BCOA refused to resume negotiations after UMWA members turned it down. Bobby Brown issued a statement implying that miners had been duped into voting against a good offer: "The majority of the miners voting apparently chose to listen to dissidents who were not involved in the bargaining process and who, for whatever reasons, misrepresented the agreement and assured its defeat."[41] Rejection, Brown said, "reflects a disturbing lack of discipline in the UMWA" and jeopardizes the "integrity" of the bargaining process. He was angry not so much about a shutdown of a couple of months—it would take that long to draw down stockpiles—but about the prospect of having to give ground on something like the royalty or subleasing clauses. Despite the contract's overwhelming defeat, the UMWA's bargaining council gave Church a full vote of confidence a week or so after the vote. Like it or not, they had to crew Sam's ship; if he was sinking, they were sunk. "Naturally, you hate to lose," Church said, "but you can't win them all. You dust your britches off and get back up. . . . I went fishing for a few days."[42]

BCOA stonewalled UMWA efforts to resume bargaining for more than three weeks. Church proposed seven changes in the original offer, including retention of the $1.90 royalty, voiding ARB decisions, elimination of the forty-five-day probationary period, a five-day workweek, and revision of the sub-contracting and subleasing language. Fat chance, Brown replied, BCOA's executive board had agreed that the UMWA's seven demands "were totally unrealistic."[43] His sage advice to Church was to resubmit the March 23 offer to the miners. Consol and some of the other bigger companies were prepared to tough out a long strike, but the smaller companies clamored for a settlement. About fifty independents met in late April to discuss the possibility of negotiating separately with the UMWA.[44] Brown's objective of eroding the UMWA's scope was less important to most BCOA members than getting back into production.

As in 1977–78, many reporters didn't grasp why miners rejected a proposal that both union and management negotiators compared favorably to the first wheel. The press had predicted that 1981 would see a contract without a strike. Coal bargaining had changed, reporters told the public. Arnold Miller, the pitiful soul who had caused the anarchy of the past, was now pasturing in West Virginia as president emeritus. Sam Church, the 250-pound brawler, would deliver the miners. When terms of the first offer were revealed, news stories mechanically reported the negotiators' view of it. Reporters rarely asked what BCOA had received in return, or which changes might stick in the craw of the ordinary miner. When miners voted overwhelmingly to continue to strike, the press and the negotiators looked like fools—the union team for totally misreading its members; the press for uncritical acceptance of negotiators' statements about a strike-free settlement and a generous offer.

To explain why press predictions landed so far from the mark, reporters—with a few exceptions—bought into one or more of four interpretations. The first was Ben Franklin's post mortem for the *New York Times*. The vote against the contract, he told his readers on April 2, "was interpreted . . . by deeply dismayed union and management leaders more as a cultural phenomenon than as a strictly labor movement phenomenon." Later, Franklin conjured up the "Appalachian enigma"—presumably a mix of cultural handcuffs, historical dialectics, and genetic programming errors—for his predominantly non-Appalachian audience.[45] Franklin chose not to say why the negotiators shared their ethno-anthropological speculations only with him, the very reporter who three years before had cut to the heart of that strike with his notions about the "Celtic ethnicity" of miners and the "Druid . . . heritage of southern mountain coal towns." Franklin absolved the negotiators and their proposal by substituting an explanation based on third-string socio-genetic musings.

A second interpretation was the "union politics" theory. Eldon Callen, union spokesman, told the *Washington Star* (April 3, 1981) that "officials at the union's Washington headquarters felt that some candidates for regional union offices had conspired to defeat the contract." Several UMWA districts were in the middle of election campaigns, but no reporter showed that miners voted against the contract to embarrass district-level allies of Church. The penalty to miners of doing so was so high and the return so hypothetical that no reasonable union member would buy this logic. Reporters who dished up this theory never bothered to find out whether it made any but the most superficial sense. Nor did they explain why spokesmen for the negotiators might blame rival politicians for their own whopping failure.

Bob Arnold of *Business Week* worked out a third theory. He wrote on April 20, 1981, that the contract went down because of the "democratic structure that reformers brought to the UMW in 1972." No union president had been able to supply the "sophisticated administrative, political, and bargaining skills" needed to negotiate contracts and gain rank-and-file approval. A *Pittsburgh Post-Gazette* editorial on April 6 scolded the miners for "flouting . . . their own leadership through the very voting rights obtained for them in the democratization of their once autocratic union." The "too-much-democracy" theory rested on some debatable assumptions—that true democracy requires the citizenry to endorse whatever its leaders propose; that democracy befuddles workers and their organizations; that rank-and-file voting represents excess democracy; and that democratic rigamarole obscured the obvious excellence of the March 23 terms.

Finally, there was the "fishing" theory. In their April 2 piece Petzinger and Hymowitz, *Wall Street Journal* reporters, quoted anonymous UMWA officials who suggested miners struck because "it's fishing season in Appalachia . . . and a lot of miners are off in Fort Lauderdale on vacations they booked months ago." The *Wall Street Journal* and *Newsweek,* among others, trolled this one without offering one worm of evidence to support it. Indeed, the only "miner" whom the press knew was fishing was Sam Church, who took a week off with his wife. Reporters failed to ask themselves why UMWA officials would belittle the judgment of the men and women who paid their salaries. Why, reporters should have asked, would a miner forfeit several thousand dollars for a couple of catfish? Why were union officials deflecting press scrutiny from their own mistakes?

The basic assumption of negotiators and reporters alike was that the March 23 offer was a good one. Miners rejected it because it wasn't. The matter was that simple and straightforward. A few reporters who had serious conversations with miners reported their objections to the provisions that threatened union jobs. Jim Ragsdale of the *Charleston* (W. Va.) *Gazette* quoted one local union president from southern West Virginia in his March 28 story: "The men are going to get a new set of false teeth and lose their jobs." But the rank and file's analysis—contract take-aways exceeded gains—was lost among the colorful quotes about itchy anglers and screwy hillbillies. As a result, the general public read about miners once again acting out their disruptive neuroses or simply being simple. Without a press agent of their own, the rank and file had no way of dispelling these damaging images.

For all of April and the first week of May, BCOA refused to resume bargaining, and returned to the table only when their lockout became counter-productive. The shutdown cut U.S. coal production by about half, but like the one in 1977–78, it was not terribly disruptive. The *Wall Street Journal* on May 13 reported that stockpiles remained high nationally after forty-eight days although a few plants reported only forty-five days' worth on hand. Utilities and steel companies began hedge buying on the spot market and were driving up prices. The centrifugal forces within the BCOA were threatening to spin some producers into orbits of their own. BCOA leaders now knew that Church, like Miller, could not ram through a bad contract. UMWA members appeared to be willing to stay out another couple of months. Church had to be given concessions to secure approval on any second try.

A second offer was announced at the end of May, which the UMWA bargaining council approved unanimously on May 29. BCOA agreed to retain royalties on both UMWA and non-UMWA coal and provide some protection against layoffs in subleasing and subcontracting arrangements. The forty-five-day probationary period was dropped. The UMWA agreed to extend the length of the contract for forty months rather than thirty-six. The new pension for widows was cut from one-hundred dollars to ninety-five dollars a month. Although the second offer retained some provisions that had doomed the first, it was significantly better on the whole. With 87,551 miners voting—about 12,000 fewer than before—two-thirds voted for it, and the strike ended on June 6, after seventy-two days.

The final contract tacked on a few new benefits—a modest dental plan and widows' pension—but on the vital matter of jurisdictional scope, the UMWA barely clung to the status quo. It weakened the UMWA's grip on mines under contract and would not make it easier to organize new ones. Safety and occupational health were neglected. The UMWA had not reestablished a health-care fund for working miners, who continued to be covered by company plans. A joint union-management study was commissioned to investigate the feasibility of converting pensions to company carriers.

What was most obvious, however, was that it had been the miners themselves who had held the line. Once again, miners were forced to act independent of their leadership—as they had in the black-lung strike of 1969, the anti-injunction strikes of the mid-1970s, and the 1977–78 strike—to protect their interests. Miners scorned Church for having tried to get them to go along with the first offer. They doubted his commitment to UMWA tradition for suggesting they might work without a contract. Some thought "Sell-Out Sam" was resting comfortably in the back pocket of Bobby Brown. Many thought Sam Church, whose career was punctuated with fisticuffs with fellow unionists, had taken a dive. Church's handling of the 1981 negotiations destroyed his political base among UMWA miners. They were no longer willing to give him the benefit of any doubt, as delegates had at the 1979 convention. The labor peace he endorsed was revealed as terms of surrender: Lay down your arms, miners were told, and trust to the generosity of the victors.

Church faced a presidential election in twenty months and hoped that time would dim rank-and-file resentment. Had the election been scheduled for the last six months of 1981, most observers felt, even a semicredible alternative candidate would have won. Church's chances were also threatened by the drastic fall in UMWA membership. Although the UMWA would not release official numbers, a reasonable guess is that only 120,000 to 130,000 working miners—of 200,000 to 230,000 nationally—carried UMWA cards in mid-1982. (At the beginning of 1984, the UMWA estimated that only 100,000 of its members were working.) The growth of the nonunion sector was coupled with devastating layoffs that swept through the coalfields after 1978. As many as 30,000 to 40,000 UMWA miners were out of work by 1982. They blamed Church, in part, for their troubles.

Other UMWA leaders took note of Church's political vulnerability. The first to break was Willard "Bill" Esselstyn, his secretary-treasurer, a big affable miner in his early thirties from western Pennsylvania. Esselstyn was almost alone among UMWA officials in standing against Church publicly. Church retaliated by installing his own financial officer in the secretary-treasurer's office, leaving Esselstyn with a title but little day-to-day responsibility. Esselstyn bridled over his personal situation as well as Church's conservative, company-oriented tilt. Although he disliked Esselstyn as much as Esselstyn disliked him, Church asked him to run as his secretary-treasurer in the fall of 1981. Esselstyn walked away from Church's offer and tried to launch a presidential bid. He never got it off the ground, but doggedly critiqued Church for almost a year preceding the election.

A few months after Esselstyn announced his candidacy, a serious challenge rolled out of western Pennsylvania in the person of Richard Trumka, a smart 33-year-old miner with a law degree from Villanova. Trumka had whipped one of Church's cronies, Walter "Fudge" Suba, by a 10-to-1 vote for a seat on the UMWA's executive board during the 1981 strike. Suba's deep involvement in the 1981 negotiations had angered his district. Trumka cut his teeth on MFD politics in the early 1970s and was one of several dozen young miners whom Miller, to his everlasting credit, brought into the middle ranks of the union hierarchy to seed the next generation of mine-worker leaders. Trumka worked in the union's legal department during the Miller years. He put in his mining time between schooling. He learned the coal business, as had his male kin, underground, at a Jones & Laughlin mine near his hometown of Nemacolin, Pa. He allied with a capable Vietnam veteran, Cecil Roberts, the thirty-five-year-old vice-president of southern West Virginia's District 17. Roberts, a native of Cabin Creek, quickly distinguished himself from run-of-the-mill union officials by his outspoken criticism of both the Miller and Church administrations. He was considered a militant unionist and had been sued for his role in a UMWA organizing rally in May 1981 that preceded attacks on company buildings and vehicles. For secretary-treasurer, Trumka and Roberts chose fifty-two-year-old John Banovic, recently elected president of District 12 in Illinois. All three candidates held political office on their own, and Trumka and Banovic had beaten incumbents who were tied to Church through the 1981 negotiations. Candidates needed to be nominated by twenty-five local unions to qualify for a ballot position: Trumka won, 449 to 283 for Church.[46]

Trumka and Church ran campaigns that reflected their differences in style and substance. Trumka, drawing on the UMWA traditions of liberal unionism and anticompany militancy, attacked Church for the 1981 negotiations and collaboration with the BCOA. "'We are pledged,' Trumka said . . . 'to cooperate with the industry only when it results in safer working conditions, higher wages, and better benefits.'"[47] He said of Church's bargaining record: "It's an indication of his type of leadership—reactionary."[48] In relations with the BCOA, Trumka promised he would follow a policy of "No Backward Step," a pledge to retain the status quo at the very least. The coal industry, unlike autos and steel, was not losing hundreds of millions of dollars, Trumka argued in the *UMWA Journal* (September 16–30, 1982). "They [the operators] made no effort to deny that they were a profitable industry . . . [and] I simply don't agree with the idea of negotiating concessions." In Trumka's view, Church's worst concession was the shift in the 1978 contract to company health insurance and the abolition of the UMWA's first-dollar medical plan for active miners. Trumka also criticized Church for failing to do enough to increase the use of coal. He proposed, for example, that the U.S. impose a special duty on imported goods that were subsidized or made by a labor force that worked under lower standards than American workers. The UMWA, he said, should have been fighting harder for federal assistance to coal-hauling rail systems, port improvements, and plans to use coal in place of other fuels. Trumka proposed contract changes that would guarantee workweeks, severance pay, supplemental unemployment insurance, and limitations on BCOA companies subleasing and subcontracting with non-UMWA companies.

Trumka hammered away at Church's dependency on "outsiders," who, he claimed, were running the UMWA and financing his opponent's campaign. He referred to Eldon and Gary Callen, two Church assistants; Paul Locigno, a right-wing consultant from the Ohio Teamsters' Pension Fund on Church's reelection staff; campaign adviser and former Nixon-Agnew worker Norman Beebee, with ties to a number of right-wing organizations; Larry Sherman, editor of the *American Labor Beacon* and author of a "Red-scare" report alleging communists in the Trumka campaign; Willie Runyon, the Baltimore ambulance impresario said to have loaned the Church campaign $75,000; Tommy Boggs, a fast-track Washington lobbyist—Continental Oil was one of his clients—whom Church appointed to the NBW's board of directors and who contributed to Church's fund raising; Nathan Landow, the Democratic party fat cat and NBW director who pulled away from Church after refusing to buy UMWA-owned NBW stock at a bargain price in return for a large contribution to Church's campaign; and Harry Huge, Church's Washington lawyer whose wife, Reba, sat on the NBW board by virtue of her wedding ring. Trumka promised to sweep the union clean of "outsiders" and run it exclusively with miners. This principle guided his own campaign. He refused to employ or accept the help of nonminer professionals. He defied conventional wisdom by raising all campaign funds—

almost $200,000—from the UMWA's ranks through donations, raffles, lotteries, and the sale of campaign trinkets. A miner-financed campaign, he said, would be beholden to no group other than coal miners.

Church ran on his record and devised a campaign strategy based on heavy spending. Church claimed credit for the good parts of the 1981 contract—the one that miners ratified after rejecting his first proposal. He touted the new pension for widows, dental insurance, and the one-hundred dollar-a-day wage rate paid to a small group of senior miners. He said he had stopped acid-rain legislation that "could have cost 80,000 jobs," and worked hard for legislation promoting synthetic fuels, coal transportation, conversion of oil-burning power plants, mine safety, and black-lung benefits.

Church impugned Trumka's personal honesty on two matters. He claimed Trumka deceived reporters regarding membership in Phi Beta Kappa, and that he did not have the full five years' mining experience needed to run for UMWA president. Trumka was not a member of the honorary academic society and was negligent in not correcting that impression among reporters. (Reporters for the *Louisville Courier-Journal* found a similar skeleton in Church's closet when they discovered, contrary to his résumé, that he had never graduated from high school.) Trumka's mining time logged depended on who was counting, and with what standards. Church himself and every other member of the executive board voted in favor of Trumka when the issue first arose after he won his board seat in July 1981. Robert T. Boylan, director of the UMWA Health and Retirement Fund, wrote on September 27, 1982, that Trumka's work record constituted seven years of service—two more than necessary—as defined by the 1974 Pension Plan. Reporters who had described Trumka as a Phi Beta Kappa graduate, like Ben Franklin of the *New York Times,* jumped on him for allowing them to purvey an untruth and hammered at his eligibility.

Where Trumka charged Church with being a tool of "outsiders," Church retorted that "Trumka is part of a left-wing near-commie plot."[49] Trumka and Roberts, he claimed, were themselves radicals who were friendly with the activists who had worked for the MFD and Miller in the early 1970s. To substantiate his charges, Church established links with Larry Sherman, who was affiliated with a research organization called Intellico and the Detroit-based *American Labor Beacon*. Sherman and about one-hundred others had until 1981 been members of the National Caucus of Labor Committees (also known as the U.S. Labor Party), a right-wing political cult led by Lyndon LaRouche. Sherman's faction split from LaRouche, but continued to embrace his basic political orientation, which interprets American politics in terms of a worldwide conspiracy of British bankers, communists, and antinuclear activists. Sherman prepared a report on the Trumka campaign for Eldon Callen, Church's principal assistant, in the summer of 1982 that charged: "An apparatus . . . of university professors, press personnel, community organizers, and communists and socialists of every stripe . . . is in place, either waiting for the call from Trumka or quietly working behind the scenes, directly or indirectly."[50] Among the conspirators, Intellico's Sherman included former Miller staff, Democratic Socialists, International Socialists, Communist Party USA, Socialist Workers' Party, Revolutionary Communist Party, the Association for Union Democracy, Joe Rauh, Jr., Chip Yablonski, the Rockefeller Family Fund, the Institute for Policy Studies, *Working Papers, Washington Post* reporter Patrick Tyler, the Council of the Southern Mountains, *Charleston Gazette* reporter John Bussie, and me. Dozens of individuals and organizations were named as part of a vast coordinated left-wing conspiracy. This was news to the conspirators, many of whom neither knew each other nor would exchange a single pleasantry if introduced. Excerpts from Sherman's report were published in the *Charleston Gazette* in early August. Although *Gazette* reporter Bussie scoffed at Intellico's conspiracy theory, his stories raised this "issue" one notch higher than innuendo. (The Trumka campaign obtained the Intellico report and leaked it to Bussie early in the campaign to avoid having it surface closer to the November 9 election.) Sherman's research acquired credibility from its specificity and general accuracy—such people and groups exist, and, if polled, would undoubtedly have preferred Trumka to Church. But none had anything to do with Trumka's campaign, even had they been inclined to help. Moreover, the conspirators held such disparate political views that cooperation among them was impossible, a fact no FBI agent or reader of the left-wing press could escape. Red-baiting was nothing new in UMWA politics, and Church did not have to prove anything for the tactic to have its effect. He merely had to raise the "issue" and force Trumka into denials. Church told one reporter: "Anytime I do anything, I go after it. I guess we'll do anything that is necessary to win as far as the campaign is concerned."[51]

Neither candidate articulated a clear vision of where the UMWA should be heading and a general plan for getting there. Both wanted to get the federal government more actively involved in promoting coal, and assumed that UMWA layoffs would solve themselves as coal demand grew. Their assumption was not necessarily true: coal demand could rise nationally with little benefit to the UMWA-sector of the industry. Federal programs promoting synthetic fuels, internal waterway improvements, and less restrictive federal leasing would benefit non-UMWA western mines far more than UMWA operations. Neither candidate addressed coal's historical inability to fit supply to demand, which contributed to layoffs, take-away contracts, lockouts, and strikes. Neither said much of anything about how to deal with a group of coal executives who were employees of oil companies and chemical conglomerates. Both agreed that the UMWA should recruit unorganized miners, but few ideas emerged during the campaign about how to do it more successfully than in the past. Neither offered many specifics about future bargaining demands. With his pinstripe suits and multisyllabic vocabulary, Trumka clearly came across as the more intelligent of the two. He also benefited from the fact that he was more in tune with the UMWA's membership—most of whom were under thirty-five—than the forty-six-year-old Church. Trumka greatly benefited from the deep resentment many miners carried over the way Church managed negotiations in 1981. The contract, more than any other factor, caused Church's defeat.

A clear difference between the two was how they organized their campaigns. Trumka built his candidacy from the bottom up, depending on self-motivated volunteer activists at each level. His decentralized approach rested on the effectiveness of grass-roots supporters and face-to-face campaigning. Trumka campaigned longer and more intensively than Church, despite being assigned duties at a remote strike in Oklahoma for much of 1982. Church, in contrast, ran his campaign from Washington rather than the coalfields. He raised most of his money from the presidents of other unions (who weren't sympathetic to insurgents) and private sector pals like *Penthouse* publisher Bob Guccione, who wanted access to the National Bank of Washington. Church raised more money than Trumka but spent it foolishly. He hired high-priced Washington political consultants and put too much faith in advertising. Red-baiting and attacks on Trumka's honesty never offset rank-and-file anger over the 1981 contract or Trumka's skillful organizing.

Trumka polled 79,000 votes, more than two to every one for Church in the November 9 election. Trumka won 23 of the UMWA's 26 districts, including all of the large ones save Alabama. He had humiliated an incumbent on the promise that he would provide tougher, more militant leadership. Where Arnold Miller's MFD had been the first rank-and-file insurgency in the labor movement to dethrone an entrenched incumbent, Trumka's has the potential for being the first to achieve more than temporary success.

The future of the UMWA is cloudy; a pessimist would say dark. The union has been in turmoil since 1969, tainted by murder and other troubles. Its last three presidents—Boyle, Miller, and Church—left in disgrace. In many ways, it is still dominated by John L. Lewis, whose larger-than-life bust glowers in the headquarters lobby. The UMWA would do well to demythologize Lewis, to set free the union's history and allow current leaders to escape his shadow. Miller's staff understood Lewis's complex legacy but tiptoed around it as if it were a ticking bomb. The 1973 and 1976 *Officers' Reports* to the conventions criticized Lewis for clamping trusteeships over districts and stifling membership participation, but avoided a frank discussion of his role in the 1950s. Trumka has tacked pieces of the Lewis history onto his incumbency to legitimize his policies. Lewis cannot and should not be Trumka's leadership model. Democratic unionism needs leadership that speaks on behalf of the members rather than leadership that tells members what to do. John L.'s brand of authoritarianism bewitched Boyle and Church, and led to their defeats. The test of

Trumka's leadership will be to carry the UMWA through some very hard years without sacrificing the basic democracy of the union and without allowing the forces within the industry to fracture the organization. For all the strife of the last fifteen years, UMWA members remain committed to their union. No exodus has occurred; no call has gone out for a new union. For bad as things have been within the UMWA, miners realize that their lives would have been far worse without its protection. The UMWA can be hurt by internal division, but it is more threatened by external forces over which it has exercised little control to date.

UMWA miners find themselves today in a troubled sector of the industry, one that is using its troubles to take power away from the union and its workers. Between 1970 and 1980, BCOA-UMWA coal production declined in both relative and absolute terms, falling from 424 to 365 million tons, while national output rose from 603 million to 824 million. In 1983, BCOA tonnage was down to about 200 million tons of 785 million mined nationally. UMWA contracts with non-BCOA companies covered another 115 million tons or so. About 30 companies left BCOA in the early 1980s to handle labor relations individually. Decentralization of bargaining will tax the union's resources and produce destabilizing contract differences from company to company. Unionized companies are increasing their subcontracting and leasing arrangements to lower their costs. The UMWA scored a few organizing successes in the 1980s, but on the whole, the growth sector of the industry—western strip mines—has resisted UMWA organizing efforts or signed with the Teamsters, Operating Engineers, or Progressive Mineworkers. The last two BCOA-UMWA contracts have weakened the union. The switch to company health insurance plans for working miners eliminated the UMWA Funds as a union-defined institution that promoted cohesion within the membership. BCOA is likely to insist that pensions be converted to company plans in future bargaining, which would further weaken union identification. UMWA negotiators were not able to place effective limits on leasing and subcontracting in 1981. It is not likely that BCOA will agree to severe limits in the future. With UMWA-organized mines producing only 40 percent of the nation's coal, contract strikes no longer create the pressure to settle that they once did. Utilities stockpile coal, buy non-UMWA coal during a strike, and buy power from unaffected grids.

The coal industry today is principally one of strip-mine producers. About 60 percent of 1983's estimated 785 million tons was strip tonnage. For every ton an underground miner digs, a surface miner produces two to eight tons, depending on conditions. If no more coal is mined in 2000 than in 1980 and the surface-to-deep ratio shifts to 80–20, the 151,000 deep miners working in 1980 would be halved while the 78,000 surface miners would increase by one-third, to 104,000.[52] The net effect of such a shift would be to cut the number of miners from 1980's 229,000 to about 180,000, 58 percent of whom would work in surface mines. That would be the first time surface miners outnumbered deep miners. The UMWA's decade-long attempt to organize western strip miners has met with little success, although a few mines voted in the UMWA in 1983 and 1984. The dilemma Trumka faces with western strip coal goes to the UMWA's core. If the UMWA fails to organize the West, it could become a remnant organization with a small number of working members burdened with a large pensioner population. If, on the other hand, the UMWA succeeds in organizing the West, it becomes a different union, an organization of western strip miners whose agenda and interests differ from—and sometimes conflict with—the mainly eastern, mainly underground miners who have been and are its flesh and blood. (The current legislative fight over limiting acid depositions from coal combustion's sulfur and nitrogen emissions spotlights the conflict. While thousands of eastern high-sulfur-coal miners are laid off, the UMWA has not endorsed federal legislation to ban or limit the use of low-sulfur western coal in the East for fear of antagonizing the western miners it is trying to organize.) The UMWA's best approach to western coal would be to seek legislation that requires environmentally safe local coal use. Eastern coal would be burned in the East but washed and scrubbed to the extent necessary to reduce the combustion emissions that cause public health damage and acid rain. The costs of retrofitting or replacing older coal-burning plants should be spread among those who benefit from environmental improvement, those who use the electricity, those who own the utilities, and those who mine the coal. A strong local-use requirement should be coupled with a plan to convert certain oil, gas and nuclear power plants to coal. Environmental protection and local-use rules would allow the UMWA to organize the West without sacrificing its eastern base. The union will change to the degree that western members replace easterners, but a creative balance may be achievable if the two sectors are not forced to compete against each other.

New underground mining technologies—including in situ gasification, robotics, and a new generation of mining machines—represent major "labor-saving" innovations that will, when commercialized, further decrease personnel requirements. Improvements in surface and underground mining productivity have enormous implications for the work force, as great as the two-thirds cut in employment from 1950 to 1969. Even modest annual productivity increases have major impacts when they are not accompanied by increases in demand. Were productivity to rise even 3 percent a year over the next twenty years, the U.S. miner would be producing more than seventeen tons for every ten he mined today.[53] If total output in 2000 is the same as 1980 and productivity rises 3 percent annually, employment would fall from 229,000 to 143,000, a loss of about 38 percent. Improvements in coal-transportation technology and electric-power transmission may work against the eastern underground sector by increasing the economic advantages of western strip-mined coal. Chronic overcapacity works against labor by contributing to persistent unemployment and downward wage pressure.

Finally, the UMWA faces the problem of demand. A sizeable portion of its membership mines metallurgical coal, the market for which has been cut almost in half over the past decade. More efficient steel-making technologies, foreign competition, and weak demand for steel will keep this portion of the UMWA's membership on pins and needles. Foreign markets for metallurgical coal appear limited, and demand for U.S. steam-coal fell in response to falling oil prices and cheaper sources of supply from low-wage countries like South Africa. Coal-based synthetic fuels are not likely to be a significant market for many years, and the potential for additional oil-to-coal boiler conversions is not very large.

That leaves the domestic utility market as the industry's hope for the future. The UMWA, concentrated as it is in the East, has been seriously affected by efforts to burn lower-sulfur coals to limit air pollution. More than 50 million tons of low-sulfur western coal move east across the Mississippi River each year, displacing almost 11,000 eastern mine workers, most of whom are UMWA members. Public concern over acid rain will make it ever more difficult and costly to increase coal consumption in densely populated areas, particularly in the East. Demand weakness in the early 1980s contributed to the loss of perhaps as many as fifty thousand UMWA jobs—almost one-third of the union's working membership in the mid-1970s.

If the UMWA continues on a business-as-usual course, it will continue to be elbowed to the periphery of the industry. Coal operators have taken one of two positions toward the UMWA down through the years: either they have tried to rid themselves of the UMWA and its contracts or they have insisted that the UMWA maintain labor stability through suppression. Although some managers have a broader outlook, no base of opinion exists within the industry to move toward a more equitable relationship. Demands for labor stability increase during periods of slack sales, which will make it difficult for Trumka to cut a new course. By 1990, the UMWA could easily be an industrial artifact—a union with a small membership and little influence even as the U.S. turns to coal. Unless the UMWA and its members do something different, their union cards are likely to be souvenir items in twenty years.

The UMWA must act boldly if it is to survive this array of labor-threatening changes. It must alter the framework of discussion; it must seek a broader role in energy decisions.

The union must find constructive ways of affecting demand decisions and excess production capacity. Market forces will not work to the UMWA's benefit even as coal consumption increases nationally. These forces are shaped by pollution laws, utility regulators, long-term contracts, utility ownership of coal suppliers, oil-company ownership of coal producers, federal coal-leasing policy, and a host of other factors that make the market free only in name. The market, historically and currently, creates destabilizing excess capacity and oversupply, though it keeps down the price of coal. The UMWA must recognize that suppliers have neither the vision nor, perhaps, the ability to change this situation. In the late 1920s, John L. Lewis reluctantly recognized the inability of the industry to cope with the problems created by its own structure and the peculiarities of demand. He began to advocate a federal role in coordinating coal supply and price, a role every other industrial democracy has adopted over the years. The problems inherent in federal energy planning and coordination are large and should not be dismissed. Public agencies make their share of mistakes, as those who have followed the history of the Tennessee Valley Authority, the Atomic Energy Commission, and the Washington Public Power Supply System must acknowledge. But a move in this direction is likely to benefit mine workers and the public at large.The industry's call for the forced conversion of oil- and gas-burning utilities to coal and export subsidies skirts the edge of public energy planning. Trumka openly acknowledged the need in a January 1984 interview:

> I think, for the sake of the country . . . federal intervention [in energy demand], federal planning . . . is the proper course of action. . . . This is the only country . . . in the world that doesn't take an active hand in planning for its energy needs but simply says—takes the short-sighted approach—'We'll let the market do it for us; we'll let the market determine whether twenty years from now we'll have enough energy or not, or what form it will take, and whether we'll be dependent on Joe Schmedlap's new country that took over in the Middle East. . . .' With our basic industries, with energy, I don't think you can simply say . . . national security is in the hands of the market and we trust it. The free market as everyone envisions it does not exist.[54]

A federal energy agency could coordinate demand among the various fuels, prevent supply gluts, and protect consumers from baseless price increases. It could smooth the inevitable transition away from hydrocarbons to more benign sources of energy. An agency of this type would make sure that the true costs of producing energy were included in its price and would better control work-place and environmental hazards. It would plan alternative employment in those energy-supply areas being phased out and provide better mechanisms for developing the economic infrastructures of areas during their productive life. It would conduct, in whole or in part, negotiations with whatever labor organization miners choose to represent them. The UMWA could initiate a discussion of a federal energy agency within a framework of seeking union participation in private and public sector management of coal.

Western Europeans have evolved a broad approach to this subject using concepts like industrial democracy, work councils, and codetermination.[55] Europeans now require 33 to 50 percent worker participation on governing boards, and have greatly broadened the scope of collective bargaining. The three countries with the "most stringent regulatory industrial democracy laws—Holland, West Germany and Sweden"[56] had the strongest economies in Europe in the late 1970s, while Italy and France—the two countries with the least stringent industrial democracy laws—did less well. The political and economic power of organized labor in the industrial democracies was much more profound than labor's power in the United States. The European model is not a road map that U.S. labor unions—and the UMWA, in particular—can follow come what may to the promised land. Differences in political cultures compel U.S. borrowers to modify European concepts. But they provide a vision and a direction the UMWA should evaluate. The codetermination/industrial democracy approach is not the socialism of John Brophy, or nationalization. It falls somewhere to the left of free-market capitalism and to the right of socialism. Most coal industries throughout the world are nationalized, either fully or partially. Mine-worker unions in Europe led postwar societies to nationalize coal because free-market policies threatened their jobs, their income, and their well-being. Nationalization came during periods of crisis—of demand slumps, high unemployment, political turmoil. If the U.S. coal industry grows evenly over the next few decades, the conditions for nationalization will not coalesce. That pattern of growth has not occurred since before 1920, and there is little evidence to suggest that it will in the future. Plans short of nationalization could strengthen America's coal industry, stabilize it, make it safer and more equitable, increase the benefits of production to mining communities, and better serve the American public. Many models exist to inspire this discussion, which will begin only when the party most threatened by its absence initiates it.

13. The Future of Coal

Fortune-tellers use crystal balls to predict the future by peering into the past. What might we learn about coal with this time-tested technology?

Coal represented about 23 percent of the seventy quads of energy we consumed in 1983. That was about the same as coal's share in 1970. Of our major sources of energy, only coal and nuclear power grew consistently in absolute terms from 1970 to the early 1980s.[1] Americans consumed about 13 percent of all their energy in 1983 as electricity, of which coal generated about 55 percent. Of the 785 million tons we mined in 1983, 80 percent—or 626 million tons—went to electric power plants.[2] That compares with only 46 percent in 1970. Coal's other markets have been declining or stagnating since the late 1960s except for *exported* steam-coal, which grew to as much as 50 million tons a year from virtually zero in the early 1970s, but then fell to 27 million tons in 1983. America could consume a lot of coal as a feedstock for synthetic gas, oil, and fuel solids. That will not happen by 2000 unless a crash program of plant construction begins immediately. Such an effort appears to be neither wise, necessary, nor likely. Therefore, as electricity goes, so goes coal.

The amount of electricity America produced increased steadily after 1951 until the early 1980s, when it skidded to a statistical halt because of consumer conservation and economic depression in the American economy. Coal's growth in this market in the late 1970s and early 1980s came from utilities reducing their oil consumption, which fell 42 percent. Oil fueled only 6 percent of electricity production in 1983. Coal will soon reach the limits of growth based on oil substitution. Coal might also grow at the expense of nuclear power, which generated about 13 percent of our electricity in 1983. The U.S. had 76 operating nuclear reactors with a total capacity of about 63 thousand megawatts at the beginning of 1984. (Another 52 reactors with 57 thousand megawatts have been granted construction permits, or have permits pending or on order.) But 25 permitted plants were cancelled in the last decade and 64 plants whose permits were under review were scrapped as well at a total cost of about $10 billion. Coal would be the windfall fuel if more nuclear plants are cancelled, decommissioned or converted. (As of early 1984, nuclear-to-coal conversion is under consideration at only one operating reactor.) As natural gas prices increase owing to deregulation, utilities will shift to cheaper fuels—coal—if possible. If annual electricity demand grows not one kilowatt by 2000, utilities will still have increased their coal purchases by several hundred million tons a year to replace these other fuels.

An even greater factor in coal's future is the rate of growth in electricity production. Here, one can choose from dozens of crystal balls. Each forecast is based on different assumptions concerning economic growth, oil prices, interest rates, and dozens of other variables. Most models project past trends into the future. If one begins to extrapolate from 1970, our energy needs would be almost twice as large in 2000 as then. If one starts in 1980, the forecast shows little growth over the next decades. If the high-growth forecasters are right, our economy may grow 4 to 5 percent annually, which would mean that coal consumption would range between 1.5 and 2.5 billion tons in 2000 compared with 728 million tons in 1981. If, on the other hand, our economy continues to drag along and Americans continue to conserve electricity, coal production might only edge up to between 1.2 and 1.3 billion tons. Coal production will rise more or less depending on the price of other means of making electricity and the degree to which our economy rests on electricity. Not one single technical change in mining, transportation, or combustion technology need occur to allow America to triple or quadruple our use of coal under current environmental standards.

Forecasters predict a startling and profound shift in how America will get its coal in the 1990s. The Department of Energy told Congress in its *Annual Report* of 1980 that eastern strip-mining had reached its apex and would decline quickly in both absolute and relative terms by 1995. Projections of doubled national coal production in 1990 to more than thirty quads (about 1.45 billion tons) had eastern surface mines producing only about 130 million tons, or 9 percent of national output (table 5), compared with 269 million tons or 40 percent in 1978. By 1995, eastern surface mining's share falls to only about 65 million tons, a little more than 3 percent. On the other hand, eastern coal production *as a whole* was projected to increase from 496 million tons in 1978 to about 930 million tons in 1995. Eastern deep mining climbs from 227 million tons in 1978 to about 865 million tons in 1995—an increase of 281 percent! Were expansion of that magnitude to come true, the eastern-based UMWA could be a major beneficiary if it is able to maintain the core of its membership base during the mid-1980s.

America is lucky enough to not have to ask whether *we can* run much of our economy on domestic energy. Our question is harder: Is coal our best way of making electricity? *Should we*—not can we—is coal's unique question. Yes or no depends as much on the attributes and price of alternative ways of making steam as it does on coal's own characteristics. It also depends on how much electricity we're talking about. Utilities will have no choice about burning more coal if electricity demand doubles and oil prices are high.

The "should-we" question has two parts. How many dollars will it cost to

Table 5. Coal Production by Region and Mining Method, 1965–1995 (Million Tons)

		History			Projections								
		1965	1973	1978	1985			1990			1995		
Region	Mining Method				Low	Mid	High	Low	Mid	High	Low	Mid	High
East	Surface	168	235	269	255	255	256	131	132	130	63	66	66
	Deep	324	289	227	418	421	424	662	672	689	838	864	895
	Total	492	524	496	673	676	679	794	804	819	902	930	961
West	Surface	11	57	154	343	339	336	612	601	597	882	883	868
	Deep	9	10	15	16	16	16	40	37	36	63	65	66
	Total	20	68	169	359	355	354	652	638	633	946	948	934
National	Surface	189	298	427	598	594	594	743	733	728	946	949	934
	Deep	338	300	243	434	437	440	702	709	725	901	929	961
	Total	527	599	670	1,032	1,031	1,034	1,445	1,442	1,452	1,847	1,878	1,895
Total Production (quadrillion Btu)		13.4	14.4	15.0	22.8	22.8	22.9	30.3	30.3	30.6	37.8	38.6	39.0

Source: U.S. DOE, *1980 Annual Report to Congress*, p. 97.

make the electricity we need? Coal is the hands-down cheapest choice among oil, gas, most renewables, and nuclear at this time.[3] The second part compares the social benefits and costs of making electricity from each energy. The social *benefits* involve employment, investor dividends, secondary business activity, community betterment, and the like. For each unit of electricity, coal—particularly, deep-mined coal—employs more men and women than other fossil fuels and nuclear, but perhaps not as many as some renewable technologies and conservation. Social *costs* are estimates of damage to worker and public health, harm to our physical environment, and disruption of communities. Coal is more harmful than the others, except if a nuclear catastrophe should occur. Some renewables harm the environment and the people who handle them, but none—with the possible exception of wood—are as dangerous as coal. Oil and natural gas are extracted, refined, and burned with less occupational and environmental harm.[4] Comparing nuclear and coal boils down to a choice between potential catastrophe and predictable harm.

Nuclear power from light-water reactors poses three major problems: (1) uranium mining exposes miners and adjacent communities to radiation; (2) no long-term or safe solution has been found for disposing of radioactive reactor waste; and (3) the specter of screw-ups haunts the nuclear plants. Nuclear power is perceived as threatening a broader cross-section of American society than coal. The transport of nuclear waste, and the 76 plants themselves, cast radiation danger and potential cataclysm directly at our middle and upper classes, who have formed the social base of antinuclear protest. Coal, on the other hand, is often seen—mistakenly—as a localized threat that primarily affects a geographically and culturally isolated group who get paid well for risking their necks. Yet, broader and more immediate danger than nuclear radiation is coal's air pollution, which blankets much of the densely populated tier of industrial states from Illinois to Maine. Coal's social costs are awesome: of 240,000 coal workers, about 140 were killed and 18,000 injured (enough to lose at least one work shift) in 1980; an estimated 460 deaths and about 4,600 injuries are associated with transporting coal from mine to utility markets;[5] an estimated 50,000 or so premature deaths are associated with coal-fired air pollution. To this toll must be added the immediate environmental disturbance of between 87,000 and 96,000 acres from surface-mining and 137,000 acres from underground mining [in 1980].[6] Eastern surface mining disturbs at least twice the amount of land as surface mining in other areas because of its thin seams. Coal mining pollutes water with silt and particulates (some of which are toxic) as well as the land used to store mining wastes and combustion sludges. Coal poses two other environmental threats—acid rain and the global buildup of carbon dioxide—that may prove so serious as to force an absolute cutback in coal combustion.

How *should* America use its coal in the future? The question of how much we mine and burn can only be answered in terms of the care with which we handle both tasks. We are technically able to double coal supply and sustain that level for a decade or two with *less* damage to workers, the general public, and our immediate environment than we now inflict on ourselves.

The one caveat to this generalization concerns the accumulation of carbon dioxide in the earth's troposphere, for which there is no known technical fix.[7] As CO_2 builds, it blocks the loss of the earth's heat into outer space. About 50 percent of the CO_2 released in fossil-fuel combustion remains in the atmosphere, although scientists disagree over the role of fossil-fuel combustion in CO_2 buildup, its consequences, and the world's ability to adapt to climatic changes. Recent forecasts predict a doubling of atmospheric CO_2 sometime after 2075. Even a no-growth scenario eventually doubles CO_2 concentration if unaccompanied by a decrease in the burning of fossil fuels. The implications of a long-term change in the CO_2 balance could be substantial, although researchers project varying degrees of severity and caution that much remains unknown. A global temperature increase would certainly result from a CO_2 doubling and bring changes in climate, water availability, sea levels, agricultural patterns, and nation-state politics. *Changing Climate*, a report issued by the National Academy Of Sciences in 1983, stressed the uncertainties and unknowns while suggesting that the consequences of CO_2 buildup are likely to be more manageable than earlier research suggested. The change and damage resulting from CO_2 will depend to a great extent on the planning and adaptions nation-states adopt before crisis-level situations develop.

As recently as 1970, the scientific community had not recognized CO_2 buildup as a consequence of fossil-fuel combustion. Have we now plumbed all the surprises tucked into hydrocarbons? Or is science still ascending an environmental knowledge curve? Can we expect new consequences to emerge in the future? The problems we are now discovering are harder and more expensive to solve than sooty smoke, rape-and-run strip-mining and mine explosions. These were addressed—if not solved—when sufficient political will was amassed to force the private sector to adopt different methods and invest in control technologies. Second-generation environmental problems tend to materialize many years down the road, rather than immediately. They may be less susceptible to technological solution. For example, some harmful smoke particles are so small that they slip through the finest filters. It is next to impossible to prevent CO_2 from going up a stack. Second-generation problems are time bombs, not grenades. Business—whose concern is immediate return—is not much interested in defusing time bombs set for 2050. Science seems to be barreling up a knowledge curve in its ability to identify, measure, and predict environmental and health impacts. The more we learn, the more serious, and impenetrable these problems appear.[8] Our political and economic institutions have much difficulty in addressing long-term social costs.

For too long, the private sector deferred or denied the social and environmental costs of using fossil and nuclear energy. A false balance sheet came into existence. On one side were the private benefits of energy; on the other, the public costs. As long as the private sector made energy decisions public costs were high, and often unpaid. Social and environmental legislation of the 1970s began to even this balance sheet, adding complexity, lowering productivity and limiting coal use. Then backlash began in the late 1970s over the quite reasonable point that controlling public costs was costing a lot of money. The false balance sheet created a false framework for evaluating each new demand for public protection. Private costs versus public benefits led the private sector to see each new protection as an burden of increasingly marginal value. Had science been able to identify the full range of environmental and social costs years ago, and had government required private business to pay them as they went, corporations would have developed more benign energies. The legacy of a false balance sheet is that it leads to a continuing resistance to rectification. Each new scientific insight faces more determined bias against paying the costs of prevention.

Apart from the CO_2 quandary, current technology allows us to mine and burn coal more safely and cleanly if we are willing to pay and plan for it. Our coal reserves are so vast that we can exclude environmentally fragile areas with little loss. Fatality-free, low-injury mining is a fact of life in some mines. It requires a mix of well-maintained equipment, management policy, worker training and participation, and sound mining procedures. Safe mining is not a matter of luck; it is a matter of will. Most coal-fired air pollution comes from older plants that lack up-to-date pollution-control systems. These plants should be abandoned, retrofitted, or replaced. Without a radical—in the sense of getting to the roots—change in how we organize mining and combustion, we will continue to use coal inhumanely and irresponsibly. We should use coal in direct proportion to how safely we mine it and how cleanly we burn it. The stricter we are, the more we can use where it is needed.

If it is only the coal America needs, private corporations are entirely capable of mining, moving, and burning it in whatever quantities are needed.[9] The coal question America faces is whether to modify the private planning system with one more responsive to public national needs before coal booms.

Why ask that question now? First, some environmental and safety questions do not appear to be resolvable in the current framework. Public regulation has succeeded to a degree. But legislators have not been able to address acid rain, CO_2 buildup, air pollution, and certain aspects of mine safety. The private sector has dug in its heels, and the limits of regulation may be approaching. Second, the new structure of the coal industry—concentrated production in the hands of noncoal parent companies—lends itself to artificial pricing. Coal is now an industry in which a handful of regional producers and utilities establish a framework of price and supply for hundreds of suppliers. Long-term contracts with price-escalator clauses insulate producers from market forces and raise the price of coal to consumers. In certain circumstances, coal-owning oil companies could limit coal in order to sell oil at four to five times the profit. Antitrust regulators gave coal a free ride during the hard years of the 1950s and 1960s, only to confront an industry now able to fend off meaningful antitrust action. Third, the private sector has an unblemished record of bad labor relations that is not likely to improve without basic changes in how supply and demand fit together. Genuine labor stability will emerge when there is genuine supply-demand stability. The system of laissez-faire as it has evolved in this century has shown itself incapable of bringing about stability of either kind. Fourth, the coal business has created colonial underdevelopment in the coal fields that is most pronounced where mining has been most intense.[10] By thwarting taxes, land-use controls, and local planning, coal suppliers have left the public infrastructure in the coalfields chronically underbuilt. Much of it is incapable of handling rapid growth. Large parts of it are decaying at a faster rate than it is being replaced. The coal business does not appear capable of rectifying these past errors or adopting a new public-interest perspective. The private sector does not generally ask how mining may bring the most benefits to the community where it occurs. That question does not belong on a businessman's balance sheet, and the public sector is not likely to insert it there in the future. Although the industry has made great improvements in the way it goes about mining and burning coal, the limits of publicly forced cost-internalization on the private sector are approaching.

Finally, private energy planning has produced a staggering and largely unexamined redistribution of income and political power in American society. High-priced energy has transferred a disproportionate amount of money from America's middle and lower classes to energy investors and corporations. The political influence of the energy industry has grown in step with its profits and campaign contributions and our economy's energy dependency. Energy suppliers have changed the balance of power within the private sector. Their needs dominate the financial markets and reduce investment capital available for others. A recent Congressional study reported:

> We have seen the redirection of . . . [America's] economic strength into one industry—energy—at the expense of the rest of the industrial base.
>
> . . . Between 1978 and 1980, those companies which comprised the *Fortune* 500 list increased their net income by 19.6 billion dollars. Of that amount, 19.2 billion dollars, or 98 percent, of the increased net income went to the 44 oil companies and a dozen oil service and supply companies which are in that list. An even more staggering statistic is that in the *Fortune* 1000 list 82 oil and gas companies obtained 96 percent of the total net income increase. This meant that the more than 900 other companies got to

share 4 percent of the total profit increase. In fact, if adjustments are made for inflation, we have a negative increase in profits—that is, we have a sharp decline in total real profits for all companies other than energy companies.[11]

Perhaps all will work out for the best if private power is simply left alone. That certainly is the message from the boardrooms and the Reagan administration. But the shaky assumption on which this optimism is based is that what is best for Exxon is good enough for America.

The argument for changing coal's status quo boils down to a simple belief that America can mine and use coal better in the future if, and only if, a basic change occurs in the planning and operation of the industry. Otherwise, the five problems noted above will persist. The energy transition gives us occasion to rethink what we want from our coal industry. We must also keep in mind that coal use will increase for a limited period of time—perhaps as long as thirty or forty years—and then fall back as new energies are adopted. The private sector can manage this scaling up and long-term scaling down as the coal business has handled these episodes in the past—badly. Public planning is likely to be better.

A national planning agency could rebalance our energy mix as efficiently as private corporations, and more equitably. It could fit coal into a broad strategy for conserving scarce and costly energies. The planning authority could develop guidelines for mining and burning coal and be empowered to close any operation that cannot or does not obey them. It would be able to invest in new ventures and fund research. It would bargain with a national union of mine workers as its European counterparts do now. It would also coordinate capital investment so that when mines are no longer needed, substitute industry will have been phased in. If CO_2 threatens a genuine global crisis, a national coal-planning authority would probably be the most appropriate way of phasing in other fuels.

Some have spoken of coal as a "bridge fuel" that will allow industrial societies to make the transition between the hydrocarbon era and renewables. Those who suggest we must *quickly* triple our coal use are really building a bridge to an obsolete, high-coal future rather than one based on slower coal growth that takes into account the need to protect miners, communities, consumers, and the environment. Left to its own design, the private sector will plan a coal future that will attempt to wash environmental and social costs out of the system. This is a short-sighted policy resulting from understandable short-term financial perspectives. The private sector is capable of making a transition to renewables, but it is not capable of planning that transition in such a way as to minimize the dislocations resulting from the abandonment of hundreds of billions of dollars of capital stock devoted to coal and other conventional fuels. Coal is a useful and appropriate bridge as long as its growth is planned and coordinated to protect its workers and the public at large from the problems it presents.

Public planning has a long history in the United States. It has, decidedly, as ambiguous a record as that which private planning has compiled. Public planners have made the same errors of omission and commission as their private sector colleagues. Had our energy industry been fully nationalized after World War II, there is every reason to believe America would have walked down the same basic energy path that Exxon, Consolidation Coal, and Commonwealth Edison traveled. Yet, many industrial societies with which we associate have recognized the benefits of national energy planning. What opportunities does public planning present for coal?

First, it can be responsive to the electorate, which, at the very least, is a more democratic way of making important decisions, and not demonstrably worse. It is, in addition, motivated by service and security criteria and not solely profit (although most nationalized coal industries are expected to make a profit). This provides a different calculus for making choices.

Second, public planning would be asked to determine the wisest long-term energy choices for America as a whole. No corporation invests its capital on what is best for the American public. Of course, public planning can err or become captive to special interests. It will be subjected to congressional and other pressures to satisfy particular constituencies. At worst, public planning would subject us to mistakes in the name of serving all Americans instead of mistakes made in the name of making a buck.

Third, when shifts of supply and demand are needed, public planning can minimize the transition pains for people and communities. When a coal mine shuts down, management now abandons the worker. Retraining, economic diversification, and gradual substitution of new energy jobs for old energy jobs are proper tasks of public planners. Several states have set up funds, based on current mineral production, to foster the development of new enterprises for the time when the mineral resource will be depleted.

Fourth, public planners should be more inclined to invest in controlling environmental and social costs. However, managers of nationalized energy industries abroad frequently substitute a production ethic for the profit motive. That easily translates into hostility toward regulators and other "nuisances." No guarantee exists that public planning will be more responsive to environmental and social considerations, but the odds are better.

Fifth, public planning should be better able to harmonize coal supply and demand, something the market has never been able to manage. This alone would contribute to stability in coal affairs and result in more efficient use of our capital, labor, and managerial resources.

Sixth, the price of coal would be arrived at on public and private criteria. Return on investment would be made predictable and windfall profits avoided.

Finally, the experience of publicly managed coal industries in Western Europe suggests that they generally have better labor relations and, in some cases, safer operations than we. This has come at a price, to be sure. Productivity is lower and inefficient pits have been subsidized to maintain employment. Miners continue to strike, but labor relations have been less disruptive in Western Europe than in America since World War II.

The forces that are making coal a transition fuel are as irreversible as those that brought coal into the age of industrialization. Coal miners may choose to fight these forces, but they will not win. Coal—as a longer and lower (in terms of annual production) bridge—is more in the interests of miners than a higher bridge, which is likely to be shorter. A rapidly achieved, high-coal future—tripled production by 2000, as suggested in the DOE's 1980 report to Congress—is likely to be reached by making massive investments in mining systems that need less labor than existing ones. If the industry reshifts to eastern underground mining, as DOE suggests, some new employment will occur. But coal operators are likely to couple rapid production increases with intensive mechanization. A high-coal future does not necessarily mean many more jobs for miners.

The transition—whether long or short—should be planned to allow coal miners to participate in the transition to a renewable future rather than be its victims. Miners should be enfranchised to be part of the planning process. Coalfield economies should be deliberately diversified, possibly with new energy enterprises. A tax on current coal production could provide capital for new businesses and labor training there. If the UMWA is sufficiently farsighted, it could gradually become a union of energy workers, of which coal miners are one part. Without some plan for adapting to and shaping change, the UMWA and coal miners will inevitably be sacrificed to it. That would be a great loss. There is much in the fiber of the union and its members that can strengthen America in the future.

Notes

Introduction

1. Richard A. Schmidt, *Coal in America: An Encyclopedia of Reserves, Production and Use* (New York: McGraw-Hill, 1979), pp. 12-15.

2. Ibid., pp. 22-23.

3. Jules I. Bogen, *The Anthracite Railroads: A Study in American Railroad Enterprise* (New York: Ronald Press, 1927), p. 7.

4. Andrew Roy, *The Coal Mines: Containing a Description of the Various Systems of Workings and Ventilating Mines* (Cleveland, Ohio: Robison, Savage, 1876) p. 25. For early history see Charles Singer et al., *A History of Technology,* vol. 3 (Oxford, Eng.: Clarendon Press, 1957); University of Maryland School of Law, *Coal and the Environment: Legal Problems of Reclamation in Maryland, Ohio, Pennsylvania and West Virginia* (Washington, D.C.: U.S. Environmental Protection Agency, November 1971), p. 4; R. Bruère, *The Coming of Coal* (New York, Association Press, 1922), p. 5; Howard N. Eavenson, *The First Century and a Quarter of The American Coal Industry* (Pittsburgh: Waverly Press, 1942) pp. 1-30; and J. Mumford, *Anthracite* (New York: Industries Publishing Co., 1925), 1915; and W.J. Nicolls, *The Story of American Coal* (Philadelphia: Lippincott, 1897).

5. Roy, *The Coal Mines,* p. 26.

6. Edward A. Wieck, *The American Miners' Association: A Record of the Origin of Coal Miner's Unions in the United States* (New York: Russell Sage Foundation, 1940), p. 32.

7. John C. Esposito, "Air and Water Pollution: What to Do While Waiting for Washington," *Harvard Civil Rights—Civil Liberties Law Review*, 5, no. 1 (January 1970):32.

8. Wieck, *The American Miners' Association*, p. 32.

9. Roy, *The Coal Mines*, pp. 29-30.

10. Ibid., p. 29. Britain's conversion from wood to coal also prompted two inventions that were central to industrialization. As her coal mines drove deeper underground, they needed more efficient and cheaper ways to pump water to the surface. The steam-driven pumps used in Staffordshire as early as 1712 laid the technical foundation for James Watt's rotary steam engine in the 1780s. As urban consumers demanded more coal, the British also constructed a network of canals and later railroads to move coal to urban areas. In this way, coal led to the widespread use of steam engines and helped to build Great Britain's iron and steel industry. (Singer et al., *A History of Technology,* 3:84.)

11. Eavenson, *The First Century*, p. 15.

12. Howard N. Eavenson, *Coal through the Ages* (New York: published for Seeley W. Mudd Fund, 1st ed., 1935), p. 43.

13. Richard Cowling Taylor, *Statistics of Coal* (Philadelphia: J.W. Moore, 1848), p. 21.

14. James M. Swank, *History of the Manufacture of Iron in All Ages* (originally published in 1892), reprinted (New York: Burt Franklin, 1965), p. 467.

15. Eavenson, *Coal through the Ages*, p. 45.

16. Taylor, *Statistics of Coal*, pp. 20-21.

17. Eavenson, *The First Century*, p. 9. Jefferson wrote:

The country on James' river, from fifteen to twenty miles above Richmond, and for several miles northward and southward, is replete with mineral coal of a very excellent quality.

In the western country coal is known to be in so many places, as to have induced an opinion, that the whole tract between the Laurel mountain, Mississippi, and Ohio, yields coal. . . . the coal at Pittsburg [*sic*] is of very superior quality. A bed of it at that place has been a-fire since the year 1765.

Thomas Jefferson, *Notes on the State of Virginia* (New York: H.W. Derby, 1961, reprinted New York: Harper Torchbooks, 1964), p. 25.

18. Taylor, *Statistics of Coal*, p. 16.

19. Wieck, *The American Miners' Association*, p. 33. The first published manual on coal mining may have been a 55-page booklet, *The Compleat Collier: Or, The Whole ART of Sinking, Getting, and Working, Coal-Mines, &c. AS Is now used in the Northern Parts, Especially about SUNDERLAND AND NEW-CASTLE* (London, C. Conyers at the Ring in Little-Brittain, 1708). Its author is known only as "F.C."

20. Small deposits of anthracite were known to exist in Rhode Island and Massachusetts "many years before the Pennsylvania anthracite was first mined," according to Taylor, writing in 1848. Taylor, *Statistics of Coal*, p. 20. This anthracite, however, is not easily burned and lies in very irregular seams.

21. *8th U.S. Census of Manufactures of the United States* (Washington, D.C., 1860), p. c/xxiii.

22. Paul J. McNulty, "The Public Side of Private Enterprise: A Historical Perspective on American Business and Government," *Columbia Journal of World Business* (Winter 1978):123.

23. Robert A. Lively, "The American System: A Review Article," *The Business History Review* (March 1955):86. Most liberal historians agree on this point. See, for example, Stuart Bruchey, *The Roots of American Economic Growth* (New York: Harper & Row, 1965); Carter Goodrich, *Government Promotion of American Canals and Railroads 1800–1890* (New York: Columbia University Press, 1960); and Louis Hartz, *Economic Policy and Democratic Thought: Pennsylvania 1776–1860* (Cambridge, Mass.: Harvard University Press, 1948).

24. Colonial Virginia gave away some of what in 1863 became West Virginia to veterans of the French and Indian Wars. Later, the Commonwealth of Virginia gave away the rest in lieu of cash payments for service in the American Revolution. Charles H. Ambler and Festus P. Summers, *West Virginia: The Mountain State*, 2nd ed. (Englewood Cliffs, N.J.: Prentice-Hall, 1958), p. 63.

25. An excellent account of the settlement of West Virginia is found in John Alexander Williams, *West Virginia: A History* (New York: Norton, 1976), pp. 5-29.

26. Bogen, *The Anthracite Railroads*, p. 76.

27. Williams, *West Virginia*, p. 26.

28. Nineteenth-century businessmen maintained steady pressure on national politicians to gain access to public land and resources. The congressional debate divided into two positions. Some believed that public land should be sold publicly and the revenue used to benefit the nation or states. Others felt that public land should be given away or sold cheap to encourage western settlement. Post–Civil War legislation authorized the federal government to sell coal land and other mineral wealth for $2.05 an acre.

29. About two-thirds of all business corporations chartered by government statute before 1800 had to do with transportation—canals, harbors, channels, toll roads, and bridges. See Annmarie Hauck Walsh, *The Public's Business: The Politics and Practices of Government Corporations* (Cambridge, Mass.: MIT Press, 1980), p. 16 and ff.

30. Bogen, *The Anthracite Railroads*, p. 185.

31. Ibid., p. 146.

32. An estimated $195 million was spent on canal construction by private and public sources from 1815 to 1860, of which 62 percent came from state and federal sources. Government

participation in mixed enterprises raised the share of public contribution to about 70 percent of the total.

33. Bogen, *The Anthracite Railroads*, p. 27.

1. Mines and Camps: The First Hundred Years

1. Underground mining had been practiced in Europe for centuries before American miners went underground. *The Compleat Collier* describes mining conditions in the British Isles:

> Collieries . . . [employ] so many thousands of the poor in those Northern parts of *England* which are maintained by it, who must otherwise of Course be Beggers, or would be starved, if Coal Mines were not carried on. . . . [pp. 9-10].
>
> . . . I myself have seen and narrowly . . . escaped that some Collieries are very subject to this fatal Surfeit [methane]. . . .
>
> And thus it is plain, that both the Officers and poor Miners are in dayly Peril and Hazard of their Lives, for a poor Livelyhood, and that they may be easily Destroyed by Ignorant and Unskilful Managers. [pp. 45-46]

2. See letter from John Grammer, Jr., January 28, 1818, cited in Eavenson, *The First Century*, p. 72.

3. Oswald J. Heinrich, "The Midlothian, Virginia, Colliery in 1876," *Transactions of the American Institute of Mining Engineers*, May 1875–February 1876, p. 309.

4. Quoted in Eavenson, *The First Century*, p. 92.

5. See C. Crozet, Principal Engineer, *Report on the Railway from the Coal-Pits to James River* (Richmond, February 6, 1828), cited in Eavenson, *The First Century*, p. 81.

Slaves mined most of the coal produced in the antebellum South, but records of Virginia mines indicate that they also employed some free blacks and whites (Ronald L. Lewis, *Coal, Iron and Slaves: Industrial Slavery in Maryland and Virginia, 1715-1865* [Westport, Conn.: Greenwood Press, 1979, pp. 68 ff]). Where the mine owner did not own enough slaves himself, he could hire them for a year; in 1800 the rental was L20 per slave in Richmond (Eavenson, *The First Century*, p. 62). In one case, a "rich old Scotchman" named Graham used a "confidential slave of his own" to supervise a slave-worked coal mine (account by Edmund Ruffin, editor of the *Farmers' Register*, 1837, quoted in Ibid., pp. 89-90.) Slave labor drove out most free labor in southern coal mining. Figures quoted by Starobin indicate that most if not all of the coal miners in Virginia, Kentucky, and Maryland were slaves. See Robert S. Starobin, *Industrial Slavery in the Old South* (New York: Oxford University Press, 1970), p.23. Black miners continued to make up a substantial portion of the coal industry's workforce for many years. About 34 percent of all coal miners in Alabama, Kentucky, Tennessee, Virginia, and West Virginia in 1900 were black. This proportion dropped to about 16 percent by 1940 and fell steadily through the 1950s and 1960s (Herbert R. Northrup, "The Negro and the United Mine Workers of America," *Southern Economic Journal* [April 1943]:314).

6. Several nineteenth-century mining manuals are available. French coal mining is portrayed in L. Simonin, *Underground Life; Mines and Miners*, translated, adapted to the present state of British mining, and edited by H. W. Bristow (London: Chapman and Hall, 1869). See also John Holland, *The History and Description of Fossil Fuel, The Collieries and Coal Trade of Great Britain* (London: Whittaker & Co., 1841).

7. Not surprisingly, 179 miners died from suffocation when, on September 6, 1879, a ventilation fire at the bottom of a shaft got out of control, sucking all the air from the underground tunnels and blocking the only passage to the surface. The Avondale disaster prompted the Pennsylvania legislature to pass an act that required all mines in Schuylkill County to have a second shaft and forced ventilation. Mines in other counties were exempt.

8. See Chris Evans, *History of the United Mine Workers of America*, 2 vols. (Indianapolis: UMWA, 1918). Arthur E. Suffern, *The Coal Miners' Struggle for Industrial Status* (New York: Macmillan, 1926); and Wieck, *The American Miners' Association*. John L. Lewis was the son of

Welsh mining families who moved to south-central Iowa in 1876. Melvyn Dubofsky and Warren Van Tine, *John L. Lewis: A Biography* (New York: Quadrangle, 1977), p. 3 ff.

9. Isador Lubin, *Miners' Wages and the Cost of Coal* (New York: McGraw-Hill, 1924), pp. 115ff.

10. In addition to the miner-entrepreneurs, there was a second group of underground workers, known as "company" men. Paid a fixed hourly rate, they provided support services for the dozens of mining teams. Company men were generally less skilled and had a lower status in the workers' hierarchy. The mine operator had "outside jobs" for children and old men, picking the slate and rock from the coal as it was sorted into uniform sizes in huge, slope-roofed buildings called "breakers."

11. The UMWA contract in 1943 banned from union mines the practice of miners' having to purchase their own supplies.

12. U.S. Department of Labor, *11th Special Report to the Commissioner* (Washington, D.C.: U.S. Department of Labor, 1904), p. 398.

13. Carter Goodrich, *The Miner's Freedom: A Study of the Working Life in a Changing Industry* (Boston: Marshall Jones Company, 1925).

14. The U.S. Coal Commission found that "when the mine is operating steadily the miner can earn a satisfactory wage before noon." See Edward Eyre Hunt et al., *What the Coal Commission Found: An Authoritative Summary by the Staff* (Baltimore: Williams & Wilkins Company, 1925), p. 260.

15. Goodrich, *The Miner's Freedom*, p. 39. Even a ratio of one supervisor to 150 miners was not atypical. See Hugh Archbald, *The Four Hour Day in Coal* (New York: W.H. Wilson Co., 1922), p. 38. Today the ratio is 1 to 8 or 1 to 10.

16. Ibid., p. 16.

17. The power that management exercised was indirect rather than direct. Section bosses, for instance, could favor or penalize through their assignment of new "places."

18. Goodrich, *The Miner's Freedom*, pp. 56–100.

19. Goodrich notes that the UMWA did not support some "natural" rights. The custom of irregularity—of quitting work when the miner felt he had earned enough on that shift—was one which the UMWA fought neither for nor against. Ibid., pp. 63–65.

20. Data from U.S. Bureau of Mines, cited in Morton S. Baratz, *The Union and the Coal Industry* (New Haven: Yale University Press, 1955), pp. 41-43. In 1925, 73 percent of the underground production was mechanically cut; by 1950, 93 percent.

21. By 1950, almost 70 percent of all underground production was mechanically loaded. Ibid., pp. 41-43. Joseph F. Joy introduced the first successful loading machine in 1922.

22. I've focused on the technological changes that had the broadest social consequences. Other aspects of mining were also being made more efficient. Dynamite and safety fuses gradually replaced gunpowder. Some mines purchased power drills to keep pace with the mechanical cutters and loaders. Electric locomotives replaced horses as a way of taking the coal to the surface in the early 1900s. In the 1950s, mechanical roof-bolting machines, which inserted metal rods into the mine roof, replaced much of the timbering in room-and-pillar mines.

23. I draw from Thompson's discussion of mine mechanization and the structure of work. Alexander Mackenzie Thompson, III, *Technology, Labor, and Industrial Structure of the U.S. Coal Industry: An Historical Perspective* (New York: Garland Publishing, 1979), pp. 138-231.

24. The UMWA tried to negotiate piece rates that were fair to all of its members. Those miners working in thick seams, close to the final consumer and aided by machines, would normally produce the most tonnage at lowest delivered cost. A complex system of wage differentials was invented to equalize income among miners in different states and with varying working conditions. It never succeeded. Thompson, *Technology, Labor, and Industrial Structure*, pp. 115–35.

25. The social implications of mine mechanization are analyzed in: Geoffrey Bauman, "Coal Miners: From Craftsmen to Industrial Workers," unpublished paper, University of Pittsburgh, May

1974; John Brophy, *A Miner's Life* (Madison, Wisc.: University of Wisconsin Press, 1964), pp. 43 ff.; Keith Dix, *Work Relations in the Coal Industry: The Handloading Era, 1880–1930* (Morgantown, W. Va.: Institute for Labor Relations, 1977); Goodrich, *The Miner's Freedom;* Curtis Seltzer, "The United Mine Workers of America and the Coal Operators: The Political Economy of Coal in Appalachia, 1950–1973" (Ph.D. dissertation, Columbia University, 1977), p. 49 ff; and Thompson, *Technology, Labor, and Industrial Structure*, pp. 17–137.

26. William N. Page, "The Economics of Coal Mining," *The Mining Industry,* vol. 3 (New York: Scientific Publishing Co., 1894), p. 151, cited in Keith Dix, "Work Relations in the Coal Industry: The Handloading Era, 1880–1930," in *Case Studies on the Labor Process*, ed. Andrew Zimbalist (New York: Monthly Review Press, 1979), p. 164.

27. U.S. Department of Labor, *11th Special Report of the Commissioner*, p. 439, quoted in Thompson, *Technology, Labor, and Industrial Structure*, p. 94.

28. Phil Conley, *History of West Virginia Coal Industry* (Charleston, W. Va.: West Virginia Educational Foundation, 1960), p. 82.

29. Thomas A. Stroup, "Cause and Growth of Unionism Among the Coal Miners," *Mining and Metallurgy* (September, 1923):467-68.

30. U.S. Department of Commerce, *Historical Statistics of the United States, Colonial Times to 1970*, Part 1, p. 607. The *frequency* of mine fatalities increased as the first generation of machines was introduced. From 1880 to 1889 when few machines were in use, coal mining averaged 2.6 fatalities annually per 1,000 workers. In comparison, the decade 1890–1899 averaged 2.8 fatalities; 1900–1909, 3.6 fatalities; and 1910–1919, 3.4 fatalities.

31. The story of coal acquisitions in Kentucky is discussed in Harry Caudill, *Night Comes to the Cumberlands: A Biography of a Depressed Area* (Boston: Little, Brown, 1962); and Henry P. Scalf, *Kentucky's Last Frontier* (Pikeville, Ky.: Pikeville College Press, 1972).

32. The federal government gave large tracts of coal-bearing land to transcontinental railroad companies to encourage rail construction. What are now known as the Burlington Northern, Inc., and the Union Pacific Corp are the second- and third-largest private coal reserve owners, with 11.4 and 10 billion tons respectively. Other railroads with substantial coal holdings include the Santa Fe; Southern Railway; Illinois Central Gulf; Missouri, Kansas, Texas; Chicago & Northwestern; Missouri Pacific; and the Southern Pacific. (The Norfolk & Western is a major coal owner in the East with 1.4 billion tons.) The federal government gave more than 94 million acres of public land to the railroads, plus an additional 37 million granted to the states to be used to encourage rail construction. The railroads continue to own a substantial amount of the coal-bearing land originally granted them. See U.S. Congress, Office of Technology Assessment, *An Assessment of Development and Production Potential of Federal Coal Leases* (Washington, D.C.: Office of Technology Assessment, 1981), pp. 227-28.

33. John D. Rockefeller, Jr., was the classic absentee coal entrepreneur. In 1913–14, miners at Rockefeller's Colorado Fuel and Iron Co. struck, and more than 50 men, women, and children were gunned down in one day. Rockefeller had conceived CF&I's antiunion policy, but was dismayed by this tragic incident. While maintaining his opposition to unions, he devised an alternative. It came to be known as "employee representation" plans, which amounted to company-dominated discussion groups. See Suffern, *The Coal Miners' Struggle*, pp. 280 ff; George S. McGovern and Leonard F. Guttridge, *The Great Coalfield War* (Boston: Houghton Mifflin, 1972); J. D. Rockefeller, Jr., *The Colorado Industrial Plan* (New York: privately published, 1916) and *Representation in Industry* (New York: privately published, 1918.)

34. Advertisement of Trustees' Sale of the Maryland Mining Company in the *Cumberland Miners' Journal*, July 2, 1852, quoted in Katherine A. Harvey, *The Best-Dressed Miners: Life and Labor in the Maryland Coal Region, 1835-1910* (Ithaca, N.Y.: Cornell Univ. Press, 1969), p. 76.

35. Winthrop D. Lane, *Civil War in West Virginia* (1921; reprint ed., New York: Arno, 1969), pp. 16 ff.

36. *Frostburg (Md.) Mining Journal*, April 15, 1882, quoted in Harvey, *The Best-Dressed Miners*, p. 81.

37. Harvey, *The Best-Dressed Miners*, p. 79.

38. Consolidation Coal Co., *Report to the Stockholders for the year 1873* (New York, 1874), p. 7, quoted in ibid.

39. Wieck, *The American Miners' Association*, p. 166.

40. U.S. Department of Labor, Women's Bureau, *Home Environment and Employment Opportunities of Women in Coal-Mine Workers' Families*, bull. 45 (Washington, D.C.: Government Printing Office, 1925), pp. 12-13; and Williams, *West Virginia*, p. 139.

41. See Jeremiah P. Shalloo, *Private Police, with Special Reference to Pennsylvania* (Philadelphia: American Academy of Political Science, 1933); U.S. Congress, Senate, Committee on Education and Labor, *Conditions in the Paint Creek District, West Virginia. Hearings before a subcommittee on Education and Labor on S.R. 37*, 63rd Cong., 1st sess., 1913; U.S. Congress, Senate, Committee on Education and Labor, *Violations of Free Speech and the Rights of Labor. Hearings*, 74th Cong., 2nd sess., 1937; U.S. Congress, Senate, Committee on Education and Labor, *Violations of Free Speech and Rights of Labor—Harlan County and Republic Steel Corporation*, 76th Cong., 1st sess., 1938; and U.S. Congress, Senate, Committee on Manufacturers, *Conditions in the Coalfields of Harlan and Bell Counties, Kentucky, Hearings*, 72nd Cong., 1st sess., May 1932.

42. Sidney Webb, *The Story of the Durham Miners 1662–1921*, (London: Fabian Society, 1921), p. 46, quoted in Wieck, *The American Miners' Association*, p. 41.

43. "The check-off for the camp physician was supposedly voluntary, but in most cases if a miner did not agree to it, he did not have a job." Janet E. Ploss, "A History of the Medical Care Program of the United Mine Workers of America Welfare and Retirement Fund" (M.S. thesis, Johns Hopkins University, School of Hygiene and Public Health, 1981), p. 16.

44. Wieck, *The American Miners' Association*, p. 42. He writes (p. 72): "The history of the system, both in this country and in Great Britain, is one of constant misuse of monopoly by its owners."

45. Some coal states outlawed or regulated scrip and company stores. At the turn of the century in Cambria County, Pennsylvania, for example, "it appears the miner was not forced to buy at the company store, but the ease of doing so, especially when he had little access to transportation, resulted in his shopping at these stores" (Bruce Thomas Williams, "Pennsylvania: An Anthropological Study" [Ph.D. dissertation, University of Pittsburgh, 1976], pp. 57-58).

46. A common pattern was to segregate native whites from blacks and recently immigrated Europeans. Sometimes miners from Great Britain were separated from other immigrants and native whites, and other camps divided into language areas, separating Italians, Hungarians, Germans, Russians, Slovakians, and others.

47. Joel I. Nelson and Robert Grams, "Union Militancy and Occupational Communities," *Industrial Relations* (October 1978): 342–46.

48. Erving Goffman, *Asylums* (Garden City, N.Y.: Doubleday, Anchor Books, 1961)

49. Wieck, *The American Miners' Association*, p. 203. Roy (*The Coal Miners*, pp. 309-10) describes the effort of about sixty Ohio miners to operate a cooperative mine during a strike in 1874. See also John R. Commons, *History of Labor in the United States*, vol. 2 (New York: Macmillan, 1926–1935), p. 111.

50. Harvey, *The Best-Dressed Miners*, pp. 74 ff; and Williams, *West Virginia*, p. 140.

51. Coal property was assessed and taxed at the local level. Because assessors had to rely on company estimates of how much coal lay under the ground, a pattern of under-assessment emerged throughout the coal fields. Even where the resource could be estimated objectively, the tax on unmined coal was only a few cents per acre. Congress levied a coal tax from 1862 until mid-1865 ($.035 per ton in 1862 to $.06 in 1864). (See Eavenson, *The First Century*, pp. 198-199, 618.)

52. Several dozen Congressional investigations into coal affairs occurred between the late 1880s and the mid-1920s. Labor "troubles" were the subject of inquiries in 1877-1888, 1912, 1913, 1917, 1919, 1921 and 1922. Coal economics were investigated in 1893, 1908, 1912, 1914, 1915, 1917, 1919, 1920, 1921 and 1922. Two federal commissions looked into both issues: the first, the U.S. Anthracite Coal Strike Commission in 1902-1903; the second, a multi-year investigation into both bituminous and anthracite, which is generally referred to as the U.S. Coal Commission. (See U.S. Coal Commission, *Report* [Washington, D.C.: Government Printing Office, 1925]; Hunt et al., *What the Coal Commission Found;* and, Ralph Baker, "The National Bituminous Coal Commission" [Ph.D. dissertation, The Johns Hopkins University, 1939]).

53. The industry's ability to produce more coal than it could sell—excess capacity—has existed in almost every year since 1890. In its simplest form, capacity increased in the early coal business when mines operated on more days than normal. From 1890 to 1940, the average number of days worked annually ranged from 142 (1922) to 249 (1918) with more than half the years averaging 200 or more. The U.S. Bureau of Mines and the National Coal Association calculated *theoretical* production capacity assuming a work year of 261, 280 and 308 days (six-day week) for 1890-1940. Production in 1920 was 569 million tons at 220 days, but might have been 675 million tons at 261 days, 725 million tons at 280 days and 796 million tons at 308 days. (See National Coal Association, *Statistical Data Concerning the Bituminous Coal Industry* [Washington, D.C.: National Coal Association, 1945], p. 100.) In this period, the excess capacity that was actually capable of being realized probably averaged about 40%. Modern coal mines often work six-day weeks using one, two and occasionally three production shifts. Current capacity can be increased by switching to a four-shift day (three production shifts, one maintenance shift), opening new coal faces, operating on more days, using old equipment in marginal production areas and opening inefficient mines. In strong markets, excess capacity allows the industry to profit quickly with minimal additional investment. When demand is weak, excess capacity produces an irresistible downward spiral of price and wage cuts.

2. "Single Stones Will Form an Arch"

1. Webb, *The Story of the Durham Miners,* p. 2. See also Sidney and Beatrice Webb, *The History of Trade Unionism* (London: Longmans, Green, 1911), and "First Report of the Commissioner for Mines, 1842, Children's Employment Commission, Great Britain" in *Coal, Part I: Social, Economic, and Environmental Aspects,* ed. Mones E. Hawley (Stroudsburg, Pa.: Dowden, Hutchinson & Ross, 1976), pp. 216-248.

2. Wieck, *The American Miners' Association*, p. 37.

3. Ibid., p. 42.

4. See ibid., pp. 42-43. The political tie between disasters and legislation is discussed in the American context in William Graebner, *Coal Mining Safety in the Progressive Period: The Political Economy of Reform* (Lexington, Ky.: University Press of Kentucky, 1976).

5. Peter Roberts, *The Anthracite Coal Industry* (New York: Macmillan, 1901), pp. 172-73.

6. Peter Lambropoulos reviewed the literature on the Molly Maguires for this book. He concluded that it is debatable whether the accused miners actually were guilty of the crimes for which they died. In one trial, the wife of the man whose testimony confirmed McParlan's charge that five miners had murdered a policeman testified that her husband was the actual murderer. At another, the state's main witness failed to identify the defendants as the murderers. In a third, Jack Kehoe was convicted of murdering a "breaker boss" despite affidavits placing him miles away from the crime. Other miners were tried and convicted of murder for which they had been acquitted after McParlan testified they had confessed to him. The official history of Schuylkill County, Pa., where these events occurred, claims that "in this county no association was organized under the name of Molly Maguires." See *History of Schuylkill County* (New York: W.W. Munsell & Co., 1881), p. 99. See also Charles McCarthy, *The Great Molly Maguire Hoax* (Wyoming, Pa.: Cro Woods, 1969) and Anthony Bimba, *The Molly Maguires* (New York: International Publishers, 1932).

7. The complete history of this union is recounted in Wieck, *The American Miners' Association*, pp. 78-79, and Andrew Roy, *A History of the Coal Miners of the United States* (New York: Trauger Printing Co., 1907).

8. The association's Constitution was reprinted in the *United Mine Workers' Journal (UMWA Journal)*, August 20, 1908. Pete Seeger has set these words to music.

9. Suffern, *The Coal Miners' Struggle*, pp. 28 ff.

10. Ibid., p. 41. Suffern emphasizes that both sides recognized that a coordinated system of wage rates would equalize costs, stabilize the industry and protect profit and wages from competition.

11. Evans, *History of the United Mine Workers, vol. 1*, p. 401.

12. Suffern, *The Coal Miners' Struggle*, describes these early efforts, as does Evans, *History of the UMWA*.

13. Suffern, *The Coal Miners' Struggle*, p. 57.

14. United Mine Workers of America, *Fifteenth Annual Convention Proceedings*, p. 37.

15. John Mitchell, *Organized Labor* (Philadelphia: American Book & Bible House, 1903), p. ix.

16. James O. Morris, "The Acquisitive Spirit of John Mitchell, UMW President (1899–1908)," *Labor History* 20, no. 1 (Winter 1979):9ff.

17. Mitchell wrung the eight-hour day from a group of operators north of the Ohio River known as the Central Competitive Field, whose spokesman, Phil Penna, had been UMWA president between John McBride and Michael Ratchford.

18. McAlister Coleman, *Men and Coal* (New York, 1943; reprint edition, Arno, 1969), p. 72.

19. Ibid., p. 74.

20. Stanley Miller, "The United Mine Workers: A Study of How Trade Union Policy Relates to Technological Change" (Ph.D. dissertation, University of Wisconsin, 1957), p. 52. Emphasis in the original.

21. United Mine Workers of America, *Nineteenth Annual Convention Proceedings*, 1908, and *Twenty-First Annual Convention Proceedings*, 1910.

22. The growth of the UMWA in these years was not easy. Its right to represent miners was always subject to challenge. The same issues—wages, fair weights on tonnage, and union recognition—had to be reaffirmed during each negotiating round. For a chronicle of these years see: Evans, *History of the UMWA*; Coleman, *Men and Coal*, Suffern, *The Coal Miners' Struggle;* U.S. Congress, Senate, Anthracite Coal Strike Commission, *Report to the President on the Anthracite Coal Strike of May–October, 1902, by the Anthracite Coal Strike Commission*, S. Doc. 6, 58th Cong., special sess., 1903; U.S. Congress, House, *Report on the Miners' Strike in Bituminous Coal Field in Westmoreland County, Pa., in 1910-1911, H. Doc. 847*, 62nd Cong., 2nd sess., 1912; U.S. Congress, Senate, *A Report on Labor Disturbances in the State of Colorado, from 1880 to 1904, Inclusive*, S. Doc. 122, 58th Cong., 3rd sess., 1905; U.S. Congress, House, Industrial Commission, *Report of the Industrial Commission on the Relations and Conditions of Capital and Labor Employed in the Mining Industry*, H. Doc. 181, 57th Cong., 1st sess., vol. 12, 1900–1902; U.S. Congress, Senate, Commission on Industrial Relations, *Industrial Relations*, Final Report and Testimony, vol. 8, S. Doc. 405, 64th Cong., 1st sess., 1916; U.S. Congress, Senate, Committee on Education and Labor, *Conditions in the Paint Creek District West Virginia, Hearings before a subcommittee of the House Committee on Education and Labor on S.R. 37*, 63rd Cong., 1st sess., 1913; Howard B. Lee, *Bloodletting in Appalachia* (Parsons, W.V.: McClain Printing Co., reprint ed., 1969); George A. McGovern and Leonard F. Guttridge, *The Great Coalfield War* (Boston: Houghton Mifflin, 1972), Winthrop D. Lane, *Civil War in West Virginia* (New York: 1921, reprint ed., Arno, 1969); Archie Green, *Only a Miner: Studies in Recorded Coal-Mining Songs* (Urbana: University of Illinois Press, 1972); Walter R. Thurmund, *The Logan Coal Field of West Virginia, A Brief History* (Morgantown, W.Va.: West Virginia University Library, 1964); and Williams, *West Virginia*.

23. The story of the 1913–14 West Virginia mine war on Paint Creek and Cabin Creek is available in several books and theses. See Lee, *Bloodletting in Appalachia;* Charles B. Crawford, "The Mine War on Cabin Creek and Paint Creek, West Virginia, in 1912–1913" (Ph.D. dissertation, University of Kentucky, 1939); Frederick A. Barkey, "The Socialist Party in West Virgina from 1898 to 1920: A Study in Working Class Radicalism" (Ph.D. dissertation, University of Pittsburgh, 1971); John M. Barb, "Strikes in Southern West Virginia Coal Field, 1912–1922" (Ph.D. dissertation, West Virginia University, 1949); Lawrence R. Lynch, "The West Virginia Coal Strike," *Political Science Quarterly* 29 (1914):626-63; Fred Mooney, *Struggle in the Coal Fields: The Autobiography of Fred Mooney*, edited by J.W. Hess (Morgantown, W.Va.: West Virginia University Library, 1967); Edwin V. Gartin, "The West Virginia Mine War of 1912–1913: The Progressive Response," *North Dakota Quarterly* (Autumn 1973): 13-27; and U.S. Congress, Senate, Committee on Education and Labor, *Conditions in the Paint Creek District, West Virginia,* hearings before a Subcommittee on Education and Labor on S.R. 37, 63rd Congress, 1st sess., 1913; Richard D. Lunt, *Law and Order vs. the Miners: West Virginia, 1907–1933* (Hamden, Conn.: Archon Books, 1979); and David Corbin, *Life, Work, and Rebellion in the Coal Fields: The Southern West Virginia Miners 1880–1922* (Urbana, Ill.: University of Illinois Press, 1981).

24. Suffern, *The Coal Miners' Struggle*, pp. 213 ff. These shutdowns were inconvenient, but not catastrophic from the operators' perspective. The time lost from labor strife was "relatively unimportant." Suffern found that man days lost during the 1899–1923 period from work suspensions, strikes, and lockouts amounted to less than 12 percent of the total time lost. Other causes of lost time—lack of demand, railroad car shortages, and mine-related factors—were more significant.

25. Stroup, "Cause and Growth of Unionism Among the Coal Miners," p. 468.

26. John Laslett, *Labor and the Left: A Study of Socialist and Radical Influences in the American Labor Movement, 1881–1924* (New York: Basic Books, 1970), pp. 192-241. Socialists began working with coal miners in the late nineteenth century. The West Virginia Miners' Union was a left-wing group that opposed Lewis's policies in the 1920s. John Brophy led the socialist Save-Our-Union Movement against Lewis in 1926. Communists organized the National Miners' Union in 1928 as an alternative to the UMWA. In the anthracite, an idealistic but non-Marxist protest movement demanding "equalization of available work" arose around UMWA insurgent Thomas Maloney in the 1930s. His group, United Anthracite Miners of Pennsylvania, fought Lewis and the companies, leaving hundreds dead. Socialists and Communists assumed leading positions among Nova Scotia coal miners who affiliated with the UMWA in 1919. J.B. McLachlan and "Red" Dan Livingstone, left-wing mine-worker leaders in Cape Breton in the 1910s and 1920s, advocated nationalization and union democracy in their Socialist politicking. John L. Lewis handled the Canadian radicals as he did those in Appalachia, the Pennsylvania anthracite, and the Midwest. He had them expelled from the union and placed their districts in trusteeship. See Paul MacEwan, *Miners and Steelworkers: Labour in Cape Breton* (Toronto: Samuel Stevens Hakkert & Company, 1976), pp. 39 ff.

27. *Minutes of the Fifteenth Annual Convention of the United Mine Workers of America*, Indianapolis, Ind., January 18–27, 1904, pp. 142-43, 151-52.

28. *Proceedings of the Eighteenth Annual Convention of the United Mine Workers of America,* Indianapolis, Ind., January 15–22, 1907, p. 420.

29. *Proceedings of the Twenty-Seventh Consecutive and Fifth Biennial Convention of the United Mine Workers of America*, Cleveland, Ohio, September 1919, pp. 868–70.

3. Too Many Mines, Too Many Miners

1. Mary Van Kleeck, *Miners and Management* (New York: Russell Sage Foundation, 1934), p. 186.

2. Most of America's growth in energy consumption occurred in the 1950–1970 period, when it nearly doubled from 34 quads to 67.3. Energy consumption grew by 13 percent from 1920

to 1930; 7 percent, 1930–1940; 43 percent, 1940–1950; 31 percent, 1950–1960; and 51 percent, 1960–1970. "Quad" is shorthand for one quadrillion Btus. One Btu (British thermal unit) is the quantity of heat required to raise the temperature of one pound of water 1° F. at or near its point of maximum density. Commonly used equivalencies include:

One ton of bituminous coal equals 24 million Btus
One quad equals 44 million tons of bituminous coal
One quad equals 172.4 million barrels of oil
One quad equals 10 billion cubic feet of gas

3. U.S. Department of Commerce, *Survey of Current Business*, July 1947 and July 1949, cited in Jules Backman, *Bituminous Coal Wages, Profits, and Productivity* (Washington, D.C.: Southern Coal Producers' Association, 1950), p. 69.

4. Use of 1920 as a base year significantly *underestimates* the steep fall in wages as a share of value. Wages represented only 49 percent of coal value of 1920 because of the extraordinary post-World War I boom that hiked the average price of a ton of coal to $3.75, $1.26 *more* than in 1919 and $0.86 more than in 1921. More typical of the early 1920s is the 70 percent + share found in other years.

5. The anthracite operators did not formally recognize the UMWA until 1920; however, a substantial part of the work force paid dues to the union and heeded strike calls. See Suffern, *The Coal Miners' Struggle*, p. 89.

6. I am indebted to Barbara Kritt, a doctoral student at the University of Michigan and research assistant at the Institute for Policy Studies in 1981, for this archival material.

7. Eliot Jones, *The Anthracite Coal Combination* (Cambridge, Mass.: Harvard University Press, 1914), pp. 62 ff. See also Arthur Donovan, "Carboniferous Capitalism: Excess Production Capacity and Institutional Backwardness in the U.S. Coal Industry," *Materials and Society* 7, no. 314 (1983).

8. Francis Walker, "The Development of the Anthracite Combination," *The Price of Coal: Anthracite and Bituminous*, in *The Annals*, vol. III (Philadelphia: The American Academy of Political and Social Science, 1924), p. 237. Walker was the chief economist of the Federal Trade Commission, which wrestled with the anthracite railroads for more than a decade. See also Jones, *Anthracite Coal Combination*.

9. Bogen, *The Anthracite Railroads*, provides detailed descriptions of the financial evolution of each of the rail lines.

10. Ibid., p. 237 ff.

11. Walker, "The Development of the Anthracite Combination," p. 238. A House of Representatives committee blamed exorbitant freight rates for eliminating independent mining companies from the anthracite trade. See U.S. House of Representatives, "The Alleged Coal Combination of the Philadelphia and Reading Railroad Company, etc.," *House Reports, No. 2278*, 52nd Cong. 2nd Sess., 1893, pp. iv–vi.

12. Ibid., p. 238 ff.

13. Hunt et al., *What the Coal Commission Found*, p. 403, quoting U.S. Coal Commission, *Final Report* (Mimeographed Report S), p. 7.

14. Walker, "The Development of the Anthracite Combination," p. 228.

15. Report of the United States Coal Commission on Anthracite, July 9, 1923, reprinted in *Coal Age* (July 12, 1923):54.

16. The U.S. Coal Commission predicted in 1923 that anthracite production would continue at about 89 million tons (the 1917 tonnage) until 1973, when depletion of the resource would force a steady decline.

4. John L. Lewis: A Paradox

1. The most recent and comprehensive biography of Lewis is Dubofsky and Van Tine, *John L. Lewis*. Their work is impressively detailed, but scants Lewis's activities in the 1950s.

2. Ibid., p. 45.

3. Tony Bubka, "The Harlan County Coal Strike of 1931," *Labor History* (Winter 1970):46.

4. A number of UMWA officials went over to the operators after leaving union jobs, and sometimes before. Daniel Weaver, a founder of the American Miners' Association in 1861, quit to take part in a mining company. He spent the rest of his life as a superintendent and mine boss in Illinois. See Wieck, *The American Miners' Association*, p. 190. Presidents Phil Penna, Michael Ratchford, and Tom Lewis went on industry payrolls after their terms. UMWA president John Mitchell accepted a high-paying job with the National Civic Federation. Anti-Lewis militant Frank Farrington was on the take from Peabody Coal Co. In the early 1970s, Joseph Brennan, head of the union's research department under Tony Boyle, became head of the Bituminous Coal Operators' Association, the industry's instrument for collective bargaining. In West Virginia, the Kanawha County Coal Operators took in UMWA District 17 president Joe Ellis after the Boyle defeat. Ellis said: "I don't think the new job will be difficult. It fits right in with my training and experience." See "Ellis Joins Operators," *UMWA Journal*, October 16–31, 1973, p. 10. Tom Pysell, a UMWA organizer under reform president Arnold Miller in the mid-1970s, later became the architect of a stop-the-UMWA campaign at an Arkansas coal mine in the early 1980s.

5. Thompson, *Technology, Labor and Industrial Structure*, discusses the implications of these changes.

6. The demand for equalization was a major source of intraunion conflict and strikes in the anthracite in the 1930s. Neither Lewis nor the operators favored equalization, although for different reasons. Both used cutbacks as a vehicle for ridding the labor force of rank-and-file activists. See Douglas Keith Monroe, "A Decade of Turmoil" (Ph.D. diss., Georgetown Univ., 1977), pp. 114 ff.

7. William Hard, "The Mind of Labor," in William Hard and Paul R. Leach, *Labor in a Basic Industry* (Chicago: Chicago Daily News, 1920), p. 13.

8. Searles's quote was issued as a press statement on January 29, 1923 (Brophy, *A Miner's Life*, pp. 170-71). Three presidents of the UMWA's anthracite districts proposed to the U.S. Coal Commission a "plan for the liquidation of the present ownership of the anthracite industry." Along with Searles and John Moore, they presented their statement, "Anthracite Accounting and Finance," to the commission in 1923. The UMWA calculated that *at the then-present rate of production*, the average cost of paying off all of the capital claims would come to about $0.28 per ton for fifty years (H.S. Raushenbush, *The People's Fight for Coal and Power* [New York: League for Industrial Democracy, 1926], pp. 28-31).

9. Brophy, *A Miner's Life*, pp. 149-75; and Sister M. Camilla Mullay, "John Brophy, Militant Labor Leader and Reformer: The CIO Years" (Ph.D. dissertation, The Catholic University of America, 1966), pp. 8-35. Brophy was a "labor progressive" who felt squeezed by theCommunist party on his left and Lewis on his right. He found his political mentors at the Brookwood Labor College and the Workers' Education Bureau, both of which he helped found. As president of UMWA District Two, he organized about twenty cooperative retail stores, a labor paper, a workers' education campaign, and a program for labor research.

10. Brophy, *A Miner's Life*, p. 218.

11. Letter from Brophy to Powers Hapgood, June 20, 1928, cited in Mullay, "John Brophy," p. 33.

12. Hard, "The Mind of Labor," p. 13.

13. Ellis Searles, editor of the *UMWA Journal*, quoted Lewis in a May 23, 1923 interview in the *New York World*. Also quoted in Joseph Finley, *The Corrupt Kingdom: The Rise and Fall of the United Mine Workers* (New York: Simon and Schuster, 1972), p. 61.

14. David J. McDonald and Edward Lynch, *Coal and Unionism: A History of American Coal Miners' Unions* (Silver Spring, Md.: Lynald Books, 1939), pp. 169, 178-79. McDonald and Lynch say the Jacksonville agreement was first abrogated in July 1925 in northern West Virginia by Consolidation Coal Company, "dominated by the Rockefeller interests." Wages fell to six dollars a

day. Three years later, the UMWA Policy Committee was forced to authorize all districts to negotiate "mutually satisfactory" contracts—that is, whatever they could get.

15. Williams, *West Virginia*, pp. 146-47.

16. Paul John Nyden, "Miners for Democracy" (Ph.D. diss., Columbia Univ., 1974), p. 424, quoting William F. Dunne, "The Government Begins Strikebreaking," *Daily Worker,* Aug. 26, 1927.

17. Ibid., p. 426, quoting *Daily Worker*, Sept. 12, 1927.

18. Anthracite's grip on its market first loosened during World War I when coal-short customers substituted other fuels. Richard Ramsey Mead, "An Analysis of the Decline of the Anthracite Industry since 1921" (Ph.D. dissertation, University of Pennsylvania, 1935).

19. Production fell from 84 million tons in 1926 to 51 million in 1940 as employment dropped from 165,400 to 91,300.

20. Monroe says Lewis "stood ready to trade the checkoff issue for arbitration at any time, and had the operators been willing to accept the status-quo, the strike would not have dragged on." Monroe, "Decade of Turmoil," p. 37.

21. Ibid., pp. 47-48.

22. One of the coal industry's historic leaders is Consolidation Coal Co., now a subsidiary of DuPont. Consolidation was chartered in the 1860s by a few small operators in western Maryland. The company quickly became the "creature of Boston and New York financiers," among whom was Jay Aspinwall, a prominent businessman (Frederick Gutheim, *The Potomac* [New York: Rinehart & Co., 1949], p. 226). Warren Delano, grandfather of Franklin D. Roosevelt, was elected a director in 1864, and James Roosevelt, Franklin's father, joined the board in 1868. (The yacht on which FDR sailed to Campobello in 1921 was owned by Van Lear Black, one of the largest stockholders in Consolidation.) Andrew Mellon and George M. Humphrey, two future secretaries of the treasury, were directors. In 1875, the Baltimore and Ohio Railroad bought a large block of the company's stock. "For thirty years thereafter very close relations existed between the railroad and the coal company, in which they exchanged coal properties for stock." Consolidation was the dominant coal supplier in 1900, producing over 18 million tons. The Pennsylvania Railroad "acquired sizable holdings in the Baltimore and Ohio" at the time that the Mellons were organizing Pittsburgh Coal (George H. Love, *An Exciting Century in Coal* [Princeton: Princeton University Press, 1955], p. 11). Under Pennsylvania president L. F. Loree, the railroad bought more stock in the coal company, to the point where Love says the railroad had control of "the greatest acreage of unmined coal in any one ownership in the Country." Former U.S. senator Clarence W. Watson of Fairmont, West Virginia, led the company between 1903 and 1928. Watson organized a stock-buyers' group that bought the railroad holdings in 1906 before passage of the Hepburn Act, which prohibited rail ownership in coal. In 1915, the Rockefellers purchased 38 percent of Consolidation after establishing themselves in the western coal business as owners of the Colorado Fuel and Iron Co. A year later, Pittsburgh Coal Company merged with the Monongahela River Consolidated Coal and Coke Company. This Pittsburgh-area producer was tightly tied into the local steel industry through Mellon, Henry R. Rea, Henry W. Oliver, and H. C. Frick, who sat as directors. Twenty years later, the two companies merged to form Pittsburgh-Consolidation Coal Co., which dominated the industry in the 1950s and 1960s. (The Rockefellers backed out of Consolidation in the late 1930s, after more than a decade without a dividend.) Consolidation's history can be found in Love's short narrative.

23. Monroe, "Decade of Turmoil," pp. 244 ff. Thomas Maloney told an insurgent rally in Wilkes-Barre on August 2, 1933: "You can talk about Philadelphia politics and rotten state, county and municipal politics but nothing was ever as corrupt as an election of the United Mine Workers of America" (pp. 244-45). Maloney's followers were denied seats at their district conventions and were purged from the organization. Monroe claims that hundreds died during the years of this rebellion.

24. The connections between the World War I experience and the coal-stabilization efforts of the late 1920s and early 1930s are discussed in James P. Johnson, *The Politics of Soft Coal: The*

Bituminous Industry From World War I Through the New Deal (Urbana: Univ. of Illinois Press, 1979).

25. U.S. Congress, Senate, Interstate Commerce Committee, *Hearings on S. 4490 to regulate, interstate and foreign commerce, require licensing, provide for consolidations and mergers, and create a Bituminous Coal Commission*, 70th Cong., 2nd sess., 1928.

26. U.S. Congress, Senate, Committee on Mines and Mining, *Hearings on S. 2935 to create a Bituminous Coal Commission, regulate interstate commerce*, etc., 72nd Cong., 1st sess., 1932. See also Coleman, *Men and Coal*, p. 21.

27. Dubofsky and Van Tine, *John L. Lewis*, p. 177. The Norris-La Guardia Act of 1932 protected unions from federal injunctions against boycotts, pickets, and strikes.

28. For information on price-fixing in coal during the 1930s, see Jules Backman, *Government Price-Fixing* (New York: Pitman, 1938); Ralph Hillis Baker, *The National Bituminous Coal Commisson: Administration of the Bituminous Coal Act, 1937–1941* (Baltimore: Johns Hopkins University Press, 1941); Waldo E. Fisher and Charles M. James, *Minimum Price Fixing in the Bituminous Coal Industry* (Princeton, N.J.: Princeton University Press, 1955); and Johnson, *The Politics of Soft Coal*. Basic sources of information on the 1930s include: Coleman, *Men and Coal;* Baratz, *The Union and the Coal Industry;* James A. Wechsler, *Labor Baron: A Portrait of John L. Lewis* (New York: Morrow, 1941); Baker, *The National Bituminous Coal Commission*; Muriel Sheppard, *Cloud by Day: The Story of Coal and Coke and People* (Chapel Hill: University of North Carolina Press, 1947); Backman, *Bituminous Coal Wages,* Waldo E. Fisher, *Collective Bargaining in the Bituminous Coal Industry: An Appraisal* (Philadelphia: University of Pennsylvania Press, 1948); Fred Greenbaum, "A New Deal for the Bituminous Coal Miners: The United Mine Workers of America and National Labor Relations Policy, 1933–1941" (M.A. thesis, University of Wisconsin, 1953); Miller, "The United Mine Workers"; Bernard Feder, "The Collective Bargaining" (Ph.D. diss., New York Univ., 1957); Saul Alinsky, *John L. Lewis: An Unauthorized Biography* (New York: Putnam's 1949; reprint edition, Random House, Vintage Books, 1970); and Charles K. McFarland, "Coalition of Convenience: The Roosevelt-Lewis Courtship, 1933 to 1941" (Ph.D. dissertation, University of Arizona, 1965).

29. United Mine Workers of America, *John L. Lewis and the International Union, United Mine Workers of America: The Story from 1917 to 1952* (Washington, D.C.: United Mine Workers of America, 1952), pp. 98-99.

30. In a number of cases, the operators settled strikes with Lewis, then disregarded their contracts. Benjamin Stolberg wrote of the 1920s: "Every victory left the union more exhausted, internally more disrupted, externally more disorganized, numerically dwindling. What the union refused to do nationally, the separate districts had to do. They had to back down on wages and conditions. The coal operators were chiseling on the contracts. . . . For the truth was that Mr. Lewis's victories were all Pyrrhic victories. They were won in the clauses of collective bargains and not in the trenches of the mines" (Benjamin Stolberg, "The Education of John L. Lewis," *The Nation*, August 8, 1936, p. 149).

31. Lewis had always been a Republican who described government control of coal as a "straight-jacket" *[sic]* that stunted initiative and enterprise. (See John L. Lewis, *The Miners' Fight for American Standards* [Indianapolis: Bell Publishing Co., 1925], p. 21.) His suspicion of federal authority was sharpened by his experience in federal courts, where the UMWA had been hampered by antitrust decisions, injunctions, and "yellow-dog" contracts (under which an employee promised to forgo union membership). Federal injunctions during the West Virginia mine wars in the 1920s even prohibited organizers from singing picket-line taunts such as: "Shoot them in the head,/ Shoot them in the feet,/ Shoot them in the dinner bucket;/ How are they going to eat?" (*New York Times*, September 26, 1925, quoted in Joanna H. Shurbet, "John L. Lewis: The Truman Years" [Ph.D. dissertation, Texas Tech University, 1975], p. 79.)

32. Backman, *Bituminous Coal Wages*, p. 64. Backman's data were based on those collected by the U.S. Bureau of Internal Revenue.

33. Commerce Department, *Historical Statistics*, part 1, pp. 590-91. Coal, on the other hand, doubled its sales to electric utilities, which in 1955 accounted for 33 percent of bituminous consumption, and maintained sales of metallurgical and industrial coals at roughly the levels of 1945.

34. The consumption of bituminous coal for all purposes by U.S. railroads dropped from 123 million tons in 1945 to 2.2 million tons in 1960. In the same period the railroads' fuel oil consumption dropped from 4.7 billion gallons to 233 million, while diesel fuel rose from 441 million gallons to 3.6 billion. *Historical Statistics of the United States*, part 2.

35. *National Bituminous Coal Wage Agreement of 1947*, p. 1.

36. In 1935, a coal miner earned $957, compared with $1,216 in manufacturing and $1,295 in iron and steel. In 1948, a miner's annual income was $3,387, compared with $3,040 in manufacturing and $3,392 in iron and steel. Data from U.S. Department of Commerce included in Backman, *Bituminous Coal Wages*, p. 29.

37. As of the mid-1940s, it was estimated that 70 percent of coal miners were covered by a company-doctor plan. Joel T. Boone, *A Medical Survey of the Bituminous Coal Industry*, Coal Mines Administration (Washington, D.C.: U.S. Department of the Interior, 1947).

38. Warren F. Draper, "The Quest of the UMWA Welfare and Retirement Fund for the Best Medical Care Obtainable for Its Beneficiaries," paper presented to the annual meeting of the American Association for the Surgery of Trauma, Hot Springs, Va., November 1, 1957. Draper was the medical director of the Fund from 1948 to 1968.

39. Dubofsky and Van Tine, *John L. Lewis*, p. 376.

40. Leslie A. Falk, "Group Health Plans in Coal Mining Communities," *Journal of Health and Human Behavior* 4 (Spring 1963):6.

41. Leslie A. Falk, "Collective Bargaining for Medical Care Benefits: A Recent Development in the USA," *British Journal of Preventive and Social Medicine* 7 (1953):88. The Department of Labor estimated that 600,000 American workers "were covered by health benefit plans established through collective bargaining" by 1945. Falk, "Group Health Plans," p. 6, quoting F. Peterson, E. Kassalow, and J. Nelson, *Monthly Labor Review* 61 (1945):191.

42. Roy, *The Coal Mines*, p. 248.

43. I draw on several studies in this discussion, among them Seltzer, "Health Care by the Ton: Crisis in the Mine Workers Health and Welfare Programs," *Health/PAC Bulletin*, no. 79, (November/December 1977); Seltzer, "The United Mine Workers of America and the Coal Operators," pp. 180 ff.; Ploss, "A History of the Medical Care Program," pp. 20 ff.; and Barbara Ellen Smith, "Digging Our Own Graves: Coal Miners and the Struggle Over Black Lung Disease" (Ph.D. dissertation, Brandeis University, 1981).

44. The original proposal did not emphasize prevention of occupational disease, only its treatment. Lewis did not demand that the industry adopt radical measures to improve safety and health conditions. The focus on treatment rather than prevention framed Fund politics throughout its history. Lewis displayed some concern over safety after World War II, but almost none over occupational health. He did win miner-elected safety committees in the 1947 contract, and urged the U.S. Bureau of Mines to take a more active role in safety. Lewis used the safety issue with a certain opportunism: he'd recite the appalling toll when he wanted something and stop when he was either placated or defeated. He appears to have become more genuinely interested in safety the more he exploited the issue.

45. *UMWA Journal*, May 15, 1946, p. 4.

46. Ibid.

47. *UMWA Journal*, April 15, 1946, p. 3.

48. The secretary agreed that UMWA miners at each mine could appoint a mine-safety committee with the right of withdrawal in especially dangerous conditions—a demand that had been on the UMWA agenda for years. The U.S. Bureau of Mines was to draft a mine-safety code that would include periodic mine inspections.

49. The Boone Report is remarkable for its portrait of coalfield conditions. The report was supposed to determine whether need justified Lewis's demand for an employer-funded medical plan. Its findings and photographs supplied an unequivocal "Yes." Boone's team of navy medical and technical personnel surveyed the housing, sanitation, hygiene, medical care, and recreation of almost seventy-two thousand miners at 260 mines the government seized. Coal counties, they reported, had substantially higher infant mortality rates than the national average. Almost half of the communities used water that was contaminated by industrial waste or untreated sewage. Housing and sanitary conditions were generally better for miners who did not live in coal camps (p. 59). The survey demonstrated the inadequacy of coalfield medical care: the vast majority of physicians were in general practice and "have not been made responsible for adequate programs of industrial hygiene and preventive medicine" (p. 92); about 70 percent of the miners were covered by a "pre-payment" plan financed by a payroll checkoff; company doctors "were not selected primarily on the basis of professional qualifications . . . but on the basis of personal friendships, financial tie-ups, social viewpoints or other non-medical considerations" (p. 123); hospitalization plans—where they existed—were usually financed by a separate checkoff, but did not cover many services; miners had no choice among hospitals—which, in many cases, company doctors had set up as for-profit institutions; and three-fourths of the hospitals surveyed were inadequate in such basics as surgery and labor rooms or X-ray facilities. The report concluded: "The present practices of medicine in the coal fields on a contract basis cannot be supported. They are synonymous with many abuses. They are undesirable and, in numbers of instances, deplorable" (p. 164). The report recommended that a "pre-payment system, with plans financed by wage deductions and predicated on a freedom of choice of physicians and hospitals by the beneficiaries, would be best" (p. 164).

50. Dubofsky and Van Tine, *John L. Lewis*, pp. 458 ff. They stress the administration's concern over the 1946 elections in taking a strong stand against Lewis.

51. Newspapers have carried headlines and stories of this type in every major coal strike.

52. Dubofsky and Van Tine, *John L. Lewis*, p. 466.

53. The Taft-Hartley Act (1947) stipulated that unions could not have exclusive power to manage welfare funds to which employers contributed. Labor and management were to have equal representation.

54. In contrast, almost all other union health plans are financed by employer and/or employee contributions per worker.

55. Initially, all medical needs were provided without charge. The Fund's comprehensive phase lasted through most of 1949. Widespread health-provider abuse—unneeded dental work, false billings, inflated fees—forced cutbacks.

56. Differences of opinion existed within the Fund over the consumer orientation of the medical services and other issues. See Ploss, "A History of the Medical Care Program."

57. Draper recruited another large contingent from the U.S. Army. Many of the Fund's progressive ideas came from Henry Daniels, Leslie Falk, Lorin Kerr, Allen Koplin, and Thomas Berret, who had worked on the migrant-farm labor health program for the Public Health Service in the late 1940s.

58. Finley, *The Corrupt Kingdom*, p. 162. Roche, the liberal daughter of a coal operator, had owned the Rocky Mountain Fuel Company in Colorado. She signed a UMWA contract in the early 1930s. Lewis set up a Delaware holding company, Lewmurken, Inc., in 1942, that invested UMWA funds in assorted enterprises, one of which was 30 percent of the Roche-family coal company. The UMWA "also loaned $1,451,104 to Rocky Mountain Fuel Company on a demand note. Even with the infusion of such an amount of union money, the Rocky Mountain Fuel Company went into receivership before the year was out. The note was carried year after year on the books in its full amount, with no chance that it would ever be repaid" (Finley, *The Corrupt Kingdom*, p. 162). The UMWA still owns this stock, which is now valuable for its Colorado land and water rights. Lewmurken made other odd investments with coal miners' money, including a $375,000 loan to the Freeport Coal Company, which maintained a 20,000-acre coal property in northeastern West

Virginia that Freeport leased to a nonunion operator. See Brit Hume, "Coal Mining: 1. The Union," *Atlantic* (November 1969):28.

5. The Lewis-Love Axis and the Great Shakeout

1. "Strategic Position," *Fortune* (June 1949):182

2. "Coal *vs.* the People," *Fortune* (October 1949):71.

3. *Coal Age* (September 1950):140.

4. "Editorial," *The Nation*, April 3, 1948, p. 367.

5. See C. L. Thompson, "Should the Coal Mines Be Nationalized," *Forum* (November 1949); Willard Shelton, "Labor: Crisis in Coal," *New Republic* (April 28, 1947); and "Editorial," *The Nation*, pp. 366–367.

6. Craufurd D. Goodwin, "The Truman Administration: Toward a National Energy Policy," in *Energy Policy in Perspective: Today's Problems, Yesterday's Solutions*, ed. Craufurd D. Goodwin (Washington, D.C.: The Brookings Institution, 1981), pp. 9 ff.

7. "Truman Administration Policies Toward Particular Energy Sources," ibid., p. 140.

8. Ibid., pp. 141-42.

9. "Lewis' Apparent Heir," *Fortune* (May 1949):190. Lewis was most comfortable and successful with summit diplomacy. As CIO leader in 1937, he "had scored his greatest collective-bargaining victory—the unionization of Big Steel—not on the picket line or via sit-down strikes, but by having breakfast with the chairman of the board of U.S. Steel, Myron C. Taylor, at Washington's Mayflower Hotel" (Nat Caldwell and Gene S. Graham, "The Strange Romance Between John L. Lewis and Cyrus Eaton," *Harper's* [December, 1961]:27).

10. "Moses Describes BCOA Relations with UMW," *UMWA Journal*, May 1, 1955, p. 11. The *UMWA Journal* quoted from Moses's speech, "The New Look," which he gave to the College of Commerce at West Virginia University on April 15, 1955.

11. See Edie Black and Fred Goff, *The Hanna Industrial Complex* (New York: North American Congress on Latin America, 1969). In the late 1960s, George Love explained the reasons for merging Consolidation Coal and Pittsburgh Coal in 1945: "And there was no technology. Nobody could afford it. No company was large enough to work on it. When we finally got the present Consolidation Coal Company together in the mid-40s, it had the assets and personnel to develop technology and do research. . . . If coal was going to be competitive with other fuels . . . you had no choice but to mechanize ("Uniting for Strength," *Nation's Business* [January 1967]:49.)

12. Thomas N. Bethell, *Conspiracy in Coal* (Huntington, W. Va.: Appalachian Movement Press, 1971), pp. 13-14.

13. "Geo. Love Offers Basis for Resuming Negotiations with Miners' Union," statement released on January 26, 1950, reprinted as *National Coal Association Industry Bulletin*, no. 2749 (January 27, 1950):2.

14. "UMWA Cited for Contempt of Court as Miners Continue Nationwide Stoppage," *Coal Age* (March 1950):141.

15. Ibid.

16. Without an "able-and-willing" clause, miners cannot legally strike except when their contract ends. In 1970, the U.S. Supreme Court said the existence of a binding arbitration provision in a contract *implies* the existence of a no-strike pledge by the union. The extension of this line of reasoning enabled coal companies to obtain federal injunctions and contempt penalties against locals and miners who stopped work for *any* reason, even those such as gasoline rationing and controversial textbooks having nothing to do with the contract. This issue came to a head in the mid-1970s when UMWA miners stopped work to protest court rulings denying them the right to strike. In September 1976, 120,000 miners struck for several weeks in protest over contempt-of-court fines of $50,000 against a West Virginia local for illegal striking and contempt citations against 213 miners. The strike forced the federal judge to back down.

17. "UMWA Cited for Contempt of Court as Miners Continue Nationwide Stoppage," *Coal Age* (March 1950):141.

18. "Court Rejects T-H Injunction Charge," *UMWA Journal*, March 15, 1950, p. 13.

19. "Coal's Future Good, Love Tells Utility Men," *Coal Age* (June 1950):191.

20. "Continuous Coal Mining," *Fortune* (June 1950):114.

21. Ibid., p. 111.

22. Ibid., p. 127.

23. Lewis, *The Miners' Fight*, pp. 108-109, 113.

24. "Lewis Gains in Wages, Welfare," *Coal Age* (October 1952):162.

25. "Lewis Blisters WSB as Miners Walk Out," *Coal Age* (November 1952):135.

26. An extensive mechanization debate occurred at the 1956 UMWA convention before delegates were to vote on the contract that Lewis negotiated. Following the debate, Lewis asked: "Do any of you boys want to make a motion back there? I want to hear an affirmative motion on this contract. If you make a negative one, I won't recognize it" (United Mine Workers of America, *Proceedings of the 42nd Constitutional Convention* [Washington, D.C.: United Mine Workers of America, 1956], p. 312). See Smith, "Digging Our Own Graves," pp. 161 ff, for a discussion of this issue.

27. "John L. at Bat," *Fortune* (March 1952):52.

28. Lewis, *The Miners' Fight*, pp. 110-11.

29. Speech by Joseph E. Moody to the National Petroleum Association on September 14, 1950 quoted in *Business Week* (September 23, 1950).

30. Ibid.

31. Bethell, *Conspiracy in Coal*, p. 17.

32. *Fortune* mentioned in 1952 that Lewis's demand of 200 days' guaranteed annual wage of 1949–50 was "traded off [for] the demand for control of the welfare fund" ("John L. at Bat," 52).

33. *Coal Age* (April 1950):135.

34. *National Bituminous Coal Wage Agreement of 1950*, (March 5, 1950): p.4.

35. At the UMWA's 1968 convention, Dr. Lorin E. Kerr, an assistant to the medical director of the Welfare and Retirement Fund, said black lung disease had been known for years and "the failure to take earlier action constitutes what may be labeled in the future as the greatest disgrace in the history of American medicine."

36. Annual reports for representative companies, such as Consolidation, Peabody, Westmoreland, Eastern Gas and Fuel, and Pittston, show a slow but steady climb in net income in the 1950s and 1960s.

37. "Contract Signed," *National Coal Association Bulletin* (March 6, 1950).

38. Bethell, *Conspiracy in Coal*, p. 8.

39. "Contract Signed."

6. Industry Peace and Labor Repression

1. "Lewis Gains in Wages, Welfare," *Coal Age* (October 1952):126.

2. Despite Lewis's record of wage increases, *Fortune* reported in August 1954 that soft-coal wages ranked seventh in 1945 and seventh again in 1953 among major industries, although they had advanced considerably from their 1939 rank of sixteenth. In 1953, coal miners' weekly wages ($85.38) were less than those of workers in construction ($91.70), in petroleum and coal products ($90.41), in crude petroleum and natural gas ($90.35), in automobiles ($87.76), in steel and iron ($87.44), and in printing and publishing ($85.67). The spread in *annual* incomes, of course, was even greater, considering the fewer number of days miners worked each year. See "Industry's Wage Score Card: Who's Up or Down," *Fortune* (August 1954):48, based on Bureau of Labor Statistics data. See also John P. David, "Worker Earnings in Bituminous Coal Since Mechanization," *Atlantic Economic Journal* (April 1974), and "Practices of the United Mine Workers of

America Within the Bituminous Coal Industry: 1950–1970," *Journal of Economics* 1 (1975).

3. "Lewis Gains in Wages, Welfare," p. 128.

4. "Lewis Blisters WSB as Miners Walk Out," *Coal Age* (November 1952):131.

5. "Moody Asks New Approach as Moses Denies Captive Rule," *Coal Age* (August 1953):129.

6. "Southern Group Weighs Cuts in Labor Costs," *Coal Age* (December 1954):117.

7. "John L. and After," *Fortune* (November 1954):72.

8. "The Gentlemen of Coal," *Fortune* (October 1955):70.

9. "Soft Coal Talks at a Standstill," *Coal Age* (December 1958):28.

10. See James Branscome, "TVA: Coal Industry Villain or Victim," *Mountain Review* (Winter 1976). Branscome quotes a May 26, 1954, letter from Ely Brandes, director of the Fuel Economics Branch of TVA, to a friend, Hyman Bookbinder, at the CIO. Brandes admits in the exchange that TVA suppliers "are paying '30–40% below the UMW scale' " (p. 3).

11. "Coal Presses Fight for Square Deal," *Coal Age* (June 1954):137.

12. *Business Week*, July 10, 1954, p. 31.

13. Ibid.

14. "Coal Presses Fight."

15. "White House Sets Coal Study as Governors Ask Action," *Coal Age* (August 1954):115. Emphasis in the original.

16. "Coal: Problems and Progress in '54, the Road Ahead in '55," *Coal Age* (February 1955):57.

17. "Expansion Proposed by Peabody," *Coal Age* (June 1955):146.

18. "Peabody, Sinclair to Combine; Island Creek, Pond Creek Merge," *Coal Age* (June 1955):112.

19. "W. Kentucky Buys Nashville Coal, Becomes Third Ranking Producer," *Coal Age* (October 1955):112. For elaboration on the West Kentucky-Nashville deal, see Caldwell and Graham, "The Strange Romance," pp. 28-29. They state that Eaton paid Nashville's Jet Potter more than three times what his company was worth—perhaps as much as $30 million, of which about $7 million was a cash down payment by the UMWA.

20. Phillip E. Griffin, *Industrial Concentration and Firm Diversification in Bituminous Coal with Special Reference to the Southeastern United States, 1950–1970* (Knoxville, Tenn.: Appalachian Resources Project, 1972), p. 114.

21. George Love, "Some Benefits of Looking Backward: Are Mergers Evil?" speech at the Wharton School, University of Pennsylvania, November 18, 1965.

22. On the Eisenhower administration's energy policy, see William J. Barber, "The Eisenhower Energy Policy: Reluctant Intervention," in *Energy Policy in Perspective*, ed. Goodwin, pp. 205-86.

23. "The Gentlemen of Coal," 69-70.

24. "$2 Miners' Raise Marked by Peaceful Negotiations," *Coal Age* (October 1956):134.

25. "Price Rises Follow Raise to UMWA Miners," *Coal Age* (November 1956):128.

26. United Mine Workers of America, *Forty-Second Annual Convention Proceedings*, October 3, 1956, pp. 308-309.

27. Ibid., p. 437.

28. Lewis expressed his opinion on union democracy in these words: "Anybody who does not think the UMWA is a democratic institution and who thinks it deprives them of their inherent and native-born rights ought to go over to the United Auto Workers. They would get all the democracy they want over there . . . A new pamphlet would be issued to them every day. They would be told how to settle the situation in Persia, Cambodia, Viet Nam, or Goa—but not much on how to get the wages the UMWA is going to get" ("Price Rises Follow Raise to UMWA Miners," *Coal Age* [November 1956]:128).

29. "Soft Coal Talks at a Standstill," p. 28.

30. *National Bituminous Coal Wage Agreement of 1950 as Amended Effective December 1, 1958*, pp. 2-3.

31. "Coal Wage Clause Ruled Illegal," *Coal Age* (October 1963):47.

32. Ibid., p. 50. Although the U.S. Court of Appeals for the D.C. Circuit remanded the NLRB's decision in mid-1965, the clause was basically inoperative with the NLRB ruling. See "Coal Wage Decision Sent Back to NLRB," *Coal Age* (September 1965):58.

33. See *Apex Hosiery Company v. Leader* (1940); *United States v. Hutcheson* (1941); and *Allen Bradley v. Local Union No. 3, IBEW* (1945).

34. "UMW Guilty of Antitrust Violation," *Coal Age* (June 1961):26.

35. *Pennington v. UMWA*, quoted in *Tennessee Consolidated Coal Company et al v. United Mine Workers of America*, 72 L.R.R.M. 2312 (1970).

36. *Tennessee Consolidated Coal Company*, 72 L.R.R.M. 2312 (1970).

37. Ibid.

38. *South-East Coal Company v. Consolidation Coal Company*, 75 L.R.R.M. 2636 (1970).

39. "UMW and Coal Company Sued," *Coal Age* (June 1966):52. Emphasis added.

40. Bethell, *Conspiracy in Coal*, p. 31. This account relies on Bethell's investigation of the South-East case.

41. *New York Times*, June 30, 1953, p. 32.

42. Brit Hume, *Death and the Mines* (New York: Grossman, 1971), p. 27.

43. "Labor Violence Continues in Preston County Drive," *Coal Age* (May 1951):164.

44. "UMW Ends 15-Mo. Drive against Elk River Coal and Lumber Co.," *Coal Age* (February 1954):139. See also "FBI Investigates Violence in Widen, WV, Strike," *Coal Age* (December 1952):126.

45. "Firm Wins $225,000 from UMWA; Pa. Injunctions Set," *Coal Age* (September 1953):135.

46. "UMWA Sues Companies and Law Officials in Ky. Drive," *Coal Age* (October 1951):136.

47. "UMWA Auto Blasted in Kentucky," *Coal Age* (February 1952):174.

48. "Ky. Organizer Shot; Picket Convicted in Widen Slaying," *Coal Age* (September 1953):174, 176.

49. "Bridge Blasted," *Coal Age* (August 1952):146.

50. Ibid.

51. "UMWA Agrees to Forget Claim for $2,000,000 in Ky. Suit," *Coal Age* (May 1953):134.

52. "Kentucky Indictment Upheld," *Coal Age* (October 1953):139.

53. Ibid.

54. "Ousted Official Sues UMWA; Arrests Follow Disclosures," *Coal Age* (April 1952):131.

55. Hume, *Death and the Mines*, pp. 26-27.

56. Ibid., p. 28.

57. Finley, *The Corrupt Kingdom*, pp. 144-45, and Hume, *Death and the Mines*, p. 28.

58. Finley, *The Corrupt Kingdom*, pp. 145-52, gives a detailed account of this campaign.

59. *UMWA Journal* (November 15, 1959):6.

60. Finley, *The Corrupt Kingdom*, p. 162.

61. Lewis invited Eaton to the UMWA's 1956 convention where he introduced him as "one of America's foremost industrialists and financiers . . . [who] if perchance, he should buy any additional coal companies in this country, and I hope he will, I am quite assured that again his official act will bring that company . . . under the industry agreement with the United Mine Workers of America," *Forty-Second Annual Convention Proceedings*, Oct. 9, 1956, p. 545.

62. Edward A. O'Neill, "Industrialist Cyrus Eaton Dies at 95," *Washington Post*, May 11, 1979. Eaton's friendship with Lewis preceded his UMWA-funded deals in the 1950s. Eaton

supported labor's claim to higher wages and better conditions as a way of saving capitalism from class-based turmoil. Alone among industrialists, he praised Lewis's behavior during World War II. The two men shared a love of poetry and great books, and their tie was one of kin as well—Lewis's mother-in-law was an Eaton. In 1958, Eaton proposed that Lewis be named to replace Secretary of State John Foster Dulles, a man he described as "an insane fanatic" (E. J. Kahn, Jr., "Communists' Capitalist, Parts I and II," *New Yorker*, October 10 and 17, 1977). The *New York Times* of September 20, 1960 reported that the UMWA had given nearly $20 million in loans to Eaton and his companies in the 1950s; the true amount appears to be about $5 million more than that.

63. Finley, *The Corrupt Kingdom*, p. 164.

64. Ibid., p. 165. Eaton and Lewis used UMWA and Fund money to buy electric utility stocks to ensure the purchase of union coal. Between 1955 and 1963, mine-worker money purchased 140,000 shares of Cleveland Electric Illuminating Company and Eaton became a director of his hometown utility. In the same period, the Fund bought 55,000 shares of Kansas City Power and Light, and loaned Eaton $27,000 to buy his own shares. Eaton exercised the Fund's proxy. The Fund also became a stockholder in Union Electric, Ohio Edison, West Penn Electric, Southern Company, and Consolidated Edison. Judge Gerhard Gesell of the U.S. District Court in Washington, D.C., found all of this to be a "clear case of self-dealing on the part of trustees Lewis and Schmidt [Henry Schmidt, the industry's representative], and constituted a breach of trust" (Finley, *The Corrupt Kingdom*, p. 167). See *Blankenship v. Boyle*, 77 L.R.R.M. 2140 (U.S. District Court, D.C., April 28, 1971).

65. Finley, *The Corrupt Kingdom*, p. 165.

66. Ibid.

67. *New York Times*, September 20, 1960 quoting UMWA financial statements.

68. Dubofsky and Van Tine, *John L. Lewis*, pp. 506-10.

69. He continued as the UMWA trustee on the Welfare and Retirement Fund until his death in 1969.

70. Griffin, *Industrial Concentration and Firm Diversification*, p. 114. Griffin's data exclude captive coal producers, all of whose coal, generally, was consumed by the parent company.

71. By the mid-1970s utilities purchased about 80 percent of their coal under long-term contracts (Charles River Associates, *Coal Price Formation*, EPRI-EA-497 Final Report [Cambridge, Mass.: 1977]). These contracts protect consumers and suppliers from changes in price, supply, and coal quality. In the 1950s, coal operators used their contracts' promise of guaranteed sales as collateral to secure financing. Many of these early contracts included only perfunctory price-escalator clauses or none at all, because suppliers believed that productivity increases would control their costs. This assumption held until the 1970s. Then, a number of suppliers—Peabody Coal, in particular—were caught between rising costs and locked in prices. Contracts negotiated after the mid-1970s contain carefully designed price-escalator clauses to protect supplier profits.

72. That strike was against the Truman administration rather than the BCOA.

7. Tony Boyle and the Black Lung Rebellion

1. See Smith, "Digging Our Own Graves," p. 187.

2. *Fortune 500*, July 1955 and July 1960.

3. *National Bituminous Coal Wage Agreement of 1950 as Amended* (effective Dec. 1, 1958).

4. Hume, *Death and the Mines*, pp. 38-39.

5. The contract included a new provision that BCOA operators pay an $0.80-per-ton royalty on coal bought from non-UMWA operators for cleaning or marketing. The National Independent Coal Operators' Association (NICOA), successor to the Southern Coal Producers' Association, attacked the discriminatory royalty. NICOA, which accounted for about 30 percent of national coal production, said the new royalty on non-UMWA coal was "making it prohibitively costly for unionized coal operators to buy, for resale, coal that has been mined by non-union workers" ("New

Bituminous Coal Contract Signed," *Coal Age* [May 1964]:26). NICOA sued the UMWA and the BCOA before the National Labor Relations Board and won the point.

6. *UMWA Journal*, January 15, 1965.

7. *Wall Street Journal*, September 24, 1965, p. 4.

8. The "qualification-to-perform-work" language was not changed. However, miners who were laid off because of mine closure could, if they requested, be placed on a hiring list of their employer's mines in that UMWA district. If his employer rehired him at another mine, a miner started at the bottom of the job scale.

9. The UMWA Constitution has prohibited wildcat strikes for years.

10. Keith Dix et al., *Work Stoppages and the Grievance Procedure in the Appalachian Coal Industry* (Morgantown, W.Va.: Institute for Labor Studies, 1972), pp. 7-8. There were twenty-two other major work stoppages in this period.

11. *Sinclair Refining Company v. Atkinson*, 370 U.S. 195 (1962).

12. *Boys Market v. Retail Clerks International*, 398 U.S. 235 (1970).

13. My discussion draws on Smith, "Digging Our Own Graves," pp. 227 ff.; Ploss, "A History of the Medical Care Program"; Nyden, "Miners for Democracy"; Hume, *Death and the Mines*; Finley, *The Corrupt Kingdom*; Seltzer, "Coal Miners and Coal Operators" and "Health Care by the Ton."

14. Hume, *Death and the Mines*, p. 31.

15. Smith, "Digging Our Own Graves," p. 230.

16. One denial letter from the Fund to a miner read: "This benefit is subject to suspension or termination at any time by the trustees of the fund for any matter, cause or thing, of which they shall be the sole judges, and without any assignment of reason therefore" (Jeanne Rasmussen, "On the Outside Looking In," *Mountain Life and Work* [September 1969]:9, quoted in Smith, "Digging Our Own Graves," p. 237).

17. Hume, *Death and the Mines*, p. 34.

18. *Blankenship v. Boyle*, p. 1101.

19. Interview with Robert Payne cited in Nyden, "Miners for Democracy," p. 710.

20. George W. Hopkins, "The Miners for Democracy" (Ph.D. diss., Univ. of North Carolina at Chapel Hill, 1976), p. 211.

21. *Black Lung Bulletin* #3, p. 2, cited in Hopkins, "The Miners for Demoocracy," p. 213.

22. Rasmussen testimony in U.S. Congress, Senate, *Hearings before the Subcommittee on Labor of the Committee on Labor and Public Welfare on the UMWA Welfare and Retirement Fund*, 91st Cong., 2nd Sess., 1971, p. 42.

23. Hopkins, "The Miners for Democracy," p. 221.

24. Interview with Robert Payne, cited in Nyden, "Miners for Democracy," p. 714.

25. F.C., *The Compleat Collier*, p. 23.

26. Friedrich Engels, *The Condition of the Working Class in England* (1892; reprint ed., London: Panther Books, 1969), p. 272.

27. "Report of the Schuylkill County Medical Society," *Transactions of the Medical Society of Pennsylvania*, 5th ser., part 2 (June 1869), p. 490; quoted in Smith, "Digging Our Own Graves," 29. Smith's dissertation on black lung disease and its politics is the best work on the subject.

28. U.S. Department of the Treasury, Public Health Service, *The Health of Workers in Dusty Trades*, bull. 208 (Washington, D.C.: Government Printing Office, 1933, part 3).

29. Ibid., p. 18.

30. Joseph E. Martin, Jr., "Coal Miners' Pneumoconiosis," *American Journal of Public Health* 44 (May 1954):581.

31. Jan Lieben and W. Wayne McBride, "Pneumoconiosis in Pennsylvania's Bituminous Mining Industry," *Journal of the American Medical Association* 183, no. 3 (January 19, 1963).

32. W. S. Lainhart et al., *Pneumoconiosis in Appalachian Bituminous Coal Miners* (Washington, D.C.: Government Printing Office, 1969).

33. Earl H. Rebhorn, "Anthraco Silicosis," *Medical Society Reporter* 29, no. 5 (May 1935):15.

34. Andrew Meiklejohn, "History of Lung Disease of Coal Miners in Great Britain: Part II, 1875–1920," *British Journal of Industrial Medicine*, 9, no. 2 (April 1952):94.

35. W. Donald Ross et al., "Emotional Aspects of Respiratory Disorders Among Coal Miners," *Journal of the American Medical Association* 156, no. 5 (October 2, 1954).

36. The respirability of an individual particle is determined by shape as well as size, but it is measured by weight. The 1969 Federal Coal Mine Health and Safety Act (PL 91-173) defined respirable dust as 2.0 milligrams of coal-mine dust per cubic meter of air. Larger particles are called nonrespirable, which means that most are caught in the upper bronchial tract. The respirable particles penetrate the lower lungs where oxygen is exchanged for carbon dioxide. Most nonrespirable particles are ejected, but some are retained and may contribute to industrial bronchitis and breathlessness. See Curtis Seltzer, "Coal Mine Health" (Washington, D.C.: Congressional Office of Technology Assessment, 1978), p. 16.

37. Interested readers should pursue Smith's discussion of the economic and ideological sources of black-lung disease ("Digging Our Own Graves").

38. The UMWA was not alone among American labor organizations in neglecting occupational health. See Curtis Seltzer, *Surveying and Analyzing the Field of Employee Rights Related to Occupational Disease* (Washington, D.C.: U.S. Department of Labor, July 1979).

39. Smith, "Digging Our Own Graves," 145-46. Roche's rationale, according to Smith, was that industry participation in financing and managing the Fund made investigations inappropriate.

40. J. Russell Boner, "New UMW President Takes Stiffer Stand on Coal Mine Automation," *Wall Street Journal*, March 15, 1963.

41. United Mine Workers of America, *Proceedings of the 45th Constitutional Convention* (Washington, D.C.: United Mine Workers of America, 1968), p. 313.

42. By taking their complaint directly to the legislature, miners did exactly as others had done years before. In the 1840s, British miners "sent agents up to London to urge upon the legislature the necessity of legislation and of government inspection for the proper security of their lives and safety" (Roy, *The Coal Mines*, p. 177). In 1870, following the Avondale disaster, anthracite miners "framed a bill providing for the proper protection of the health and safety of persons employed in the anthracite coal mines . . . , and they sent a committee of intelligent miners to the State Capital to urge upon members of the legislature the necessity of the passage of the bill" (ibid., p. 181).

43. Before 1972, coal producers had to pay only $1.35 per $100 of sales. From 1953 to 1972, coal's share of total state tax revenue was less than 10 percent, even though its share of gross state product was generally double that. See Seltzer, "Miners and Operators," pp. 730-32, based on data supplied by the Business Research Division of the C&P Telephone Company, April 1976 (on coal's share of gross state product); and West Virginia Tax Department, *Thirty-Fifth Biennial Report* (Charleston, W.Va.: Tax Department), pp. 22-23.

44. Initially, some of the BLA's allies got the jitters over a strike with which they were associated. Kaufman called it a "mistake." Hechler said: "I don't believe that endless, bitter strikes and walkouts will either solve the problem or result in good legislation." The UMWA denounced the strike, as did the coal operators. See Hume, *Death and the Mines*, p. 134.

45. *Washington Post*, February 27, 1969, p. A-4.

46. Jim Ragsdale, "Arnold Miller Recalls Days Gone By in Fight for Black Lung Legislation," *Charleston* (W. Va.) *Gazette*, January 13, 1980.

47. *Raleigh Register and Post Herald*, March 1, 1969, pp. 1,8, cited in Smith, "Digging Our Own Graves," 346-47.

48. Interview with black-lung activist recalling the anonymous legislator's statement to him, cited in Smith, "Digging Our Own Graves," p. 350.

49. Quoted in Robert G. Sherrill, "The Black Lung Rebellion," *The Nation*, April 28, 1969, p. 530.

50. Hume, *Death and the Mines*, p. 9.

51. See Seltzer, "Coal Mine Health"; Congressional Office of Technology Assessment, *The Direct Use of Coal: Problems and Prospects of Production and Consumption* (Washington, D.C.: Office of Technology Assessment, 1979), pp. 266 ff.; and James Weeks, "Data for the U.S. Standard for Permissible Exposure to Coal Mine Dust" (Washington, D.C.: OTA, November 1978).

52. The federal dust-sampling program is intended to indicate the level of compliance with the 2 mg. standard. Dust conditions are indeed better now than they were before 1970, but the sampling program is badly flawed. The samples are infrequent, and there is no guarantee that dust levels are in compliance on days the samples are not taken. For the federal standard to be effective, compliance must be universal and consistent over time.

Most investigations of the federal dust-sampling program in the 1970s found numerous weaknesses. The General Accounting Office noted "many weaknesses in the dust-sampling program which affected the accuracy and validity of the results and made it virtually impossible to determine how many mine sections were in compliance with statutorily established dust standards" (U.S. Congress, General Accounting Office, *Improvements Still Needed in Coal Mine Dust Sampling Program and Penalty Assessments and Collections* [Washington, D.C.: GAO, December 31, 1975], p. i). The National Bureau of Standards confirmed GAO's findings: "When the miners and mine operators perform and supervise the sampling and when the weighings are made in the normal manner, . . . the uncertainty [of accurate measurements] is *estimated* to be as large as 31 percent or 0.63 mg /m^3" (National Bureau of Standards, *An Evaluation of the Accuracy of the Coal Mine Dust Sampling Program Administered by the Department of the Interior, a Final Report to the Senate Committee on Labor and Public Welfare* [Washington, D.C.: Department of Commerce, 1975], p. iii).

Sharp surveyed the dust-sampling practices in East Kentucky underground mines in 1977. His interviews and federal records

> suggest that many dust samples are being collected incorrectly. In some instances, extra dust has been added; but in substantially more cases, the samples are too low. One explanation for the inaccurate sampling, of course, lies in the fact that the mineowners control sampling in their own mines. Since the penalty for exceeding federal standards may be to close the mine, coal operators have a strong incentive to err in the direction of samples showing less than the actual dust levels [Gerald Sharp, "Dust Monitoring and Control in the Underground Coal Mines of Eastern Kentucky," Master's thesis, University of Kentucky, November 1978, p. 4].

What is needed is a continuous sampling system that immediately warns miners of high dust concentrations. Such a system would allow miners and supervisors to reduce dust as soon as the limits are exceeded. As it is, samples are taken and the results are returned weeks later. Continuous-sampling technology is currently available, but is costly.

53. Data from the Social Security Administration, 1984. Jerry Landauer of the *Wall Street Journal* correctly criticized the Black Lung Benefits Act of 1972 as a bail-out to an "industry that has for the most part refused to accept its full workmen's compensation burden." President Nixon signed both the 1969 and 1972 bills, despite reservations, in the face of political pressure from both liberal and conservative coal-state politicians. Landauer, "Who Will Pay for Black Lung?" *Wall Street Journal*, June 21, 1972.

54. The 1977 Amendments is common shorthand for two acts, both of which were actually enacted in 1978. The Black Lung Benefits Reform Act addresses eligibility standards, while the Black Lung Benefits Revenue Act established the Black Lung Disability Trust Fund.

55. United Mine Workers of America, *An Analysis of the Condition of the Black Lung Disability Trust Fund* (Washington, D.C.: UMWA, circa 1981).

56. The UMWA would have made a mistake to seek compensation through collective bargaining. Had the union succeeded, it would have created a great cost disadvantage for union operators. The disability "benefit" would have reduced the wages-and-benefit packages for working

miners. One black-lung activist has summed up his impression of the first few years of the BLA political activity: "The Black Lung Association was always a kind of mythical organization. It was a Wizard of Oz operation. There were pockets that were good, that were genuine, like Mingo County. But a lot of it was smoke, a lot of the mass action, the publicity, it was orchestrated by outsiders in Charleston. We could shut down the mines with a half dozen guys. It was sort of guerrilla tactics." (Interview with mine-worker activist included in unpublished notes of Barbara Smith's interviews in preparation for her dissertation.)

8. The Rise of Miners for Democracy

1. Trevor Armbrister, *Act of Vengeance* (New York: E.P. Dutton & Co., 1975), p. 70. Armbrister provides an interesting journalistic account of the 1969 Yablonski campaign, but provides little insight into its significance.

2. Titler explained Yablonski's years as an official in these terms: "You don't know that foreign element of coal miners from Russia and Yugoslavia and the like up there in Pennsylvania. I don't mean there's anything wrong with being foreign, but they stick together and stick behind their man. It's not like down in the Kentucky fields or someplace where everybody's an Anglo-Saxon." (*Charleston Gazette*, March 18, 1970, p. 7, cited in Hopkins, "The Miners for Democracy," p. 109-10.)

3. Suzanne Crowell, "Leadership Contest Shakes Up UMWA," *Southern Patriot* (October 1969):1.

4. U.S. Congress, Senate Committee on Labor and Public Welfare, *Hearings on United Mine Workers Election 1970* (Washington, D.C.: GPO, 1971), p. 133; and transcript of debate between Louis Antal and Michael Budzanoski, WQED-TV, Pittsburgh, December 4, 1970, cited in Nyden, "Miners for Democracy," pp. 495-96.

5. "One Non-Union Coal Miner Is A Threat to All Members, Yablonski Warns," *UMWA Journal*, June 1, 1968, p. 11.

6. Recording of Yablonski speech distributed by Boyle administration on a record, "Coal Miners for Boyle-Titler-Owens," 1969.

7. Armbrister, *Act of Vengeance*, p. 58.

8. Jeanne Rasmussen, "The Miners: What Happens Now," *Mountain Life and Work*, (February 1970):6-7

9. Hume, *Death and the Mines*, p. 173.

10. Ibid., p. 178.

11. Boyle indicated to the industry's trustee that neutral trustee Josephine Roche, who was not present at the time, supported the increase. She did not, but the 2–0 vote stood. The Fund could not afford the higher benefit without additional revenue.

12. Rasmussen, "The Miners," pp. 6-7.

13. Joe McGinniss, "The Yablonski Murders: A Bitter Struggle for Life in the Mines Slides into Savagery," *Life*, January 23, 1970, p. 37.

14. John Bronson, "Yablonski Case Still Affects UMW," *Pittsburgh Post-Gazette*, January 1, 1980.

15. This discussion draws on Hopkins, "The Miners for Democracy"; Nyden, "Miners for Democracy"; Hume, *Death and the Mines;* Finley, *The Corrupt Kingdom;* Smith, "Digging Our Own Graves"; and my own writing and observation.

16. Rauh letter to Secretary George Schultz (January 13, 1970) cited in Hopkins, "The Miners for Democracy," p. 66.

17. Rauh's law firm donated space and administrative services. The Kaplan Foundation gave the Washington Research Project enough money in early 1971 to allow the MFD legal team to step up its efforts. Jay Rockefeller, then West Virginia's secretary of state, sent $3,000 for legal work on the autonomy suit. Rauh estimated legal costs to have been a minimum of $300,000, some of which

he billed to the UMWA in 1973 after the MFD victory. See Hopkins, "The Miners for Democracy," pp. 288, 323; and Nyden, "Miners for Democracy," p. 509.

18. Letter from Edgar James to Yablonski and Feldman from MFD collection, cited in Hopkins, "The Miners for Democracy," p. 174.

19. Nyden, "Miners for Democracy," p. 526.

20. Hopkins notes that a MFD press release doubled the number of participants, and *The Miner's Voice* "conveyed the impression that the dissidents had conducted an effective meeting" ("The Miners for Democracy," p. 177).

21. Ibid., p. 199.

22. Immediately following the Yablonski murders, Rauh said: "The fight against the United Mine Workers was not over. It was just beginning when Joe Yablonski was killed" (*New York Times*, January 7, 1970).

23. Chip Yablonski initially supported Tom Pysell, a local union president from Ohio, who became a UMWA organizer. Ten years after the Wheeling Convention, Pysell was working for Garland Fuel, an Arkansas coal company, against a UMWA organizing campaign there.

24. Hopkins, "The Miners for Democracy," p. 327.

25. Ibid., p. 360. Seven years later, Patrick claimed that he, Miller, and Trbovich had decided the evening before the nomination that Trbovich would be nominated for the president, Miller as vice-president, and himself as secretary-treasurer. "If anyone else was nominated for those positions, they would decline in favor of the other person. We took a solemn pact and shook hands on it. Trbovich and I went to bed, but Arnold . . . campaigned like hell along with . . . Chip Yablonski . . . and Rick Bank. . . . It is a sordid story" ("The UMW: 'It was Brother Against Brother,' " *Coal Age* [June 1979]:19).

26. Hopkins interview with Ed James cited in Hopkins, "The Miners for Democracy," p. 357.

27. After his election, Miller stated his position on strip mining:

> . . . the platform I ran on last year [1972] . . . took the position that if you couldn't restore the land to some useful purpose mining should not be allowed. . . . I think surface mining is acceptable in certain areas as the only feasible way to get the coal out such as the anthracite fields. . . . There is some flat land in western Kentucky and Illinois that they certainly have demonstrated that you can go in and take the coal out, recontour the land and restore it to some useful purpose. . . . But you get down here [West Virginia], and particularly in the area where I live . . . they have no intentions of doing anything but going in and destroying the land. . . . I'm opposed to it [strip mining] wherever it's done in an irresponsible manner. ("Roads May Decide UMW Headquarters," *Charleston* (W. Va.) *Gazette*, April 6, 1973.)

28. *Coal Patrol*, June 1, 1972, p. 1. Bethell's candid account of the Wheeling Convention drew sharp attack from certain MFD activists. The MFD campaign had notions about loyalty and criticism that were not much different from Boyle's. Bethell offered a public apology in a subsequent *Coal Patrol* (which he edited), but remained the object of lingering doubt even after Miller appointed him to be director of the UMWA's Research Department.

29. Jo Ann Levine, "Will Miners' Union Reform Slate Win?" *Christian Science Monitor*, November 28, 1972. Miller seems to exaggerate on both counts. Even if his numbers are cut by two-thirds, the tally—133 bathhouses and 20,000 miners—is impressive.

30. Rockefeller came to West Virginia in 1964 as a social worker with the intention of beginning a political career. Although he never involved himself directly in the mine-worker ferment, he was known to be sympathetic to the rebels. He and Miller were friends, although Rockefeller took few risks on behalf of the insurgent miners. Congressman Hechler and state senators Si Galperin, Warren McGraw, and Robert Nelson and former state senator Paul Kaufman supported MFD enthusiastically.

31. MFD acquired a gloss of left-of-center radicalism through its association with the young

nonminers who worked in the campaign and assumed important staff positions in the UMWA after winning. Miller embraced some of their radicalism when he worked for Designs for Rural Action in the early 1970s. His basic instincts were those of a social democrat and advocate of public regulation. Like the miners with whom they worked, the radical staff had always been "outs." Their support for rank-and-file unionism emerged from their own struggle against powerlessness. They were suspicious of power and leadership. Their ideas about participatory democracy and decentralized power stemmed from campus politics in the 1960s and their own experience as organizers of poor people in Applalachia. Like the MFD leaders, they wrote their lines as the play unfolded.

32. "The Truth About MFD Campaign Money," *UMWA Journal*, September 15, 1972, p. 29.

33. See Hopkins, "The Miners For Democracy," p. 436.

34. Ibid.

35. James Ridgeway, "Politics Mine-Worker Style," *New Republic*, November 4, 1972, p. 18.

36. The MFD benefited from a sympathetic press. Reporters watched lawsuits pile up against Boyle and figured he was guilty of at least some of the charges. Boyle's pompous, know-nothing arrogance grated on reporters who had covered the Yablonski campaign. They were impressed with Miller's simple directness. The MFD's staff were quick-witted and iconoclastic—qualities reporters liked. Some MFD staff were journalists and knew what made good copy. They befriended the press, whereas Boyle feared and baited it. Sympathetic journalists failed to probe MFD with the same scalpels they used on Boyle. Jim Ridgeway's handling of the MFD was typical of liberal journalists and beat reporters on national dailies: "The Miners for Democracy is a spare but impressive political organization which grew out of Yablonski's 1968 *[sic]* campaign against Boyle. . . . (Actually, the Miller campaign is nearly broke. Total assets are about $5000. There is no foundation support of any kind. The bulk of support comes from $5 and $10 contributions by coal miners.)" ("Politics Mine-Worker Style," pp. 16-17.) When the MFD candidates failed as union administrators in the mid-1970s, the press reported their troubles with the vengeance of consumers who had been duped into buying cars with no warranties.

9. Union Reform and Industry Turmoil

1. Thomas N. Bethell, "The UMW: Now More Than Ever," *The Washington Monthly*, March, 1978, pp. 19-20.

2. Most of the big nonunion stripping operations offer higher wages and comparable medical benefits to undercut UMWA appeal. The UMWA contract generally offered a better pension plan and a wide range of job protections—grievance procedure, safety committees, training, seniority, safety rules (e.g., the individual's right to refuse imminently dangerous work; the right of the safety committee to remove workers from imminent dangers) and local practices and customs. The UMWA is also the only institutional voice for *all* coal miners in national politics, contributing to the political campaigns of prominer candidates and lobbying for prominer legislation and administrative rules.

3. Testimony of Ed Ziolkowski, vice-president and general manager, Energy Development Company, Hanna, Wyoming, before the President's Commission on Coal, published transcript of a public hearing in Denver, Colorado, October 26, 1978, p. 132.

4. Office of Technology Assessment, *The Direct Use of Coal,* pp. 111-21.

5. While the Miller administration developed effective new programs in communications, safety, research, and political action, Miller shied from an education program for the rank and file and a training program for future leaders when Rick Bank and Tom Bethell proposed them. Research director Bethell initiated a couple of training projects, but the absence of staff and funding handicapped his makeshift efforts. The 50,000 young miners who entered the workforce in the 1970s picked up knowledge of union procedures and contracts as best they could (UMWA, *It's Your*

Union. Pass It On. Officers' Report to the United Mine Workers of America 47th Constitutional Convention [Washington, D.C.: UMWA, September 1976], p. 70).

6. Ward E. Sinclair, "Miners for Democracy: Year One at the UMW," *Ramparts* (June 1974):38.

7. "Labor: New Vigor in the Pits," *Time*, April 30, 1973, p. 85.

8. UMWA's *President's Report* to the 46th Constitutional Convention, December 1973.

9. Tom N. Bethell, "1974: Contract at Brookside," *Southern Exposure* (Spring/Summer 1976):115-16.

10. Ben A. Franklin, "2 Years After His Election, Miller Is Facing Criticism," *New York Times*, January 25, 1975.

11. Ibid.

12. FSO evolved into more than a service center when Gail Falk, a UMWA lawyer, made herself the link between UMWA headquarters and black-lung activists associated with the various BLA chapters. She translated the feelings of black-lung victims and their widows into UMWA legislative and regulatory proposals and took certain precedent-establishing cases into court. The more FSO represented grass-roots BLA interests, the more Miller viewed it with distrust. Some black-lung activists saw FSO as Miller's vehicle to coopt their militancy and channel it into litigation and legislation. Following a multistate wildcat strike over black-lung legislation, Miller closed FSO in 1976.

13. In Miller's first four years, the UMWA claimed to have organized 120 mines employing 5,500 miners who produced about 35 million tons annually. (UMWA, *It's Your Union. Pass It On*, p. 53).

14. The best source on UMWA conventions is the published proceedings. See also Curtis Seltzer, "How Much Can a Good Man Do," *Washington Monthly* (June 1974).

15. Reversals came on several issues, two of which were the eligibility of pensioners in voting for local officers and the disbanding of bogus locals (those with fewer than ten active miners). Miller wanted to disfranchise pensioners because, he felt, they no longer understood mining conditions. Bogus pensioner locals had been a source of political strength for Boyle. They were illegal, and Miller had been disbanding them. Pensioners felt they were being cut out of a union they had built.

16. Union employees were freed from the obligation of contributing to incumbent politicians. Officials were forbidden to use union funds to advance their political goals.

17. Miller supported a resolution allowing the executive board to suspend autonomy for any of the following reasons: (1) to prevent or correct corruption; (2) to assure compliance with collective-bargaining agreements; (3) to restore democratic procedures or protect the democratic rights of members; and (4) "to otherwise enforce the UMWA's Constitution and carry out the legitimate objectives of this Union." Ex-Boyle supporters opposed Miller's position in speeches that would have won cheers at MFD rallies in 1972. The vote boiled down to a matter of trust in the fairness of Miller and the executive board. A district-by-district caucus returned a verdict in favor, 821 to 674.

18. Sinclair, "Year One," p. 41.

19. One MFD veteran recalls Chip Yablonski asking him the night before the nominating vote at the 1972 convention whether Miller could "be controlled." Other versions have Yablonski using the word "handled." Yablonski threw his support to Miller following this assurance.

20. After spending $2 million on the year-long strike, the UMWA secured a contract for 180 miners only after one was murdered. The UMWA assumed that a victory in the much publicized strike in Harlan County would tumble other nonunion mines in East Kentucky like dominoes. The theory proved wrong. A central part of the UMWA's strike strategy was to harass Duke Power Co., owner of the Brookside mine, through national publicity and on Wall Street's capital markets. See Bethell, "1974: Contract at Brookside," and Bob Hall's interview with Bernard Aronson, UMWA press director and strike strategist, in that article.

21. Tom N. Bethell, "It's Our Time to Catch Up," *UMW Journal*, October 16–31, 1974, p. 10. Bethell described the negotiations from his position of UMWA research director:

I somehow assumed that the operators would come to the bargaining table equipped with a battery of ideas and proposals that would address the fundamental causes of the problems in labor relations that already loomed so large; we prepared detailed proposals of our own and drafted position papers to answer theirs. Theirs, however, didn't exist. Sometimes they put forward practical reasons why one of our proposals wouldn't work, but if their long experience had allowed them to glean any insights into improving labor relations, they left them at home (Thomas N. Bethell, "The UMW: Now More Than Ever," p. 20).

22. BCOA calculated the number of idle days due to wildcat strikes as:

Year	Workdays Lost
1970	593,100
1971	565,000
1972	515,600
1973	529,200
1974	1,023,800
1975	1,417,400
1976	1,950,300
1977	2,273,600

225 days (average work year) × 130,000 workforce = 29,250,000
possible annual work days. Multiplied by 3 years = 87,750,000 days
Days lost for 1975–1977 = $\frac{5,641,300}{87,750,000} = 6\%$

23. The Bureau of Labor Statistics publishes data on work stoppages in the bituminous mining industry. Since 1932, the *fewest* number of workdays idle was 90,700 in 1961, which still represents more than half the work force on strike one shift during the year. See Keith Dix et al., *Work Stoppages and the Grievance Procedure in the Appalachian Coal Industry* (Morgantown, W. Va.: Institute for Labor Studies, West Virginia University, 1972).

24. U.S. Department of Labor, Bureau of Labor Statistics, *Labor Relations in the Fairmont, West Virginia Bituminous Coal Field*, by Boris Emmet, bull. no. 361 (Washington, D.C.: Government Printing Office, 1924), p. 8. In some early contracts, mine operators who failed to collect wildcat penalties from strikers were themselves fined $4 per worker to be paid to a charity the UMWA and operators designated. In the late 1960s, the UMWA and the BCOA agreed to deduct $10 from the $120 Christmas bonus for each month during the previous year in which miners failed to work all scheduled days. See Dix et al., *Work Stoppages*, pp. 53-54.

25. Before Lewis gained control in the 1920s, wildcat strikes frequently leaped state lines in response to wage cuts. See Arthur E. Suffern, *Conciliation and Arbitration in the Coal Industry of America* (New York: Macmillan, 1926).

26. Curtis Seltzer, "West Virginia Book War: A Confusion of Goals," *The Nation*, November 2, 1974.

27. Brett and Goldberg's findings suggest that "wildcat strikes are related to management practices in dealing with employee relations matters." Where management met regularly with the UMWA mine committee and made a good-faith effort to resolve grievances, miners struck infrequently. Strike-plagued mines had no regularly scheduled labor-management meetings, communication was poor, and grievances were not handled expeditiously. See Jeanne M. Brett and Stephen B. Goldberg, "Wildcat Strikes in the Bituminous Coal Mining Industry: A Preliminary Report" (unpublished manuscript undertaken for the National Science Foundation, circa 1978), p. 65.

28. From 1970 to September 30, 1977, companies obtaining the most restraining orders were: 145, Pittston Co.; 131, Bethlehem Steel; 119, Westmoreland Coal; 113, Armco Steel; 111, American

Electric Power; 91, International Telephone and Telegraph (Carbon Fuel); 84, Continental Oil; 81, Lykes-Youngstown Corp.; 62, Amherst Coal; and 61, Occidental Petroleum. Two of the top three coal producers in southern West Virginia in 1976—Eastern Gas and Fuel and U.S. Steel—sought injunctions infrequently, preferring to settle disputes through other means. See Barbara Smith, "Coal Miners and the Federal Courts: Wildcat Strikes in West Virginia, 1972-1977" (unpublished manuscript, February 1978).

29. The UMWA adopted a ten-point antiwildcat program in the mid-1970s and expelled two strike leaders associated with the Miners' Right to Strike Committee. Lewis consistently opposed rank-and-file-initiated strikes. In 1951 his executive board set up penalties for wildcat strikes. District presidents levied the fines, which generally fell between $100 and $500. The board reaffirmed the penalty procedure in 1956, but ten years later found it necesary to ask employers to impose a $5-per-day fine for each day an individual struck. Most BCOA members did not cooperate, and the scheme was cancelled in 1967 (Dix et al., *Work Stoppages*, p. 54).

30. Although some left-of-center observers believe wildcat strikes are always useful, Daniel Marschall comes closer to their protean nature.

> Wildcat strikes have positive and negative aspects. They enable miners to win demands that otherwise could not be achieved, or would take months passing through a grievance procedure. They also give miners a sense of their own power, and build rank-and-file organization outside of union structures.
>
> . . . But wildcats are inherently limited, and to a certain extent destructive of the strength and cohesiveness of the union. By their nature, they are oriented towards immediate issues and do not present a coherent, positive alternative to union procedures. Their impact on companies is lessened by the fact that . . . companies ordinarily compensate for "down-time" in their yearly production estimates. Wildcats cut miners' salaries, and decrease industry contribution to the Funds. Most importantly, wildcats foster divisions within the union. (Daniel Marschall, "The Miners and the UMW: Crisis in the Reform Process" *Socialist Review*, no. 40/41 [July–October 1978]:109-10.)

31. Radicals who went into the mines within a year or two of the MFD victory were not tied to the Miller administration. A few worked locally for the MFD, and one was Miller's bodyguard for a short time. The miner Communists had little political faith in the left-of-center staff, who had never been miners and were not Communists. Miller's opponents painted both groups with the same red brush, despite their differences.

32. Vice-president Trbovich openly joined the anti-Miller board faction by the end of 1975, as did other former supporters of Miller and the MFD.

33. The exceptions were the two strikes shaped by the Miners' Right to Strike Committee. In both cases, most strikers never acknowledged the committee's leadership or embraced its political philosophy.

34. William H. Miernyk, "Panic Over Coal Future in America Unjustified," *Charleston* (W. Va.) *Gazette*, March 15, 1978.

35. "President of Cannelton Discusses Coal Issues," *Charleston* (W. Va.) *Gazette*, April 12, 1978.

36. This language was incorporated in the text of the antistaff speech Trbovich distributed at the convention, but is not to be found in the *Proceedings*. His text was circulated among the delegates.

37. Address by Mike Trbovich, *Proceedings, 47th Consecutive Constitutional Convention, UMWA*, pp. 59-61.

38. A detailed discussion of the convention can be found in Paul F. Clark, *The Miners' Fight for Democracy: Arnold Miller and the Reform of the United Mine Workers* (Ithaca, N.Y.: Cornell University Press, 1981), pp. 75-90. During Miller's successful re-election campaign in 1977, his running mate, Sam Church, said in response to a reporter's question about the meaning of a Miller

Administration slogan, "Safety—Or Else": "Well, I don't know. It could mean a lot of things. It's a slogan." ("UMW Contenders Play Safety Tune," *Mine Safety Week*, May 2, 1977, p. 3.)

39. "Patterson's Action Draws Criticism," *Charleston* (W. Va.) *Gazette-Mail*, May 7, 1977.

40. *UMWA Journal*, July 16-31, 1977, p. 12.

41. He carried only one district outside this area, Nova Scotia.

42. First-dollar coverage is not the same as "free" medical care. Miners paid for first-dollar coverage at the bargaining table by forgoing other benefits and accepting less pay.

43. Discussion of the Funds' successes and failures can be found in Suzanne Jaworski Rhodenbaugh, "Death by Computer and Contract: The UMWA Health and Retirement Funds," *Crossroads* (July–August 1978):22–26; and Seltzer, "Health Care by the Ton," *Health/PAC Bulletin* (November–December 1977):1-8, 25-33.

44. UMWA Welfare and Retirement Fund, *Annual Report*, 1973–74, p. 99.

45. Suzanne Jaworski Rhodenbaugh, "Letter to the Editor," sent to forty coalfield newspapers, July 6, 1977.

46. Ben A. Franklin, "Years of Conflict in Coal Fields Set Stage for a Strike," *New York Times*, November 27, 1977.

47. Tom N. Bethell, "Fiasco at the Funds," *Coal Patrol*, no. 33, (August 5, 1977):3. U.S. Steel witheld more than $6 million from the Funds over a legal matter.

48. Ibid., pp. 2-3.

49. William Kissick and American Health Management and Consulting Corp., "West Virginia Primary Care Study Group: Problems of Reimbursement," Department of Community Medicine, West Virginia University, update, November 1978.

50. Virginia Gemmell and Jane Ray, *Physician Loss in Central Appalachian Coalfield Hospitals and Clinics*, draft report (Washington, D.C.: Appalachian Regional Commission, 1978).

10. Wrong War, Wrong Place: The 1977-78 Strike

1. The UMWA's research was so shoddy that its public relations material could not even get straight the name of the federal mine safety agency, which it referred to as the Mine Engineering and Safety Administration. The Mine Safety and Health Administration had once been called the Mining Enforcement and Safety Administration. See "UMWA Bargaining Update," November 1, 1977, p. 3.

2. Another 1,500 to 2,000 small coal companies adapted to the terms of the BCOA contract although they played no part in its negotiations. The UMWA negotiated separate contracts with the anthracite operators, western surface-mine companies, and mine-construction firms.

3. The coal industry was familiar with rebellious miners and meddlesome government, but to these traditional sources of industrial angst, was added women in the 1970s. Not a single woman worked in American coal mines in 1970. A decade later, more than 3,000 were at work in the mines, about 1 percent of all coal workers. Coalfield women had always shared mining's economic and emotional consequences. They made do without income when the men struck. For each of the hundred thousand plus mining fatalities since 1900, there was a widow, a mother, a sister. When Mother Jones barnstormed through strike after strike during the UMWA's first thirty years, she made a point to organize the women, knowing that male solidarity was only as strong as the unity between the striker and his wife. In the 1970s, women played central roles in grass-roots coal politics—the black-lung movement, the Brookside strike, the Farmington widows, and anti-strip-mining protests, to name a few. Several professional women had breached the UMWA before the 1970s. The daughters of both Lewis and Boyle were UMWA staff. Suzanne Richards, the lawyer who was Boyle's administrative assistant, ran day-to-day UMWA affairs in the late 1960s. Lewis retained the redoubtable Josephine Roche as the Fund administrator and trustee for twenty years. While the MFD

and its staff was almost exclusively white and male, several women worked for the reformers and later as UMWA lawyers, lobbyists, and administrative staff. Women went into the mines for bread-and-butter reasons. See "Women Coal Miners, Special Issue," *Mountain Life and Work* (July/August, 1979). The money was the best available for unskilled or semiskilled labor in coal communities. Employers resisted them at first. The coal industry still has no women at the policy-making level and few female professionals. (Coal management is, almost to a person, white, male, and Christian. In more than a decade of covering the coal industry, I have never seen or heard of a black mine operator or superintendent. At the 1977 Knoxville convention of the National Independent Coal Operators' Association, the organization of small independent suppliers, its Washington lawyer, John Kilcullen, began his remarks on excessive federal regulation with a "nigger" joke.) Miners, like other males, resented the intrusion into "their" world as well as the competition for jobs. Each woman has had to prove herself to the veterans, as do all new miners. Some males like to bully or harass the women. Those who have had the fewest problems are those who take their work seriously and stand their ground. A number of women have reported that supervisors harass them, either by assigning nasty jobs or offering easy jobs for easy sex. A key breakthrough came when a young, disarmingly soft-spoken lawyer, Betty Jean Hall, sued 153 coal companies in May 1978 over employment discrimination. Gradually, one by one, the companies have worked out hiring plans with Hall's Coal Employment Project, which also provides training, prods all parties to be alert to women's special safety-and-health issues, and organizes conferences. Most women are still too new to the work force to have moved up the job ladder into supervisory positions, the union, or government service. Women were elected delegates to UMWA conventions since 1973. One miner, Barbara Angle, is turning into a novelist of strength and perceptiveness.

4. The cash-penalty proposal against strikers was a page out of coal's past. Here is one precedent. Mark A. Hanna, the political kingmaker of the 1890s, owned extensive coal interests, including the Panhandle mine in Painter's Run near Pittsburgh. Hanna's manager, Thomas E. Young, informed his miners in December 1896 that henceforth they were to promise to refrain from striking for a period of twelve months. The miner would have $.10 set aside from each ton credited to him—⅙ of his tonnage rate—that would be returned at the end of the year only if Panhandle had no strikes. The miner forfeited his money if he struck regardless of cause or provocation.

5. "A Coal Company President Discusses Labor Stability," *Coal Age* (February 1978).

6. "Roads May Decide UMW Headquarters," *Charleston* (W. Va.) *Gazette*, April 6, 1973.

7. Clark, *The Miners' Fight for Democracy*, p. 118; and David B. Hecker, "Internal Politics Splits Mine Workers' Convention," *Monthly Labor Review* (January 1977):60.

8. UMWA, *Proceedings of the 47th Constitutional Convention*, p. 495.

9. Helen Dewar, "A Coal Walkout No Longer Poses an Instant Crisis," *Washington Post*, November 20, 1977. Ben A. Franklin in a November 27 story announced that Miller "seems unalterably wedded . . . absolutely determined" to win some form of the right to strike.

10. "Miller Fears Pension Program May Suffer," *The Charleston* (W. Va.) *Gazette*, February 9, 1978.

11. No one ever spelled out clearly and in detail what was guaranteed. Insurance companies and individual operators settled disputes over coverage.

12. "Marshall Outlines Role in Strike," *The Charleston* (W. Va.) *Gazette*, March 26, 1978.

13. Huge had also lobbied Miller to endorse Carter early in the 1976 presidential primary campaign.

14. While advising Miller on bargaining strategy, Huge was making $50,000 a year as the UMWA trustee at the Funds. Huge was sensitive to criticism that he gave only token service to the Funds while conducting a successful law practice. "[The Funds] get more than a day a week—a lot more, and they get quite a bit of expertise," he told the *Coal Daily* on November 1, 1977. So blatant did Huge's conflict of interest become that he was forced to resign from the Funds during the negotiations.

15. Terrence Smith, "Test of Crisis Management," *New York Times*, March 12, 1978. Miller also retained two Washington lawyers, Ronald G. Nathan and Philip J. Mause, to help in the bargaining. Neither knew much about coal negotiations or coal miners.

16. Wayne Horvitz, "1977–1978 Coal Negotiations: Background," mimeograph (Washington, D.C.: U.S. Federal Mediation and Conciliation Service, June 1980), p. 3.

17. Peabody is primarily a surface-mining company, with only about 30 percent of its output originating in underground mines. About 25 percent of its production is not covered by the UMWA-BCOA contract. Hills inclined toward either company-by-company bargaining or separate contracts for underground and surface miners.

18. Katina Cummings, "Decentralized Bargaining in the Bituminous Coal Industry? Emerging Shifts in Power Relations" (M.S. thesis, Massachusetts Institute of Technology, December 1980), p. 149.

19. Johnston's observation is quoted in Cummings, "Decentralized Bargaining," p. 76.

20. "Miller Fears Pension Program May Suffer," *Charleston* (W. Va.) Gazette, February 9, 1978.

21. William J. Eaton, "Why Carter Leaned on Coal Firms—And How it Worked," *Miami Herald*, February 26, 1978.

22. Roderick Hills, speech to the Edison Electric Institute, Houston, Texas, April 30, 1978.

23. Ann Hughey, "Huge to Quit UMW Post When Contract Is Signed," *Charleston* (W. Va.) *Gazette*, October 29, 1977.

24. A *Wall Street Journal* editorial said: "The more we look at it, the more we think that if we were a coal miner we'd have voted against the proposed contract too" ("The Miners Have a Point," *Wall Street Journal*, March 7, 1978). The *Journal* figured the offer had more wages than benefits, which meant that after taxes and inflation were added "there is no small chance that they [miners] will end up going backward." BCOA operators lit up the *Journal*'s switchboard. A week later, a chastised *Journal* wrote, "We . . . had to stand up and eat crow like a man." Based on how "the producers figure[d]" the wage-and-benefit offer, the *Journal* said the "distribution is not out of line with settlements in other industries" ("Eating Crow [A Correction]," *Wall Street Journal*, March 14, 1978). The *Journal* missed the miners' point twice: they rejected the offer not over their after-tax pay but because of the takeaway nature of the contract as a whole.

25. Social commentators and reporters have routinely painted Appalachian people—and miners, in particular—as peculiar, sadistic, violence-prone half-wits. This picture has tangibly affected public perception and public policy. It all started with the Hatfield-McCoy feud that the national press featured as a kind of comedy among white savages. See Williams, *West Virginia*, pp. 99 ff.

An early commentator on the industrialization of Appalachia, Malcolm Ross, described mountaineer miners as lacking instruction in the "intricate conventions of morality." Their fundamentalist religious beliefs, he wrote in 1931, foster a fatalism that allowed them to be killed and maimed in the mines with little protest. "Traditions of dismal ignorance have made it a sardonic joke to single them out as the 'purest Anglo-Saxon blood in the United States,' for their present slovenly habits of body and mind are no cause for racial pride. They are children, totally unprepared to meet the struggle of modern civilization" (Malcolm Ross, *Machine Age in the Hills* [New York: Macmillan, 1933], p. 76).

Social scientists contributed "objective" scholarship. Arnold Toynbee wrote in 1946: "The Appalachian 'mountain people' today are no better than barbarians. They have relapsed into illiteracy and witchcraft. They suffer from poverty, squalor, and ill health. They are American counterparts of the latter-day white barbarians of the Old World—Rifis, Albanians, Kurds, Pathans, and Hairy Ainus; but whereas these latter are belated survivals of an ancient barbarism, the Appalachians present the melancholy spectacle of a people who have acquired civilization and then lost it" (Arnold Toynbee, *A Study of History* [London: Oxford University Press, 1946], p. 149).

Three decades after Ross poisoned the well against bias-free interpretations of Appalachian miners, social scientists and journalists spun out derivative theories when the War on Poverty engulfed the mountains. Jack Weller compared Appalachians to middle-class Americans and found them badly wanting (Jack Weller, *Yesterday's People: Life in Contemporary Appalachia* [Lexington: University of Kentucky Press, 1966]). He described Appalachians as fatalistic, resigned, incapable of collective action, incorrigibly individualistic, and uninterested in changing the conditions of their lives. (Fewer than five years later, these same political cripples had undertaken successful movements for union democracy, electoral reform, and environmental protection.) Casual journalistic imagery spread these characterizations and made victim-blaming acceptable, as in this *Time* story on the 1968 Farmington disaster that management negligence caused: "Yet the grimy, coughing coal miner is inured to the dangers of his calling. He is stoic about the deaths of friends and relatives and accepts as 'part of the game' the high possibility that he may be crushed, burned, suffocated or drowned in his own Cimmerian tomb. His very fatalism has helped to perpetuate the conditions under which miners work and die." ("Disasters: Too Late for 78," *Time*, December 6, 1968, p. 32.) These media images have become commonplace over the years because miners and Appalachians lack a watchdog group, such as the Anti-Defamation League or the NAACP, on the national scene. See Seltzer, "The Media vs. Appalachia: A Case Study," *Mountain Review*, (May 1977).

26. "My own feelings on the merits of the Taft-Hartley vs. seizure changed during this period [following the rejection of the second contract]. The liability I saw to Taft-Hartley was that traditionally an injunction ordered workers to resume production under the terms of the last contract. During a period of high inflation, it seemed unfair to order miners back to work under a 1974 contract. . . . When I discovered [that it was possible] . . . to pay wages higher than those in the 1974 contract . . . it helped persuade me that Taft-Hartley would be useful in this unique situation." ("Marshall Outlines Role in Strike," *Charleston [W. Va.] Gazette*, March 26, 1978.)

27. Byron E. Calame, "Coal Miners May Walk Out in December, But Impact Won't Be What It Used to Be," *Wall Street Journal*, October 6, 1977.

28. Helen Dewar, "A Coal Walkout No Longer Poses an Instant Crisis," *Washington Post*, November 20, 1977.

29. Bureau of National Affairs, *Daily Labor Report*, no. 33, A-11, February 13, 1978.

30. Martha Bryson Hodel, Associated Press, "Coal Supply Falls, Foes of Pact Grow," *Washington Post*, February 10, 1978.

31. "Coal Walkout's Impact Mounts on Wide Front," *Wall Street Journal*, Feb. 10, 1978.

32. Steven Rattner, "Carter Orders Plans to Send Coal to Areas Hit by Miners' Strike," *New York Times*, February 12, 1978.

33. Art Pine, "The Impact: Economy on the Brink of Setback," *Washington Post*, March 6, 1978.

34. Reginald Stuart, "Cities Turn Off the Lights as They Run Out of Coal," *New York Times*, February 16, 1978.

35. "Entering the Doomsday Era: Shutdowns and Blackouts Loom as the Coal Strike Rumbles On," *Time*, February 27, 1978.

36. J.P. Smith, "Easing the Impact: Dim Prospects," *Washington Post*, February 13, 1978.

37. William Robbins, "West Virginia Areas Made to Cut Power," *New York Times*, March 9, 1978.

38. U.S. Congress, General Accounting Office, *Improved Energy Contingency Planning Is Needed to Manage Future Energy Shortages More Effectively* (Washington, D.C.: GAO, October 10, 1978). The National Electric Reliability Council (NERC) published a study in November 1977 that argued electric utilities could survive a three-month nationwide coal strike through the adoption of conservation measures and the "re-dispatching of coal and oil-fired generation" within and among the nation's power systems. (National Electric Reliability Council, "A Study of Interregional Energy Transfers for Conservation of Coal Winter 1977/78," November, 1977, pp. 2-3.) Following

the strike, NERC reported: "Throughout the most prolonged fuel crises ever encountered by the electric utility industry, the reliability of the overall bulk power network was never in jeopardy. No blackouts were experienced, and except for the voluntary and state-mandated curtailments, the electric service of all utility customers continued uninterrupted ." NERC showed how 15 billion kilowatt-hours of electric energy were shipped into coal-deficient power systems. Electric-power transfers and the shipment of non-union and western coal prevented the strike from seriously affecting coal-dependent utilities. (Idem., "The Coal Strike of 1977-1978: Its Impact on the Electric Bulk Power Supply in North America", undated, pp. 8-9.) The Energy Department helped to coordinate these measures and was aware of their success. For that reason, Energy never sought to invoke its mandatory power transfer and coal allocation authority—further evidence that the White House knew no national emergency existed.

39. United Press International, "Weather—Not Coal Strike—To Blame for Steel Slump," *Times-West Virginian*, March 16, 1978.

40. GAO, *Improved Energy Contingency Planning*, pp. 36-37.

41. John A. Ackermann, "The Impact of the Coal Strike of 1977–1978," *Industrial and Labor Relations Review* (January 1979):184-85.

42. Ibid., p. 40.

43. *Newsweek* (November 20, 1978) reported that President Carter invoked executive privilege to keep locked in the White House a report by his Council of Economic Advisers that "assured the President that no coal emergency existed." GAO wrote its report without access to certain documents the White House withheld.

44. Telephone interview with George Getschow, summer 1978.

45. "After 109 Days," *Washington Post*, March 26, 1978.

46. Edmund Faltermayer, "What the Coal Strike Has Obscured," *Fortune*, April 10, 1978.

47. Urban C. Lehner, "Coal Strike: A Needless Panic?" *Wall Street Journal*, March 28, 1978.

48. Ben A. Franklin, "Spur-of-Moment Meeting Led to Tentative Coal Accord," *New York Times*, March 16, 1978.

49. Another way of measuring wages is by annual income, which varies according to the number of days worked. Rarely have miners worked more than 225 days a year. A miner who worked forty-nine weeks (without overtime or escalators) would have earned $17,963 in the first year of the March 14 contract (245 days × $73.32). But owing to demand weakness, the typical UMWA member logged only about 215 days, or $15,764.

50. The Council on Wage and Price Stability announced that the total compensation increase to the BCOA would be 37 percent over three years, but the "direct inflationary impact will be limited by the small size of coal shipments relative to the overall economy. . . . The 37 percent rise in compensation would translate into about 0.08 percent on the overall price index over the three-year life of the contract and 0.05 percent in the first year" (Council on Wage and Price Stability, "The Inflation Impact of the 1978 National Bituminous Coal Wage Agreement," news release, April 14, 1978, p. 5).

51. "Coal Productivity Gains Seen as Offsetting Inflation in Contract," *Bluefield* (W. Va.) *Telegraph*, April 14, 1978.

52. A miner from Company A who picketed miners at Company B might not be fired if his identity were concealed or Company A chose not to force the issue.

53. Curtis Seltzer, "Miners: An Unhealthy Contract," *Health/PAC Bulletin* (1979):58.

54. The first contract offer also referred to "trust documents" at the Funds. When a copy of the mysterious document made its way to the coal fields, miners learned that new limitations were written into the fine print.

55. Anne A. Scitovsky and Nelda McCall, "Coinsurance and Demand for Physician Services: Four Years Later," *Social Security Bulletin* (May 1977). Don Conwell, administrator of the

New Kensington clinic, called the new contract regressive. "What this new contract is doing is putting up a barrier to preventive medicine, because no one's going to pay $7.50 to a doctor until they get really sick. . . . The irony of it is that the UMW with its health funds started the whole thing of preventive medicine 20 years ago" (Bob Morris, "All Sides Anxious to See New Pact," *The Charleston* (W. Va.) *Gazette*, March 16, 1978).

56. Other unions tried to even the sides by giving money to the miners. But the Miller administration did not distribute any contributions until after the contract was ratified. Farmers sent food convoys directly to the coalfields.

57. " 'Miller Has No Business Being the UMW President,' " *Wheeling* (W. Va.) *News-Register*, March 27, 1978.

58. Ibid.

59. The bargaining council formally rejected the first proposal on February 7, and an "improved" offer on February 18.

60. Donald Cook, chairman of the coal-burning American Electric Power Company, said, "The new democracy is the new anarchy." "Coal Negotiations," *Editorial Research Reports* 25 (October 1974):818, quoted in Cummings, "Decentralized Bargaining," p. 40.

61. "Awash in Coal," *Wall Street Journal*, May 8, 1978.

62. Wayne Horvitz, "What's Happening in Collective Bargaining?" *Labor Law Journal* 29 (Aug. 1978): 464.

11. The Interwar Years: A Phony Peace

1. Office of Technology Assessment, *Development and Production Potential of Federal Coal Leases*. When these contracts were negotiated in the mid-1970s, the advantages of western coal seemed persuasive, but increases in rail rates have left many utilities believing they made the wrong choice.

2. *Triangle* quoted in "Coal: UMW, BCOA Say Industry Needs Good Labor Relations" (Beckley, W. Va.) *Register and Post-Herald*, January 1, 1980.

3. Testimony of John D. Rockefeller IV, President's Commission on Coal, published transcript of a public hearing in Charleston, W. Va., October 20, 1978, p. 8. Rockefeller was quoting Executive Order 12062, May 26, 1978.

4. Only Wirtz might have been expected to cut to the heart of the problem. But he never forced the commission to dig into the roots of coalfield conflict.

5. Miller wanted to appoint a pal of his Charleston girlfriend to the commission. Rockefeller refused to allow "his" commission to be made into a laughingstock. Church suggested Friedman.

6. Stephen B. Goldberg, "Report to the President's Commission on Coal: Collective Bargaining Study," consultant report to President's Commission on Coal, July 10, 1979.

7. President's Commission on Coal, *Recommendations and Summary Findings* (Washington, D.C.: U.S. Government Printing Office, March 1980), p. 13. A companion report, *Staff Findings*, concluded: "The origins and duration of that [111-day] strike are best understood in the context of the substantial changes" accompanying a prior ten-year period of transition and "the pressures that were placed on the UMWA and the BCOA" (*Staff Findings*, p. 49). The commission never made explicit what it thought the coal industry had made a transition from, or to. Saying that "instability in the coal industry comes from pressures on the actors" is like saying day changes into night because the earth rotates on its axis. The staff concluded: "Effective and ongoing communication at the top level between union and management is the central need" (*Staff Findings*, p. 57). The problem in 1977–78 was *not* a lack of communication at the top. Rather, what was being communicated was unacceptable to miners at the bottom.

8. Commission, *Recommendations*, p. 11.

9. Ibid., p. 17.

10. The Boone Report justified the UMWA's demand for a comprehensive industry-financed

health-care system. Unlike this famous predecessor, the commission's reports produced not a single identifiable change in public or private policy related to coalfield living conditions.

11. Michael S. Koleda, "Introduction," *The American Coal Miner: A Report on Community and Living Conditions in the Coalfields* (Washington, D.C.: President's Commission on Coal, 1980), p. ix. A commission questionnaire of 1,299 miners found an average expected income in 1979 of $19,422.33.

12. Commission, *Recommendations*, pp. 5-6. Earlier, the commission proposed an oil-backout plan for 2.3 million barrels of oil per day by 1990, coupled with a proposal for a synthetic fuels industry capable of producing another 2.3 million barrels a day. The total $159.2 billion package, the commission suggested with great tact, "should be shared by all Americans," that is, subsidized by taxpayers. See President's Commission on Coal, "Acceptable Ways to Hasten the Substitution of Coal for Oil, An Interim Report" (Washington, D.C.: President's Commission on Coal, July 1979), p. 4.

13. Ann Hughey, "New Rain on Coal's Parade," *Forbes*, April 14, 1980.

14. Stonie Barker, Jr., of Island Creek Coal Co., chairman of the BCOA; Nicholas Camicia of Pittston Co., representing the American Mining Congress; and Robert Quenon of Peabody Coal Co., representing the National Coal Association.

15. "UMW, Coal Industry Launch Program to Encourage Greater Coal Use in U.S.," *Energy Users' Report*, March 6, 1980, pp. 9-10.

16. Obviously, coal can help America use less imported oil. Any major, publicly funded study of oil-backout strategies should carefully evaluate the best mix of substitute fuels and conservation rather than assume coal is the wisest way of spending billions in tax dollars. Advocates of energy conservation at the Energy Productivity Center of the Mellon Institute, for example, had their own ideas about how to spend hundreds of billions of dollars. They wanted to invest in energy conservation measures so that by 2000 A.D. the U.S. would need to import no foreign oil at all. See Douglas Martin, " 'Least-Cost' Energy Plan Stirs Debate," *New York Times*, April 6, 1981.

17, Rockefeller ran for reelection in 1980 against former two-term Republican governor Arch Moore, who defeated him in 1972. Rockefeller had to please the state's coal interests so they would not contribute to Moore's campaign. Jay, who had metamorphosed into coal's political cheerleader, beat Moore with 54 percent of the vote by spending more than $11.5 million of his own money.

18. Commission, *Recommendations*, p. 8.

19. Ibid., p. 19.

20. U.S. Department of Energy, "An Assessment of National Consequences of Increased Coal Utilization: Executive Summary," vol. 1 (Washington, D.C.: U.S. Department of Energy, February 1979), p. 77. A figure of 50,000 premature deaths is frequently used.

21. Robin Toner, "Church Puts Weight Behind Coal Industry," *Charleston* (W. Va.) *Daily Mail*, February 19, 1980.

22. "UMW, BCOA bury the hatchet on safety." *Mine Regulation & Productivity Report*, October 5, 1979, pp. 3-4.

23. Ibid., p. 4.

24. "Unions: Woes Multiply for Arnold Miller," *Business Week*, April 3, 1978.

25. Jim Ragsdale, "Union in Upheaval with Palace Intrigue," *Charleston* (W. Va.) *Gazette*, August 9, 1979.

26. Jim Ragsdale, "Miller Splits with Church over Ambitions," *Charleston* (W. Va.) *Gazette*, November 6, 1979.

27. When Miller was first approached in July 1979 with this proposition, he said, "I don't want no hokey titles." See Jim Ragsdale, "IEB to Consider Placing Miller on Sick Leave," *Charleston* (W. Va.) *Gazette*, November 15, 1979.

28. Robin Toner, "Miller Resigns Post, Offers Sam Church 'Best Wishes,' " *Charleston* (W. Va.) *Daily Mail*, November 16, 1979.

29. Jim Ragsdale, "Miller Seeking Power to Fire District Officers," *Charleston* (W. Va.) *Gazette*, July 25, 1979.

30. Ben A. Franklin, "In the Bitter Aftermath, Whither the U.M.W.?" *New York Times*, March 26, 1978.

31. Jim Ragsdale, "New Union President Former Miller Foe," *Charleston* (W. Va.) *Gazette*, November 17, 1979.

32. The assault charge was dropped because it was Church's first offense in the District of Columbia.

33. Jim Ragsdale, "New UMW President Gives Call for Unity," *Charleston* (W. Va.) *Gazette*, December 11, 1979.

34. Ben A. Franklin, "Big Coal, U.M.W. Want More of Energy Pie," *New York Times*, May 11, 1980.

35. Clark, *The Miners' Fight for Democracy*, p. 142.

36. Evelyn Ryan, "UMW Officers Now 'Wait and See,' " (Morgantown, W. Va.) *Dominion-Post*, January 6, 1980.

37. Church appointed as his vice-president Wilbert Killion, who had run unsuccessfully for secretary-treasurer with Boyle in 1972. Church said Killion was never closely identified with the Boyle administration and was "loyal to the organization." See "Ex-Boyle Man Named UMW Vice President," *Charleston* (W. Va.) *Gazette*, February 14, 1980.

38. Church originally asked for a legal assessment without limit.

39. *Newsweek*, for example, said of this period: "Church became the union's point man with industry and government officials, who found that he had a far better grasp of the issues than Miller. . . . [Since] he was elevated to the top post . . . Church has moved to restore authority and financial health to the UMW" ("Sam Church's Bully Pulpit," *Newsweek*, April 28, 1980, p. 67).

40. John P. Moody, "UMW's Church: New John L. Lewis," *Pittsburgh Post-Gazette*, December 19, 1979.

41. Ben A. Franklin, "Miners' Union Gains Respect in Industry," *New York Times*, January 28, 1980.

42. Ibid.

43. Wages, Pension Funds Have Priority," *Charleston* (W. Va.) *Daily Mail*, February 20, 1980.

44. "Church, Brennan Discuss UMWA-BCOA Relations," *UMWA Journal*, December 1979, p. 7.

45. Ibid.

46. Ben A. Franklin, "Big Coal, U.M.W. Want More of Energy Pie," *New York Times*, May 11, 1980.

47. "Image Doesn't Sparkle in Poll for Mine Group," *Wall Street Journal*, October 28, 1981.

48. Jim Ragsdale, " 'Rapist' Portrayal, Regulations Upsetting State Coal Operators," *Charleston* (W. Va.) *Gazette*, November 13, 1981.

49. Ben Z. Hershberg, "Miners, Owners Work in Concert to Sing Coal's Praise," *Louisville* (Ky.) *Courier-Journal*, October 28, 1981.

50. Patrick E. Tyler, "Corporate Deception and Intrigue Plague a Shaky Financial Citadel," *Washington Post*, October 12, 1980.

51. See Patrick E. Tyler, "The Tenure of a Dynamic Duo: A Chronicle of Troubled Loans," *Washington Post*, October 14, 1980.

52. Ibid.

53. Ibid.

54. Helen Winternitz and Timothy M. Phelps, "D.C. Bank in Turmoil Under Union Leaders' Growing Influence," *Baltimore Sun*, May 19, 1980.

55. Runyon edified his fellow bank directors by playing a coal-is-good record that Church cut in Nashville.

56. Helen Winternitz and Timothy M. Phelps, "Union Ties Mark Credentials of New Additions to NBW Board," *Baltimore Sun*, May 19, 1982.

57. Tyler, "Corporate Deception and Intrigue." Federal examiners "criticize" loans they believe present potential problems or those in which problems have appeared.

58. Thomas Love, "NBW Lists as High Risk $8.7 Million in Loans to Bank Insiders," *Washington Star*, March 19, 1980.

59. Patrick E. Tyler, "How $5,000 Check Saved Area's Largest Theater Chain," *Washington Post*, October 13, 1980.

60. Ibid.

61. Winternitz and Phelps, "D.C. Bank in Turmoil." Owing to this predicted cash-flow shortfall, NBW loan officers asked Goldman to put up $4 million in collateral. " 'The credit people at the NBW were kicking and screaming [about the Goldman loan]. Finally, they were told, you make that loan, just make it as palatable as possible. To make a loan that's bad when it [goes] on the books, much less down the road is incompetent, stupid and asking—just asking—for problems,' the (anonymous) bank official said."

62. Merrill Brown, "Lyons Quits NBW Board, Owes Bank $3.6 Million," *Washington Post*, March 19, 1980.

63. Jernberg shared his opinion of Church with the UMWA delegates in Denver: "[Sam Church] has shown to me over the last couple of years that I've known him that he has a positive insight into every facet of union leadership. I think you see in the making here a modern John L. Lewis." See remarks of Dale Jernberg, *Proceedings of the 48th UMWA Constitutional Convention*, December 17, 1979.

64. Patrick E. Tyler, "Officials' Gas Well Deals Plunge Bank into Further Crisis," *Washington Post*, October 13, 1980.

65. Patrick E. Tyler, "National Bank of Washington: A Financial Citadel in Hot Water," *Washington Post*, October 12, 1980.

66. Patrick E. Tyler, "Solving the Sensitive Dilemma of Joseph Danzansky's Bank Loans," *Washington Post*, October 14, 1980.

67. In another deal NBW officers helped campaign aides of 1978 District of Columbia mayoral candidate Sterling Tucker "disguise a $50,000 line of credit as a series of small contributions from individuals to the Tucker campaign." Bank and campaign records showed that at least $12,000 of this amount was borrowed. Campaign laws may have been violated because the loans to individual borrowers were "made under circumstances different from those described to election officials who approved them." See Patrick E. Tyler, "Bank Provided $50,000 Credit for Tucker's Mayoral Campaign," *Washington Post*, October 15, 1980.)

68. Patrick E. Tyler, "Laurel Track Loan Had the Right Things Going for It," *Washington Post*, October 16, 1980.

69. A Maryland grand jury would investigate Wyatt and charge him with bribing a state health official. See Timothy M. Phelps and Helen Winternitz, "Links to Bank Helped Track Owner Get Unusual Loan," *Baltimore Sun*, May 20, 1980.

70. Ibid.

71. Tyler, "Laurel Track Loan."

72. The most intriguing aspect of NBW's loan to Laurel was Shamy's statement to a NBW officer quoted in the *Sun* that "he had paid a $150,000 [not $250,000] 'finder's fee' " to obtain the loan. The *Sun*'s check of NBW records determined where all but $165,000 of the $4.5 million went. Had Church, Runyon, or any other NBW director received any part of a fee, it would violate federal law. On the other hand, Mandel or Wyatt could legally accept a fee. See Phelps and Winternitz, "Links to Bank."

73. Ibid. A federal jury in Baltimore found Shamy guilty of mail fraud, wire fraud, and racketeering in connection with using Laurel's treasury to pay $160,000 in personal debts. Shamy was convicted in 1977 of ordering the track's old clubhouse burned and was later charged with

awarding a $2.4 million rebuilding contract to a company in which he held a concealed interest (Saundra Saperstein, "Laurel Raceway Co-Owner Found Guilty of Fraud," *Washington Post*, December 15, 1979).

74. NBW made the second loan despite the knowledge that Laurel's top officers "had been indicted on 13 counts of fraud and racketeering." See Patrick E. Tyler and Dale Russakoff, "Indicted Race Track Officials Lent $325,000 by D.C. Bank," *Washington Post*, June 1, 1980.

75. Tyler, "Laurel Track Loan."

76. Ben A. Franklin, "Mine Workers' President Defends His Role in Bank," *New York Times*, May 25, 1980.

77. Karen Kaplan, "Church Answers Allegations, Explains Endorsements," *Bluefield* (W. Va.) *Daily Telegraph*, May 25, 1980.

78. "UMW Leader Emphasizes Lacy Wright Endorsement," *Welch* (W. Va.) *Daily News*, May 27, 1980.

79. Patrick E. Tyler, "Mine Union May Lose D.C. Bank," *Washington Post*, October 25, 1980.

80. Ibid.

81. Landow and Nesline had their fingers in many pies. Nesline has been described as an "internationally known gambling figure for 40 years . . . [and has] long been identified in local police files as the suspected 'godfather' of bookmaking and other gaming action in and around Washington" ("Gambling Kingpin Is Linked to Prominent Investors' Casino Deal," *Washington Post*, January 26, 1978). Nesline's FBI rap sheet shows twelve arrests and convictions beginning in 1934, including a conviction for carrying a pistol without a license, an acquittal on a homicide charge, and acquittal on federal gambling charges. He has served about three years in jail. He was convicted of income tax evasion in U.S. District Court in May 1981 for failing to pay $96,000 in taxes from 1971 to 1979. Nesline was fined $10,000 and placed on probation for five years after being sentenced to a year and a day. In that case *(U.S. v. Joseph A. Nesline)*, the Justice Department described Nesline as "widely reported to be the dominant figure on the local gambling scene for decades." Nesline was jailed in Miami in December 1981 for civil contempt when, after being granted immunity, he refused to testify before a grand jury investigating a bribery case. One Landow-Nesline deal was a proposed hotel-casino on St. Maarten which involved Edward Cellini, brother of Dino Cellini, a former associate of Meyer Lansky. Landow also asked Lester Matz, a skiing buddy, to help locate funding for the St. Maarten venture. Matz is the Maryland construction engineer who admitted paying kickbacks to Spiro T. Agnew and other Maryland politicians for state contracts (ibid.). Another Landow-Nesline plan was to open a hotel-casino in Atlantic City jointly with Resorts International. This deal fell through when the partners couldn't arrange the $85–$100 million financing. Landow involved Smith Bagley, heir to the Reynolds tobacco fortune and close friend of Jimmy Carter, in the venture because he thought Bagley could open Arab pocketbooks, specifically that of the Pahlavi Foundation. Nesline, at the time of the *Post* article, rented a two-bedroom apartment in the Promenade, a Landow building, in the name of his girlfriend for $480 a month, a cut rate. The two spent social time together in Monte Carlo and Washington (ibid.). The organized-crime section of the Montgomery County, Maryland, police department kept tabs on Landow in the late 1970s, and one search of Nesline's apartment turned up written documents and art objects tying the two together.

82. Information on Intellico, a splinter group from Lyndon LaRouche's U.S. Labor party, is taken from Douglas Foster, "Teamster Madness," *Mother Jones*, (January 1982); Alan Crawford, "Lyndon LaRouche's goon squads," *Inquiry*, February 15, 1982; Dennis King, "Hypocrites! Antidrug Cult Linked to Mob Cronies," *Highwitness News*, December 1981; and various communications with Chip Berlet of *The Public Eye*, a journal of the National Lawyers Guild. For information on Presser and Boffa, see Jonathan Kwitny, *Vicious Circles: The Mafia in the Marketplace* (New York: Norton, 1979). Church's meeting with Boffa was observed during a 1982 investigation by *Baltimore Sun* reporter Helen Winternitz, who shared that information with me.

12. The 1981 Strike and UMWA's Struggle for Survival

1. Thomas Petzinger, Jr., "Overcapacity in Coal Industry Is Blamed for Weak Prices; Shakeout Is Under Way," *Wall Street Journal*, September 25, 1979. Several factors create excess capacity. First, it is relatively cheap and easy for marginal suppliers to get into production. Second, big mines can open more sections within a short period of time. Third, most mines can increase production by 15 to 30 percent simply by adding a second or third production shift and scheduling weekend work.

2. "A Startling Jump in Coal Output," *Business Week*, November 17, 1980, p. 122.

3. Speech by E. Morgan Massey, president of A. T. Massey Coal Company, at the 12th Annual Institute on Coal Mining Health, Safety and Research at Virginia Polytechnical Institute, 1981, quoted in *The National Independent Coal Leader*, September 1981. Massey told the mine-safety professionals how one of his companies, Martin County Coal Corporation, handles safety. "Instead of a safety department, we set up a training department—the objective being to teach coal mining, to teach production and to teach the proper operation and maintenance of the new machinery with safety being a constraint rather than an objective of the endeavor. Martin County Coal added a production bonus of $0.50 per ton which was not constrained in any way by the safety record."

4. E. Morgan Massey explained this cause-and-effect relationship with simple clarity: "Most of my 30-plus years in this business has been involved with marketing coal and attempting to get that black stuff mined, cleaned and loaded into railroad cars with as few people as possible. That is called productivity and . . . if you do not have some fair amount of it, you do not stay around this business very long" (ibid.).

5. Douglas Martin, "Talking Business with Bailey of Conoco: Bright Hopes for U.S. Coal," *New York Times*, October 28, 1980.

6. Simon Dresner, "Consol Capital Investments Begin Paying Productivity Dividends," *Coal Industry News*, November 2, 1981, reprinted from *Conoco, '81*.

7. H. Thomas Kaib and James Lawless, "Oil Companies Control Coal: Plot Feared," (Cleveland, Ohio) *Plain Dealer*, April 21, 1980.

8. Anthony J. Parisi, "The Coal Stocks, Too, Are Heating Up," *New York Times*, January 20, 1980. MAPCO is not a BCOA member.

9. Dean Witter Reynolds, quoted in *U.S. News & World Report* (William C. Bryant, "Taking Stock: As Demand for Coal Rises," September 29, 1980).

10. OTA, *Direct Use of Coal*, p. 113 ff. A *Business Week* survey of twenty-six oil-owned coal companies in September 1979 revealed that they expected to be producing between 40 and 50 percent of total U.S. consumption by 1985, without any new coal acquisitions ("The Oil Majors Bet on Coal," *Business Week*, September 24, 1979, p. 105).

11. Ibid., p. 115.

12. On a Btu basis, the coal reserves of Exxon and Phillips Petroleum each exceed "the entire proved crude oil reserves of the United States." Together, they own the coal equivalent of 66.6 billion barrels of oil—almost 2.5 times America's proved oil reserve of 27.1 billion barrels (Joy Manufacturing Company, *Annual Report*, 1981, p. 3).

13. Office of Technology Assessment, *Development and Production Potential of Federal Coal Leases*, p. 373.

14. Energy companies expanded coal operations after the 1978 strike. They acquired coal producers. In 1979, Sun Oil bought Elk River Resources for $300 million; SOHIO took over Dahlgren-Moores Prairie Coal; Diamond Shamrock acquired Falcon Seaboard. After buying Kewanee Industries in 1977, Gulf Oil subsequently bought a 50 percent interest in Republic Steel's Alabama reserves and mines. Gulf paid $325 million for Kemmerer Coal, which produced more than 4 million tons at a Wyoming surface mine. Cities Service and Mobil both began developing major Wyoming mines. Standard Oil of Indiana bought two coal properties in 1980 for $183 million. The Fluor Corporation, a giant engineering and construction company that was also part owner of Peabody Coal—the nation's largest—bought St. Joe Minerals Corporation in 1981 for $2.9 billion.

St. Joe owned Omar Mining and other coal properties. Getty Oil paid $70 million for Energy Fuels Corporation, a Colorado outfit producing 3.6 million tons. The biggest independent producer in West Virginia—Amherst Coal, which mined 1.9 million tons in 1980—was sold in a 1981 stock swap for $201 million to Diamond Shamrock, a Dallas-based energy and chemical conglomerate. SOCAL, however, could not gain control of AMAX, coal's third biggest producer, even though it owned 20 percent of its stock. The star-crossed SOCAL bid for AMAX puzzled many. AMAX appears to have invited the merger offer after SOCAL indicated it wanted to up its share to between 30 and 50 percent. The AMAX board split over SOCAL's $4 billion offer, and SOCAL chairman, George M. Keller, said afterward: "I really find no thrill in the domestic coal business for us." (It certainly hadn't appeared that way when SOCAL made its bid.) Finally, Du Pont outbid Seagram and Mobil for Conoco in 1981 amid promises that Consolidation Coal would not be sold to pay for the acquisition. Seagram ended up owning a large bloc of Du Pont stock. Du Pont's long history of antiunionism is not at odds with the labor policies of Conoco and Consolidation Coal executives ("AMAX Management Invited Merger Offer That Led to $4 Billion Bid, SOCAL Says," *Wall Street Journal*, May 6, 1981; "SOCAL Lifts Outlays for '81 Budget 29% from Previous Goal," *Wall Street Journal*, August 12, 1981; and Steve Mufson, "Conoco's Chairman Says Coal Subsidiary Won't Be Sold After Merger with DuPont," *Wall Street Journal*, August 7, 1981).

15. Beginning in 1974, the UMWA bargained separately with western producers. AMAX ended its UMWA contract at its massive Belle Ayre mine in Wyoming. SOCAL invested in AMAX during this battle, and some observers believe that its resources helped AMAX survive the strike. See John F. Schnell, "The Impact on Collective Bargaining of Oil Company Ownership of Bituminous Coal Properties," *Labor Studies Journal* (Winter 1979):213 ff.

16. Thomas Petzinger, Jr., and Carol Hymowitz, "Anxiety Is Growing in the Coal Industry over Conoco Unit's Tough Labor Stance," *Wall Street Journal*, March 27, 1980.

17. Bob Howard, "Coal Industry: A New Era for Coal Threatens Miners, Part I," *In These Times*, December 12–18, 1979, p. 12.

18. Bob Robinson, "UMW's Church Faces Tough Challenge in Strike," *Charleston* (W. Va.) *Gazette-Mail*, April 6, 1980.

19. John P. Moody, "Church to Coal Operators; Negotiate, Stop Pressure," *Pittsburgh Post-Gazette*, March 28, 1980.

20. Bob Hall, "Consol's Hard Line Turns Small Dispute into Weeks-Long Wildcat Walkout," *UMWA*, May 1980, p. 6,

21. Keith Dix, quoted in Ibid.

22. Gil Hazelwood, "Agreement Reached in Wildcat Strike," *Charleston* (W. Va.) *Gazette*, April 14, 1980.

23. Hall, "Consol's Hard Line," p. 1.

24. Ibid., p. 4.

25. Robinson, "UMW's Church Faces Tough Challenge."

26. "Coal Industry's Chief Bargaining Group Is Restructured: Major Firms Get Control," *Wall Street Journal*, March 20, 1980.

27. Ibid. The decision of the 1979 UMW convention to authorize a selective strike fund—which everyone knew was aimed at Consolidation Coal—may have played a part in Consol's decision.

28. Petzinger and Hymowitz, "Anxiety Is Growing in the Coal Industry."

29. Ibid.

30. The two were Pittston Co., which also owns Brinks and an oil distributing company, and Peabody Coal, which is owned by a five-company partnership. The other CEOs came from Continental Oil (Consolidation Coal), Occidental Petroleum (Island Creek Coal), AMAX, Inc. (AMAX Coal), U.S. Steel, Bethlehem Steel, American Electric Power Co., Standard Oil of Ohio (Old Ben), Algoma Steel (Cannelton Industries), and St. Joe Minerals (Omar Mining). Of the top 30

companies in the BCOA, 8 were owned by oil companies, 10 by steel companies, and 7 by other noncoal companies. Only 5 were primarily coal producers (Joseph P. Brennan, "Coal Industry Labor Issues," speech to the Fuel Supply Seminar of the Electric Power Research Institute, St. Louis, Missouri, October 13, 1982).

31. Cummings, "Decentralized Bargaining," p. 159. Cummings's thesis is a useful analysis of recent changes in the structure of coal-industry bargaining.

32. If they left the BCOA, they had to be prepared to negotiate separately with the UMWA, meet substantial pension obligations, and withstand a strike in which strikers received benefits from working miners. BCOA's disfranchised members felt that continued affiliation was the lesser of the evils.

33. Ann Hughey, "One Year After the Big Strike, Coal Issues Are Different" *Charleston* (W. Va.) *Gazette-Mail*, March 25, 1979, p. 4E. Church's concern for Jay's feelings knew no bounds. The *UMWA Journal*, whose copy Church read before printing, reported a gathering of more than 500 persons to commemorate the 1913 Ludlow massacre in which militiamen sympathetic to the Colorado Fuel and Iron Co., whose principal owner was John D. Rockefeller, Jr., murdered more than 50 UMWA strikers and family members. The Ludlow strike is synonymous with the name Rockefeller, but the *Journal*'s account of the memorial service tactfully omitted the family name. "Ludlow Killings Remembered," *UMWA Journal*, July, 1980, p. 12.

34. "Surprise Strike," *Time*, April 13, 1981; "Why Coal Peace Is So Elusive," *Business Week*, April 20, 1981; and "Tough Mine Union Chief," *New York Times*, March 24, 1981.

35. David Moberg, "Miners' Rights Are Not for Sale," *In These Times*, April 8–14, 1981.

36. The UMWA did not try to collect the royalty before 1974. When it did, Consol, U.S. Steel, and some other companies refused to pay, leaving it to the courts to decide. The royalty brought in a total of about $152 million since its beginning in 1964. Most of that had been collected after 1976 when BCOA companies began marketing a lot of non-BCOA coal.

37. *Coal Outlook*, March 30, 1981.

38. "After a Short Strike, Labor Peace for Coal," *Business Week*, April 6, 1981, p. 28.

39. Ibid., pp. 28-29.

40. BCOA's attack on the cost-of-living adjustment (COLA) was repeated in other unionized industries in these years. COLAs raised labor costs in step with inflation, which from management's view made them expensive and unpredictable. Management found other objections—COLAs rewarded all workers uniformly rather than favoring the most productive; less money was available for other items, including bonus-for-production plans; the Consumer Price Index—on which COLAs are figured—overstates the inflation rate; prices could not be increased with the same quarterly regularity as COLA-affected labor costs. Merlin Breaux, Gulf Oil's vice-president for labor relations, who played a hand in coal's 1978 negotiations, told the *Wall Street Journal* that COLAs remove from management "the right to trade off what labor claims is automatically coming to them" (Amal Nag, "Cost-of-Living Clauses in Labor Pacts Come Under Attack by Firms," *Wall Street Journal*, August 5, 1981). When labor saw a COLA as a basic entitlement, management couldn't get anything in return for it. The oil industry did not have COLAs and two oil men, Breaux and Brown, ended coal's.

41. "Membership Rejects Tentative Agreement, Negotiators Return to Bargaining Table," *UMWA Journal*, April–May 1981, p. 4.

42. Robin Toner, "Church Sees BCOA Not in Mood for Accord," *Charleston* (W. Va.) *Daily Mail*, April 17, 1981.

43. Andrew Blum, "BCOA Decries Last Union Offer as Unrealistic," *Charleston* (W. Va.) *Gazette*, April 21, 1981.

44. Robin Toner, "UCE May Seek Separate UMW Pact," *Charleston* (W. Va.) *Daily Mail*, April 23, 1981.

45. Ben A. Franklin, "The Appalachian Enigma Rules in Coal Strike Settlement,"

Charleston (W. Va.) *Gazette-Mail*, June 21, 1981. For a more thorough discussion of press coverage of the 1981 contract, see Curtis Seltzer, "The Pits: Press Coverage of the Coal Strike," *Columbia Journalism Review* (July–August 1981):67-70.

46. Two other candidates announced for UMWA president during the prenomination period. Fred Carter, a retired miner with black lung from southern West Virginia, sought to build a campaign on his black-lung advocacy and plans to organize new mines. The other candidate was none other than Arnold Miller, who, in a fit of fantasy, told reporters in June that he was the only candidate who could perform the president's job well. His announcement was greeted more with pity than anger or contempt. Miller won one local nomination.

47. John P. Moody, "Trumka View to Unseat UMW Chief," *Pittsburgh Post-Gazette*, November 28, 1981.

48. "Church Faces Tough Opponent in UMW Race," *Coal Age* (October 1982):11.

49. James Warren, "Watch Out for Rich Trumka," *Washington Post*, August 8, 1982.

50. Intellico/Larry Sherman, "Preliminary Status Report on Trumka/Roberts Campaign Support," photocopy manuscript, circa summer 1982, p. 1.

51. "Church Faces Tough Opponent," p. 11.

52. Employment data come from U.S. DOE, *Coal Production, 1980*, p. 76. This rough calculation understates the employment loss because most of new surface-mine production will come from the huge, tremendously efficient open pits in the West whose productivity is much higher than eastern contour or trench surface mines. Not included are another fifteen thousand or so mine-construction workers and those who work around the mines.

53. The annual productivity gain is not compounded.

54. Personal interview with author, January 15, 1984.

55. The literature on codetermination, worker participation, and European industrial democracy is voluminous. Some perspectives include: Robert J. Kuhne, "Co-Determination: A Statutory Re-Structuring of the Organization," *Columbia Journal of World Business* (Summer 1976):17-25; Ted Mills, "Europe's Industrial Democracy: An American Response," *Harvard Business Review* (November–December 1978):143-52; Hugh Neuburger, "Codetermination—The West German Experiment at a New Stage," *Columbia Journal of World Business* (Winter 1978):104-10; Michael Poole, "Industrial Democracy: A Comparative Analysis," *Industrial Relations* (Fall 1979):262-71; A. W. J. Thomson, "Trade Unions and the Corporate State in Britain," *Industrial and Labor Relations Review* (October 1979):36-54; and Malcolm Warner and Arndt Sorge, "The Context of Industrial Relations in Great Britain and West Germany," *Industrial Relations Journal* (March/April 1980):41-49.

56. Mills, "Europe's Industrial Democracy," p. 144.

13. The Future of Coal

1. Consumption of natural gas remained at about the 20-quad level in those years while petroleum peaked in 1978 and then declined (U.S. Department of Energy, *1981 Annual Report to Congress, Volume 2, Energy Statistics* [Washington, D.C.: DOE, May 1982], p. 7).

2. The United States consumed 697 million tons and exported 78 million in 1983. Of the tons we consumed, electric utilities used 90 percent. The Department of Energy reported a total of 814 million tons of coal supply in 1983, distributed as 785 million tons mined, 1 million tons imported, and 27 million tons as changes in consumers' stocks, losses, and unaccounted for (U.S. Department of Energy, telephone interview, March 1984.)

3. While the fuel cost of nuclear energy is less than coal, its capital costs and fixed charges are substantially higher. Not included in many cost comparisons is the large federal subsidy to nuclear energy compared with the lower subsidy to coal. While economists disagree over how to calculate these costs, many utility managers now agree with Komanoff that the generating costs of *new* nuclear plants are higher than those of an equivalent coal plant. Komanoff estimated generating

costs of $0.09 per kwh for a nuclear plant in 1985 (in 1978 dollars) compared with $0.06 per kwh for coal (Charles Komanoff, "A Comparison of Nuclear and Coal Costs," testimony before the State of New Jersey Board of Public Utilities, Docket No. 762-194, phase 3, October 9, 1978).

4. Some fuels, such as liquified natural gas, are relatively safe and environmentally benign on a day-to-day basis, but a serious accident can turn them into human or environmental disasters. Experts, of course, disagree over how accidents of this kind should be handled in the calculus of social costs and benefits.

5. OTA, *Direct Use of Coal*, p. 327. OTA estimated national transportation casualties based on data computed for one 1000-MW coal-fired power plant. See MITRE Corp./Metrek Division, *Accidents and Unscheduled Events Associated With Non-Nuclear Energy Resources and Technology*, February 1977, p. 51.

6. The estimate of land disturbed by surface mining includes 65,797 acres in the Appalachian region, 19,800 in the central region, and 1,600 to 10,240 in the western region; for underground mining the regional breakdown is 107,329 acres Appalachian, 21,875 central, and 8,225 western. These calculations assume 44 million tons of coal per quad for eastern and central coal, 59 million tons for western surface-mined coal. The number of acres disturbed per quad was calculated from OTA, *Direct Use of Coal*, p. 246. Data on 1980 coal production by mining method and region of origin came from U.S. Department of Energy, *Coal Production, 1980* (Washington, D.C.: May, 1981), p. 7. Disturbance of surface-mined land does not necessarily imply irreparable short-term or long-term damage. Some surface-mined land can be reclaimed; other parcels can be reclaimed in name only. The degree of long-term environmental damage that surface mining causes depends on what mining method is used, local topography and weather conditions, and the seriousness with which the mine operator pursues reclamation.

7. Council on Environmental Quality, *Global Energy Futures and the Carbon Dioxide Problem* (Washington, D.C.: Government Printing Office, January 1981): J. Hansen et al., "Climate Impact of Increasing Atmospheric Carbon Dioxide," *Science*, August 28, 1981; and U.S. Senate, Committee on Governmental Affairs, *Carbon Dioxide Accumulation in the Atmosphere, Synthetic Fuels and Energy Policy: A Symposium*, 96th Cong., 1st sess., 1979.

8. For example, Dow Chemical Co. and Swedish and Dutch chemists raised the issue of dioxin, the toxic contaminant in Agent Orange, being emitted from coal-burning power plants, municipal incinerators, and diesel engines. Dow found a level of one or two parts of dioxin per trillion in combustion by-products. Some scientists believe dioxin accumulates in living organisms, which would make even small exposures of general concern. Dioxin is shorthand for a large class of chemicals of which the most toxic are 2, 3, 7, 8 tetrachlorodibenzo-p-dioxin (known more commonly as TCDD or dioxin). Lab animals have died from exposures to TCDD of as few as 50 parts per trillion, but the lethal dose for humans has yet to be determined (Adrian Peracchio, "Contaminant Dioxin Is Found in Coal Plants' Emissions," *Washington Post*, August 17, 1979).

9. Geologists at the U.S. Geological Survey and Bureau of Mines divide coal deposits into two broad categories: resources, which are undiscovered and uneconomic to mine at current prices; and reserves, which are economically minable and proven. America's estimated coal resources are overwhelming—almost 4,000 billion tons. Of that, about 1,760 billion are considered a recoverable reserve. However, not all of the recoverable coal can *be* recovered; most, in fact, will be left in the ground. The U.S. Bureau of Mines estimates that only 8 percent—136 billion tons—of the recoverable reserve can actually be extracted, 46.9 billion by stripping and 89 billion by deep mining. These reserves are divided by rank as follows (in millions of tons): anthracite 2,334, bituminous 68,340, subbituminous 56,227, and lignite 8,895.

Our ability to mine "recoverable" coal is limited by land-use conflicts (seams underlying residential areas), geological constraints (faults and thinness in the seam; water problems), laws that prevent mining in certain areas (e.g., national wildernesses, public forests, land surrounding preexisting oil and gas wells) and the inherent limitations of our recovery methods. *At any given mine site*, surface mining extracts between 70 and 90 percent of the recoverable reserve and underground mining takes between 40 and 65 percent.

Several immutable geological facts of life confront us. First, more than half of what we can mine is bituminous coal, and most of the rest is subbituminous. Second, underground mining is the only way to get *two-thirds* of our total recoverable reserve. Third, our largest single deposit lies in deep bituminous seams. Finally, the amount of strippable bituminous coal—13.5 billion tons—is comparatively small.

By 1985, America is likely to be producing 1 billion tons annually. At that rate, our coal will last 136 years.

Strippable Reserves of Bituminous Coal and Lignite in the United States, U.S. Bur. Mines. Inf. Cir. 8531, 1971, quoted in Richard Schmidt, *Coal in America: An Encyclopedia of Reserves, Production and Use* (New York: McGraw-Hill, 1979), p. 62.

10. OTA, *Direct Use of Coal*, pp. 290–331.

11. U.S. Congress, House of Representatives, *The Changing Distribution of Industrial Profits: The Oil and Gas Industry Within the Fortune 500, 1978–1980, Report of the Committee on Energy and Commerce*, 97th Cong., 1st sess., December 11, 1981, pp. v-vi.

Index